STABILITY IN NONLINEAR CONTROL SYSTEMS

STABILITY IN NONLINEAR CONTROL SYSTEMS

BY

Alexander M. Letov

TRANSLATED FROM THE RUSSIAN

BY J. GEORGE ADASHKO

1961

PRINCETON, NEW JERSEY

PRINCETON UNIVERSITY PRESS

L. C. Card 59-5599

Printed in the United States of America

TABLE OF CONTENTS

TABLE OF CONTENTS

TABLE OF CONTENTS

AUTHOR'S FOREWORD TO THE AMERICAN EDITION

The American edition of my book 'Stability in Nonlinear Control Systems' is in the main an exact translation of the USSR edition of 1955. The revisions are confined to discovered misprints and to a slight editing of the text and addition of a few literature references.

This edition does include, however, an additional chapter, based for the most part on a paper I delivered to the International Congress on Automatic Control at Heidelberg in 1956.

AUTHOR'S FOREWORD TO THE RUSSIAN EDITION

This book contains the results of certain investigations of stability* and of the degree of stability of several nonlinear control systems with one or two regulators.

The problems considered here concern what is called absolute stability, i.e., stability under unbounded perturbations, with regulators of arbitrary nonlinear characteristics. The characteristics are defined precisely only to the extent of their belonging to a certain class of functions.

The first investigation of stability of this kind is contained in a note by A. I. Lur'e and V. N. Postnikov, devoted to the solution of one particular problem.** The system they consider belongs to one widely used class of control systems, a general study of which was first begun by Lur'e, who formulated the problem of absolute stability of such systems and submitted for its solution a method which he carried through to the algorithm stage.***

Lur'e's methods can be developed, generalized, and extended to cover a whole series of broader and more complicated class of control systems. This book is an attempt at such a development and generalization, and at the development of various methods for solving problems in the absolute stability of control systems.

The author's desire to obtain a certain generalization of Lur'e's results was stimulated by three factors.

The first was the constant efforts of Boris Vladimirovich

---

* A. M. Lyapunov, Obshchaya zadach ob ustoǐchivosti dvizhenia (General Problem of Stability of Motion), Gostekhizdat, 1950.

** A. I. Lur'e and V. N. Postnikov, Concerning the Theory of Stability of Control Systems, PMM [Prikladnaya matematika i mekhanika (Applied Mathematics and Mechanics)], Vol. VIII, No. 3, 1944.

*** A. I. Lur'e, Nekotorye nelineǐnye zadachi teorii abtomaticheskovo regulirovaniya (Certain Nonlinear Problems in the Theory of Automatic Control), Gostekhizdat, 1951.

Bulgakov to perfect methods for the mathematical analysis of control systems. He paid particular attention to the Lyapunov direct method and, as a teacher, frequently called my attention to this method.

The second factor was the publication of N. G. Chetaev's monograph,* in which all the general problems of the theory of stability were successfully solved by Lyapunov's direct method. It uncovered the analytical and geometrical possibilities of this method, which were illustrated convincingly by many specific examples.

The third and final factor was the articles on the stability of control systems, by A. I. Lur'e, alone and in collaboration with V. N. Postnikov, published in the journal "Prikladnaya matematika i mekhanika" (Applied Mathematics and Mechanics).

Lur'e demonstrated convincingly in his articles the benefits that ensue from the application of Lyapunov's direct method to the solution of the first principal problem of the theory of automatic control, and thereby influenced the author's own approach to this method.

The basic source material for the present book were, on the one hand, the aforementioned monograph and the articles by Lur'e, and on the other the author's own articles, published in 1948-1954 in the journals 'Prikladnaya matematika i mekhanika' and 'Avtomatika i telemekhanika' (Automation and Telemechanics).

These articles incorporate many suggestions made by Lur'e upon reading the manuscripts. Many valuable comments were also made by him concerning the contents of this book, and I take this occasion to express my deep gratitude for continuous interest in my work.

The book contains eleven chapters and an introduction. The introduction treats the main premises of stable motion, as developed by A. M. Lyapunov. They are included for the benefit of readers unfamiliar with Lyapunov's monograph. Principal attention is paid here to the Lyapunov direct method, which serves as the basis for the entire book.

Chapter I is devoted to the equations of control systems and the definition of absolute stability. In Chapters II and III the original equations of the control systems are converted to different canonical forms, which are employed in the solution of the principal problems.

Chapter IV treats the principal results obtained by Lur'e for an inherently-stable control system with a single regulator.

In Chapter V are considered different forms of stability criteria

---

* N. G. Chetaev, Ustoichivost' dvizheniya (Stability of Motion, Gostekhizdat, 1946.

for control systems of the class treated in Chapter IV. Particular attention is paid to one of the methods of formulating simplified criteria. These criteria are of practical value, for they make possible the analysis of equations expressed in symbolic form, for systems with many degrees of freedom. In particular, the simplified criteria help explain the role of small, so-called 'parasitic' parameters of control systems.

The analysis of inherently unstable control systems is the subject of Chapter VI. Two methods of investigation are developed here, one belonging to Lur'e and the other to the author.

The problem of stability of program-control processes is formulated in Chapter VII, where a particular solution is obtained for the problem.

In Chapter VIII we discuss the calculation of the degree of stability of a control system with one regulator. Two methods are given here for estimating the degree of stability. Such an estimate enables us to define the quality of a control system.

The absolute stability of a control system with two regulators is discussed in Chapter IX.

In Chapter X we study two related special topics in stability theory, first-approximation stability and stability under continuous disturbing forces.

Finally, in Chapter XI, we investigate the stability of unsteady motion in any (finite or infinite) time interval, and propose one method of solving this problem.

Numerous examples are used throughout to illustrate the sequence of the computations involved in the applications of the methods as they are developed.

The book makes no pretense of being complete. In particular, there is no mention of the valuable work by B. V. Bulgakov and A. A. Andronov with their students, who employ a different point of view.

The main purpose in publishing this book is to acquaint the many readers engaged in the field of automation with the possible effective utilization of Lyapunov's direct method, and to solve the major problems in the theory of automatic control. The book is intended for undergraduates and graduate students specializing in automatic control and applied mechanics.

In view of the acute shortage of literature devoted to the direct application of the Lyapunov direct method to the theory of automatic control, the author hopes that this book will be found useful.

The author thanks A. N. Rubashov for much help in preparing this book for publication and for many valuable comments.

STABILITY IN NONLINEAR CONTROL SYSTEMS

INTRODUCTION

1. STATEMENT OF THE STABILITY PROBLEM

The modern theory of automatic control, no matter how presented, is based on a single strong foundation, A. M. Lyapunov's theory of the stability of motion.*

Consequently, by way of introduction, we shall give a brief treatment of certain known premises of this theory, in a form and approach necessitated by the scope of this monograph.

Readers familiar with the Lyapunov theory as presented in the original papers can immediately proceed to Chapter I; readers not familiar with the theory can for the time being confine themselves to a reading of this Introduction.

Modern automatic control methods represent, in the majority of cases, rather complicated electromechanical devices, consisting of regulated objects and regulators. The purpose of the regulator is to maintain continuously in the regulated object either a certain steady state or a state that varies in a prescribed manner. Consequently, the control process consists of utilizing the regulator to prevent deviations from the desired state of the regulated object, deviations that could occur in the regulated object as a result of certain disturbances of its operation.

One of the fundamental problems in the theory of automatic control is the time variation of the control process. The mathematical analysis of this problem is as follows:

Corresponding to each automatic control system is a definite system of differential equations of the form

---

* See the author's foreword to this book, and also: I. G. Malkin, Teoriya ustoĭchivosti dvizheniya (Theory of Stability of Motion), Gostekhizdat, 1952, and G. N. Duboshin, Osnovy teorii ustoĭchivosti dvizheniya (Principles of the Theory of Stability of Motion), Moscow State University Press, 1952.

(1) $$\frac{dx_k}{dt} = X_k(x_1, \ldots, x_n) \qquad (k = 1, \ldots, n) .$$

Here $x_1, \ldots, x_n$ are variables describing the state of the system, and X_k are certain functions of these variables, defined in some fixed region G in the space of the variables $x_1, \ldots, x_n$. This space is called the phase space* and we shall denote it by E_n.

In this space, (1) defines the components X_k of a velocity vector v of the motion of a certain point M, called the representative point. From the physical point of view (1) should be considered as the mathematical form of expressing these laws to which the controlled systems are subject.

The properties and features of these laws are either fully or approximately (but with good accuracy) represented by the character of the functions $X_k(x_1, \ldots, x_n)$. The region G where the functions $X_k(x_1, \ldots, x_n)$ are defined is that portion of the space E_n over which the action of the above physical laws extends.

Let the quantities $x_{10}, \ldots, x_{no}$ denote the initial values of the variables $x_1, \ldots, x_n$. They define uniquely the initial state of the control system at $t = 0$. Corresponding to each system of initial conditions $x_{10}, \ldots, x_{no}$ is a solution

(2) $$x_k = x_k(t, x_{10}, \ldots, x_{no})$$

of (1). It is assumed that such a solution exists for all values of $t > 0$ and is unique.

Solution (2) describes the motion of the automatic control system, as determined by its initial state, and assumes that this motion is unique and corresponds to the essence of the majority of physical laws.

In such systems, steady-state processes are described by the so-called obvious solutions of (1). These solutions

(3) $$x_1 = x_1^*, \ldots, x_n = x_n^*$$

are the roots of the equation

(4) $$X_k(x_1, \ldots, x_n) = 0 \qquad (k = 1, \ldots, n).$$

---

* In the general case, (1) may contain the time t in explicit form. We confine ourselves, however, to an analysis of only the specific case given here.

These are included in the family of solutions (2) and are determined by the initial values $x_{10} = x_1^*, \ldots, x_{n0} = x_n^*$.

(4) describes the static behavior of control systems. The fundamental problem in statics is the determination of solution (3) and a study of its structure as a function of the various constant parameters of the control system.

We usually consider also cases in which there exists only one solution (3) fully corresponding to a definite steady state process in the control system.

One of the fundamental problems in the theory of automatic control is whether or not the obvious solution (3) corresponds to any one physically feasible steady-state process. This problem can be resolved by investigating the solution (3) for stability.

Physically realizable steady states correspond uniquely to the so-called stable solutions (3), while physically unrealizable ones correspond to the unstable solutions (3). Consequently, in the mathematical treatment, the problem of the correspondence between the solution (3) and a physically realizable steady state of the control system is the problem of stability (or instability) of solution (3).

Hereafter we shall find it convenient to deal with equations derivable from (1) by making the following change in variables:

$$(5) \qquad x_k = x_k^* + y_k \qquad (k = 1, \ldots, n) \quad .$$

In Lyapunov's terminology, the equations obtained after such a transformation are called the equations of disturbed motion; they are of the form

$$(6) \qquad \frac{dy_k}{dt} = Y_k(y_1, \ldots, y_n) \qquad (k = 1, \ldots, n) \quad ,$$

in which

$$(7) \quad Y_k(y_1, \ldots, y_n) = X_k(x_1^* + y_1, \ldots, x_n^* + y_n) \qquad (k = 1, \ldots, n) \quad .$$

Formula (5) determines the transformation of the shift of the origin to a point with coordinates $x_1^*, \ldots, x_n^*$. As a result, the obvious solution (3) of (1) corresponds to the obvious solution

$$(8) \qquad y_1^* = 0, \ldots, y_n^* = 0$$

of (6). In Lyapunov's terminology, this solution describes the undisturbed motion of the control system.

Let the variables $y_1, \ldots, y_n$ assume at $t = 0$ some arbitrary initial values $y_{10}, \ldots, y_{n0}$, of which at least one does not vanish; these values are called disturbances. Corresponding to each given system of such disturbances is a unique and continuous solution

$$y_k = y_k(y_{10}, \ldots, y_{n0}, t) \qquad (k = 1, \ldots, n) \tag{9}$$

of (6); this solution describes the disturbed motion of the automatic-control system.

Were we to know all the solutions (9), we would know all the disturbed motions of the system. But for the general case it is impossible in practice to obtain all these solutions, thus complicating the rational choice of control parameters. A qualitative investigation is therefore necessary to permit a survey of the entire family of the disturbed motions (9) and, without resorting to integration, to ascertain whether these motions tend, as $t \longrightarrow \infty$, to the undisturbed motion (8) independently of the initial values $y_{10}, \ldots, y_{n0}$.

Lyapunov's theory of the stability of motion permits an estimate of the properties of the disturbed motions of interest to us, without resorting, in final analysis, to integration of (6). The theory thus points a way towards rational construction of regulators.

If solution (8) is stable at a particular setting of the regulator, the regulated system will itself, without interference from the outside, choose the mode of the undisturbed motion corresponding to this solution. But if solution (8) is unstable, such a steady-state mode turns out to be physically unrealizable.

Thus, the definition of stability of motion (8), given by Lyapunov, acquires great practical value in the study of important problems of modern technology.

Let us proceed now to a definition of this stability.

We shall observe and compare at each instant of time t any disturbed motion (9) with the undisturbed motion (8), by studying the differences

$$x_k - x_k^* = y_k - y_k^* \qquad (k = 1, \ldots, n) \quad ,$$

whose initial values are

$$x_{k0} - x_k^* = y_{k0} - y_k^* \qquad (k = 1, \ldots, n) \quad .$$

DEFINITION. An undisturbed motion (8) is called stable with respect to the quantities y_k if for any specified positive number ε, no matter how small, there exists another positive number $\eta(\varepsilon)$, such that for all disturbances of y_{ko}, satisfying the conditions

$$|y_{ko}| \leq \eta \qquad (k = 1, \ldots, n) \tag{10}$$

the disturbed motion (9) will satisfy the inequalities

$$|y_k(t)| < \varepsilon \qquad (k = 1, \ldots, n) \tag{11}$$

for all $t > 0$.

The undisturbed motion (8) is called unstable, however, if there exists a value of ε such that given a value of η, no matter how small, it is possible to find y_{ko} satisfying conditions (10), so that at least one inequality (11) is not satisfied for a certain $t > 0$.

This definition can be given a geometric character, if we bear in mind that the whole manifold of disturbed motions (9) begins near the origin defined by (8). With this as our purpose, we can formulate the definition in the following manner: The undisturbed motion (8) is called stable with respect to the quantities y_k if for any specified positive number A, no matter how small, it is possible to choose another number $\lambda(A)$, such that for all disturbances y_{ko}, satisfying the condition

$$\sum_{k=1}^{n} y_{ko}^2 \leq \lambda \tag{12}$$

the disturbed motion (9) will satisfy the inequality

$$\sum_{k=1}^{n} y_k^2(t) < A \tag{13}$$

for all $t > 0$. In the contrary case the undisturbed motion is called unstable.* The set of points satisfying condition (13) will be called the A-vicinity, while the set of points satisfying inequality (12) will be called the λ-vicinity of the obvious solution (8).

Inequality (12) limits the totality of the initial disturbances of the system; inequality (13) limits the character of the course of its

---

* N. G. Chetaev, Ustoĭchivost' dvizheniya (Stability of Motion), Gostekhizdat, 1946.

disturbed motion. In all cases in which these inequalities are satisfied we say that the disturbed motion of the system converges to the undisturbed motion.

Defined in this manner, however, the character of the convergence of disturbed motion to the undisturbed motion (or, which is the same, the character of the stability of the undisturbed motion), can be twofold: either the undisturbed motion will be stable in the usual sense, i.e., inequalities (12) and (13) are satisfied, or, in addition, the following equalities will hold:

$$\lim_{t \to \infty} y_k(t) = 0 \quad .$$

In the latter case we say there is asymptotic stability of the undisturbed motion.

One premise which is of fundamental importance in the problem of stability must be pointed out. This is that the number λ, which enters into the formulation of stability according to Lyapunov, can always be defined. The method of calculating λ was given by Lyapunov in a proof of one of the stability theorems. The magnitude of this number depends substantially on the form of the so-called Lyapunov V-functions and the parameters of the systems of equations that describe the disturbed motion. An example of such a calculation was given by N. G. Chetaev. Having calculated λ, it is possible to establish the size of the region

$$\sum_{k=1}^{n} y_{ko}^2 < \lambda \quad ,$$

in which stability of the undisturbed motion is assured.

As already mentioned, in the case of asymptotic stability the following equalities are satisfied:

$$\lim_{t \to \infty} y_k = 0 \quad .$$

If these equalities occur for all the y_{ko} that belong to G, we speak of asymptotic stability in the large. If, however, the fulfillment of these equalities requires that y_{ko} be sufficiently small, one speaks of asymptotic stability in the small.

In certain cases it is necessary to choose the parameters of the regulator such that the investigated steady-state mode of the regulated system be known to be unstable, i.e., that it be physically unrealizable.

Such, for example, may be the flight of an airplane in spin.* The study of the stability of motion permits solving this complicated and very important problem with the aid of the known theorems of Lyapunov and Chetaev on stability.

2. LYAPUNOV'S DIRECT METHOD

The Lyapunov direct method reduces to the construction of such functions V of the variables $y_1, \ldots, y_n$, whose total derivatives with respect to time have, in accordance with (6), certain properties that assure stability.

Each V-function is defined in a certain region G', specified by the inequality

$$\sum_{k=1}^{n} y_k^2 < L \ , \tag{14}$$

where L is a certain constant. It is assumed that L can assume any positive value. Then the region G' will be contained in the region G, if L is sufficiently small, or else will contain the region G (or will coincide with it), if L is sufficiently large.

The function V will be called sign-invariant if, except for the null values, it will assume everywhere in G' the values of only one sign. A sign-invariant function that assumes zero values only at the origin will be called sign-definite, or, if it is desired to call attention to its sign, positive-definite or negative-definite.

Thus, for example, of the two functions

$$V_1 = y_1^2 + y_2^2 + y_3^2 \quad \text{and} \quad V_2 = (y_1 + y_2)^2 + y_3^2$$

the function V_1 is sign-definite, while the function V_2 is merely sign-invariant.

If V is a sign-definite function, then the equation V = C = const. represents a one-parameter family of closed surfaces. When the parameter C is decreased, each such surface contracts to the origin, and in the limit, as C goes to zero, it contracts into a point, namely the origin. These surfaces intersect all paths that lead from the origin to infinity.

---

* N. G. Chetaev, Concerning Stability of Motion, Izv. AN SSSR, Div. of Technical Sciences, No. 6, 1946.

Along with the functions V we shall consider their total derivatives with respect to time, namely

$$\frac{dV}{dt} = \sum_{k=1}^{n} \frac{\partial V}{\partial y_k} \cdot \frac{dy_k}{dt} \quad . \tag{15}$$

LYAPUNOV'S FIRST THEOREM. A disturbed motion is stable if its differential equations are such that it is possible to find a sign-definite function V, having a derivative that, by virtue of these equations, is sign-invariant and of sign opposite to that of V, or vanishes immediately.

The functions V, which satisfy the conditions of this theorem, are called Lyapunov functions. To prove the theorem, let us assume that in a given problem we know some positive-definite Lyapunov function for the region G'. Let us calculate its total derivative (15) in accordance with (6). We have

$$\frac{dV}{dt} = \sum_{k=1}^{n} \frac{\partial V}{\partial y_k} Y_k (y_1, \ldots, y_n) \quad . \tag{16}$$

Let us consider the surfaces $V = C$ $(C > 0)$ and some point on one of these surfaces. As is shown in Euclidian multi-dimensional geometry, the quantities $\partial V/\partial y_k$ are proportional to the direction cosines of the normal n to such a surface:

$$\frac{\partial V}{\partial y_k} = + \sqrt{\sum_{k=1}^{n} \left(\frac{\partial V}{\partial y_k} \right)^2} \cos (ny_k) \qquad (k = 1, \ldots, n) \quad . \tag{17}$$

(The positive direction of the normal is considered to be that of the outward normal.)* According to (17), expression (16) assumes the form

$$\frac{dV}{dt} = + \sqrt{\sum_{k=1}^{n} \left(\frac{\partial V}{\partial y_k} \right)^2} V_n \quad ,$$

where V_n denotes the projection of the velocity of the representative point M on the normal to the surface $V = C$:

---

* If V is a negative-definite function, it is necessary to consider the surfaces $V = C$, where $C < 0$, and the inward normal direction is then assumed positive.

$$V_n = \sum_{k=1}^{n} Y_k \cos (ny_k) \quad . \tag{18}$$

In accordance with the conditions of the theorem, this projection is either negative everywhere in G' or vanishes, since $dV/dt \leq 0$. Consequently, the point M moves in the phase space E_n along trajectories that intersect the family of surfaces $V = C$ inward, i.e., it moves from surfaces with greater values of C to surfaces with smaller C (for $V_n < 0$), or else remains during the entire time on some surface (if $V_n \equiv 0$). Let A be some specified positive number. Let us choose the numbers λ and C such that the λ-vicinity of the obvious solution (8) lies entirely inside the surface $V = C$, which in turn would belong to the A-vicinity. Let us then place there the representative point M for $t = 0$. Then, by virtue of the above ($V_n \leq 0$), the point M will not reach the surface $V = C$ for any value $t > 0$, and consequently, it will not leave the A-vicinity of the obvious solution. But the latter is exactly a statement of the stability of the obvious solution (8).

LYAPUNOV'S SECOND THEOREM. If the differential equations of disturbed motion are such that it is possible to find a sign-definite function V, whose derivative would, by virtue of these equations, be a sign-definite function of sign opposite to V, then the disturbed motion is asymptotically stable.

The proof of the second theorem is identical to the proof of the first theorem. However, by virtue of the sign-definite nature of the function dV/dt, the representative point M cannot remain in this case on any surface $V = C$ ($C \neq 0$), for the quantity V_n can vanish only at the origin of the phase space. Consequently, when $t \longrightarrow \infty$ the disturbed motion will converge to an undisturbed motion, with

$$\lim_{t \to \infty} y_k = y_k^* \qquad (k = 1, \ldots, n) \quad .$$

Thus, the undisturbed motion will be asymptotically stable.

The above argument calls for one important remark. Recently* attention was called to the fact that the surfaces $V = C$ are closed only if C is sufficiently small. For example, the function**

* N. P. Erugin. Certain General Problems in the Theory of Stability of Motion, PMM (Prikladnaya Matematika i Mekhanika, Appl. Math. and Mech.), Vol. XV, No. 2, 1951; N. P. Erugin, Concerning One Problem in the Theory of Stability of Automatic Control Systems, PMM, Vol. XVI, No. 5, 1952.

** E. A. Barbashin, N. N. Krasovskiĭ. On the Stability of Motion as a Whole, Dokl. AN SSSR, Vol. XXXVI, No. 3, 1953.

$$V = y_1^2 + \frac{y_2^2}{1 + y_2^2}$$

defines a family of closed curves $V = C$ only if $C \leq 1$; for any $C > 1$ the curve $V = C$ consists of two branches, which have no common points (Figure 1). Consequently, the given V function can serve as a Lyapunov

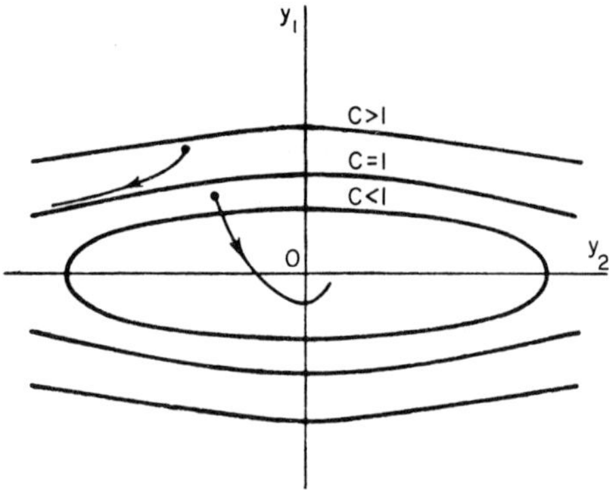

FIGURE 1

function only for an investigation of stability in which the disturbances are limited by the condition

$$y_{10}^2 + \frac{y_{20}^2}{1 + y_{20}^2} < 1 \quad .$$

If this condition is not satisfied, the representative point can go outside the limits of the curve $V = C$ through the zone of discontinuity of its corresponding branches.

Another example is given by the function

$$V(y_1, y_2) = \int_0^{y_1} \varphi(y_1)\, dy_1 + y_2^2 \quad ,$$

where $\varphi(y_1)$ satisfies the conditions:

$$\varphi(0) = 0, \quad y_1\, \varphi(y_1) > 0, \quad y_1 \neq 0 \quad .$$

Let us plot the curve

$$\int_0^{y_1} \varphi(y_1)\, dy_1 + y_2^2 = C \quad .$$

It naturally must be closed if C is sufficiently small. But, if the function $\varphi(y_1)$ is such that the integral

$$\int_0^{y_1} \varphi(y_1)\, dy_1$$

tends to a certain limit a as $y_1 \longrightarrow \infty$, then the curves $V = C$ will be closed only for values $C < a$. Consequently, whenever the constant C is not sufficiently small, it is necessary to verify whether the curves $V = C$ are closed.

Closure of the curves $V = C$ is assured if, in addition to the above, the Lyapunov function approaches infinity as

$$\sum_{k=1}^{n} y_k^2 \longrightarrow \infty \quad .$$

The latter means that no matter how large a number N may be, we can always choose a number L so large, that at

$$\sum_{k=1}^{n} y_k^2 > L$$

the function V will assume values $V > N$. In this case we say that the function V is unbounded.

The Lyapunov theorem concerning stability proved above will be unconditionally correct if the function V that enters into these theorems is unbounded.*

3. INVESTIGATION OF STABILITY WITH FIRST-APPROXIMATION EQUATIONS

In the majority of problems in the theory of control the functions Y_k can be expanded into a power series that converges in a certain H-vicinity of the origin,

* E. A. Barbashin, N. N. Krasovskiǐ. Concerning the Stability of Motion as a Whole. Dokl. AN SSSR, Vol. XXXVI, No. 3, 1953.

(19) $$\sum_{k=1}^{n} y_k^2 < H \quad ,$$

provided the constant H is sufficiently small.* In these cases (6) can always be rewritten as

(20) $$\frac{dy_k}{dt} = a_{k1}y_1 + \dots + a_{kn}y_n + F_k(y_1, \dots, y_n) \qquad (k = 1, \dots, n) \quad ,$$

where a_{ki} $(k, i = 1, \dots, n)$ are the constants of the linear portion of the expansion and the functions F_k do not contain terms of order less than the second.

The stability of solution (8) is frequently estimated by considering only the so-called first-approximation equations

(21) $$\frac{dy_k}{dt} = a_{k1}y_1 + \dots + a_{kn}y_n \qquad (k = 1, \dots, n) \quad .$$

Since it is not immediately obvious that this can be done, the need arises for investigating those cases in which an idea concerning the stability of solution (8) can be obtained by considering the first-approximation equation (21). This important problem in mechanics was first solved by Lyapunov; it is indeed this solution that resulted (after I. A. Vyshnegradskiĭ) in the development of the theory of automatic control.

To prove the Lyapunov first-approximation theorems let us introduce new variables (22)

(22) $$z_s = \sum_{\alpha=1}^{n} c_\alpha^{(s)} y_\alpha \qquad (s = 1, \dots, n) \quad .$$

If the transformation constants $c_\alpha^{(s)}$ can be chosen to satisfy

(23) $$\lambda_s c_\beta^{(s)} = \sum_{\alpha=1}^{n} a_{\alpha\beta} c_\alpha^{(s)} \qquad (\beta, s = 1, \dots, n) \quad ,$$

then (21), rewritten in the new variables, becomes simply

---

* This limitation comprises the substantial difference between the first-approximation method and the direct method of stability investigation, explained above.

(24) $$\frac{dz_s}{dt} = \lambda_s z_s \qquad (s = 1, \ldots, n) \; .$$

Here λ_s are the transformation parameters, which are determined as the roots of the equation

(25) $$\Delta(\lambda) = |a_{\alpha\beta} - \lambda\delta_{\alpha\beta}| = 0 \; ,$$

where

$$\left.\begin{aligned} \delta_{\alpha\alpha} &= 1 \\ \delta_{\alpha\beta} &= 0 \end{aligned}\right\} \qquad (\alpha \neq \beta) \; .$$

The solution of the above problem depends on the roots of this equation. We consider here only the case in which the roots are simple. Then (20) can be written in the very simple form

(26) $$\frac{dz_s}{dt} = \lambda_s z_s + \Phi_s(z_1, \ldots, z_n) \qquad (s = 1, \ldots, n) \; ,$$

in which the functions Φ_s, like the functions F_s, contain no expansion terms of order less than the second.

(26) has the obvious solution

(27) $$z_1^* = 0, \ldots, z_n^* = 0 \; .$$

Since the transformation (22) is linear and one to one, it is possible to state that the stability (instability) of the solution (27) also determines uniquely the stability (instability) of solution (8).

Let us consider the positive definite function

$$V = \frac{1}{2} \sum_{k=1}^{n} z_k^2 \; .$$

It determines half the square of the distance between the representative point and the origin. For simplicity we shall also assume that the λ_s are real numbers. Then the total derivative of the function V, calculated in accordance with (24), will be

$$\frac{dV}{dt} = \sum_{k=1}^{n} \lambda_k z_k^2 \; .$$

On the other hand, according to (26) we find

$$\frac{dV}{dt} = \sum_{k=1}^{n} \lambda_k z_k^2 + \sum_{k=1}^{n} z_k \Phi_k \quad .$$

If we confine ourselves to an examination of such disturbed motions in the H-vicinity as are characterized by sufficiently small values of the variables z_k, then for these values the sign of the function dV/dt will be determined only by the expression

$$\sum_{k=1}^{n} \lambda_k z_k^2 \quad ,$$

since the expression

$$\sum_{k=1}^{n} z_k \Phi_k$$

contains no terms below the third order of smallness.

Consequently, for $\lambda_k < 0$ we find that dV/dt is a negative definite function in the H-vicinity. But then, according to Lyapunov's theorem, the undisturbed motion (27), and consequently also the motion (8), is asymptotically stable.

In the case in which among the roots λ_k there is at least one positive root, the undisturbed motion (27), and consequently also the motion (8), is unstable, as can be proved by an analogous method.

These results can be extended to include also that case in which among the roots λ_k there are complex conjugates or multiple ones. In all such cases the stability (instability) of the undisturbed motion takes place at

$$\text{Re } \lambda_k < 0 \ (\text{Re } \lambda_k > 0) \qquad (k = 1, \ldots, n) \quad .$$

Thus, the following two Lyapunov theorems are valid:

FIRST THEOREM. If the real parts of all the roots λ_k of the characteristic (25) of the first approximation are negative, then the undisturbed motion (8) is asymptotically stable, independently of the terms F_k above the first order of smallness.

SECOND THEOREM. If among the roots λ_k of the characteristic (25) there is at least one with a positive real part, then the undisturbed motion (8) is unstable independently of the terms above the first order of smallness.

Also possible is an intermediate case, when among the roots λ_k of (25) there are such having a zero real part, while the remaining roots have a negative real part. In all these critical cases the stability of the undisturbed motion (8) cannot be determined by investigating the first-approximation equations.

As was shown by Lyapunov, in critical cases the stability (instability) of undisturbed motion is determined by the form of the nonlinear functions F_k, and then it becomes necessary to consider (6) in their original form.

It should be kept in mind that an investigation of the critical cases is of great interest for the solution of a whole series of important applied problems, including, in particular, the problem of the stability of longitudinal motion of an airplane.

An important application of the Lyapunov stability theory to critical cases of dynamic and automatic-control systems was developed by N. N. Bautin* and A. I. Lur'e.** They have shown that the question of stability of systems in critical cases is connected with the determination of the unsafe and safe portions of the boundary of their stability region.

Methods of solving the problem of the stability of undisturbed motion were developed by Lyapunov himself, and also by N. G. Chetaev, I. G. Malkin, K. P. Persidskiĭ, G. V. Kamenkov, and others, and were extended to include more complicated systems of equations of undisturbed motion.

4. THE HURWITZ THEOREM

It is clear from the above that to solve the problem of stability of motion in non-critical cases it is important to obtain the necessary and sufficient conditions under which the real parts of the roots of the characteristic (25) become negative.

These conditions were formulated by Hurwitz in the form of the

---

* N. N. Bautin, Povedenie dinamicheskikh sistem vblizi granits oblasti ustoĭchivosti, (Behavior of a Dynamic System Near the Boundaries of the Stability Region), Gostekhizdat, 1949.

** A. L. Lur'e, Nekotorye nelineinye zadachi teorii avtomaticheskovo regulirovaniya, (Certain Nonlinear Problems in the Theory of Automatic Control), Gostekhizdat, 1951.

following theorem: Let there be an equation of the n'th degree

(28) $$a_0\lambda^n + a_1\lambda^{n-1} + \dots a_{n-1}\lambda + a_n = 0 \quad ,$$

in which all a_k are real numbers, and $a_0 > 0$. Let us set up the Hurwitz determinants

(29)
$$\Delta_1 = a_1; \quad \Delta_2 = \begin{vmatrix} a_1 & a_0 \\ a_3 & a_2 \end{vmatrix}, \dots$$
$$\dots, \quad \Delta_n = \begin{vmatrix} a_1 & a_0 & 0 & \dots & 0 \\ a_3 & a_2 & a_1 & \dots & 0 \\ \cdot & \cdot & \cdot & \cdot & \cdot \\ a_{2n-1} & a_{2n-2} & & \dots & a_n \end{vmatrix} \quad ,$$

in which all $a_i = 0$ if $i \geq n$. The Hurwitz theorem is formulated in the following manner: in order for all roots of (28) to have a negative real part, it is necessary and sufficient that all the Hurwitz determinants be positive.

There are also other known criteria for the negative real parts of the roots of (28), for example the Routh, Nyquist, or Mikhaĭlov criteria, etc. We shall not stop to consider these.

5. CHARACTERISTIC NUMBERS

To investigate the stability of unsteady motion, Lyapunov formulated the theory of characteristic numbers, which was further developed and applied in the works by Chetaev, Persidskiĭ, Malkin, and others.

In this book the concept of the characteristic number will be applied to the investigation of the problem of the quality of control (formulated in a certain sense).

Thus, we shall consider the real functions $f(t)$, defined at $t \geq 0$. We shall say that $f(t)$ is a bounded function if $|f(t)|$ remains less than the certain finite limit m for all $t \geq 0$. To the contrary, if $|f(t)|$ can be greater than any specified number $m > 0$ for a corresponding choice of $t > 0$, then the function $f(t)$ will be called unbounded. Finally, any bounded function, which goes to zero in the limit as t increases without bounds will be called vanishing.

Let us consider a certain function $\varphi(t)$. The number λ_0 is called, according to Lyapunov, the characteristic number of the function

$\varphi(t)$ if, for all $\varepsilon > 0$, the function $\varphi(t)e^{(\lambda_0+\varepsilon)t}$ is unbounded, and the function $\varphi(t)e^{(\lambda_0-\varepsilon)t}$ is vanishing.

Let us give a few examples. Thus, any constant $\varphi(t)$ has a characteristic number $\lambda_0 = 0$. The same pertains to any function $\varphi(t) = t^m$, since

$$\lim_{t \to \infty} t^m e^{-\varepsilon t} = 0$$

and

$$\lim_{t \to \infty} t^m e^{+\varepsilon t} = \infty \quad ;$$

consequently, any polynomial of t has a characteristic number $\lambda_0 = 0$.

Next, for $\varphi(t) = t^t$ the function $f(t) = \varphi(t)e^{\lambda t}$ is unbounded, and for $\varphi(t) = t^{-t}$ it is vanishing for all λ; therefore, in the former case we have $\lambda_0 = -\infty$, and in the latter $\lambda_0 = +\infty$.

Lyapunov derived many important properties of characteristic numbers. We shall need here only some of these, on which we shall dwell below.

Let us return to (21) and consider the system of functions that define the following fundamental system of solutions of the equations

$$(30) \qquad \begin{matrix} y_{11}, & \dots, & y_{n1} \;, \\ \cdot\;\cdot\;\cdot & \cdot\;\cdot\;\cdot & \cdot\;\cdot\;\cdot \\ y_{1n}, & \dots, & y_{nn} \;. \end{matrix}$$

Here the first number is that of the function and the second number is that of the solution.

It is assumed that these solutions are linearly independent. Let us consider, say, the s'th solution of (30). Each function y_{ks} of this solution has a corresponding characteristic number λ_{ks}. The lowest among the characteristic numbers λ_{ks} is called the characteristic number λ_s of the s'th solution. Choosing from among the numbers λ_s the smallest one, we obtain the characteristic number of the total system of functions (30), comprising the solution of the system (21).

If we are able to find a method of calculating the characteristic numbers of the given (21), we can obtain a unique estimate of the stability or instability of the solutions of the first-approximation equations. All positive characteristic numbers represent stable solutions, while an unstable solution is obtained if only one of the characteristic numbers is negative.

6. STABILITY UNDER CONSTANTLY APPLIED DISTURBING FORCES*

We studied above the stability of motion of any system under the condition that the disturbances are instantaneous initial deviations of the system from the equilibrium position, determined by the solution (8).

Frequently, however, many systems are subjected to the influence of disturbing forces which cannot be evaluated in the majority of cases. We shall therefore assume that these forces are describable by random functions $R_s(t, y_1, \ldots, y_n)$, and that the equations of the disturbed motion of the system are of the form

(31) $$\dot{y}_s = Y_s(y_1, \ldots, y_n) + R_s(t, y_1, \ldots, y_n) \qquad (s = 1, \ldots, n) \quad .$$

The only assumption concerning the functions R_s will be that they have a sufficiently small modulus for all values of the variables y_k and t, defined in the region G. In all other respects the functions R_s remain arbitrary.

In the case of unbounded disturbing forces the statement of the stability problem has no physical meaning, since in such cases these forces always become known and therefore should be taken into account in the formulation of the functions Y_s.

Under these assumptions, we shall again be interested in the undisturbed motion (8) of (6), a motion which we shall say is stable under constantly applied disturbances, provided that for all specified positive numbers A, no matter how small, there are found two other positive numbers $\lambda(A)$ and $\varepsilon(A)$ such that for all disturbances y_{ko} satisfying the condition

(32) $$\sum_{k=1}^{n} y_{ko}^2 \leq \lambda \quad ,$$

and for all disturbing forces R_k satisfying the condition

---

* N. G. Chetaev, "On the Stability of Trajectories of Dynamics," Scientific Notes (Uchenye Zapiski) of the Kazan' University, Vol. 4, No. 1, 1931; N. G. Chetaev, Concerning Stability of Motion, Izv. AN SSSR, Division of Technical Sciences, No. 6, 1946; N. A. Artem'ev, Realizable Motions, Izv. AN SSSR, Mathematics Series, No. 3, 1939; G. N. Dubushin, Concerning the Problem of Stability of Motion with Respect to Constantly-Acting Disturbances, Trudy GAISh, Vol. XIV, No. 1, 1940; N. G. Malkin, Stability Under Constantly-Acting Disturbances, PMM, Vol. VIII, No. 3, 1944.

(33) $$\sum_{k=1}^{n} |R_k(t, y_1, \ldots, y_n)|^2 < \varepsilon \quad ,$$

in the region G, the disturbed motion y_k, determined by (31), will satisfy the inequality

(34) $$\sum_{k=1}^{n} y_k^2 < A \qquad (k = 1, \ldots, n)$$

for all $t > 0$. In the contrary case the undisturbed motion is called unstable.

According to this definition, in the case of stability of undisturbed motion under constantly applied disturbing forces we are assured only of the fact that any disturbed motion of the system will take place in the A-vicinity of the origin.

The system will not approach its undisturbed state as $t \longrightarrow \infty$.

Lyapunov's methods can also be used to investigate the stability of the amplitude motion under the action of disturbing forces. Furthermore, in the cited literature there are general theorems according to which the asymptotic stability of a system at $R_k = 0$ also guarantees in many cases its stability under constantly applied disturbing forces R_k that differ from 0, provided the latter are sufficiently small in modulus.

CHAPTER I: EQUATIONS OF CONTROL SYSTEMS. STATEMENT OF STABILITY PROBLEM

1. EQUATIONS OF REGULATED OBJECTS

We shall study henceforth regulated objects whose disturbed motion is described by linear differential equations of the form

$$(1.1) \qquad \dot{\eta}_k = \sum_{\alpha=1}^{m} b_{k\alpha}\eta_\alpha \qquad (k = 1, \ldots, m) ,$$

where η_k are generalized coordinates, and $b_{k\alpha}$ are constant parameters of the regulated object.

All these objects are classified on the basis of an analysis of the roots of the equation

$$(1.2) \qquad \Delta(\rho) = |b_{\alpha k} + \rho\delta_{\alpha k}| = 0 ,$$

where

$$(1.3) \qquad \delta_{\alpha k} = \begin{cases} 0 & \text{if} \quad k \neq \alpha , \\ 1 & \text{if} \quad k = \alpha . \end{cases}$$

We shall say that

(1) The control system is inherently stable if

$$\operatorname{Re} \rho_k > 0 \qquad (k = 1, \ldots, m) ,$$

(2) The control system is neutral with respect to the coordinates $\eta_1, \ldots, \eta_s$, (when the matrix $b_{k\alpha}$ is in canonical form), if

$$\operatorname{Re} \rho_1 = \ldots = \operatorname{Re} \rho_s = 0, \ \operatorname{Re} \rho_{s+\alpha} > 0 \qquad (\alpha = 1, \ldots, m - s) ,$$

(3) The control system is inherently unstable, if $\operatorname{Re} \rho_k < 0$ for at least one value of k.

In an examination of particular problems it is necessary to adhere strictly to this classification, since for each control system, according to the class to which it belongs, different analyses will be proposed in accordance with the class to which it belongs.

Let us assume that the regulated object (1.1) is subjected to the action of a regulating organ. Let μ be the coordinate of the regulating organ, and n_k the constant parameters that characterize the measure of the action of the regulating organ on the coordinate η_k. In this case, the equations of the disturbed motion of the control become

$$\dot{\eta}_k = \sum_{\alpha=1}^{m} b_{k\alpha}\eta_\alpha + n_k\mu \qquad (k = 1, \ldots, m) . \tag{1.4}$$

2. EQUATION OF THE ACTUATOR

Each actuator will be treated as a mechanical (electromechanical) system with one degree of freedom. We assume that the disturbed motion of such an actuator is described by a total differential equation of second order, of the form

$$V^2\ddot{\mu} + W\dot{\mu} + S\mu = f^*(\sigma) \quad , \tag{1.5}$$

where the quantities S, W, and V^2 are, generally speaking, known functions of the variables μ, $\dot{\mu}$, and σ.

It is customarily assumed that for a known class of actuators the quantities S, V^2 are constant; the first quantity characterizes the so-called load reaction of the regulating device, and the second the combined inertia of the regulator and of the actuator.

In the particular case when the actuator is an ordinary hydraulic motor, fed by an incompressible liquid, it is appropriate to assume $S \approx V^2 \approx 0$. In this case the function

$$\dot{\mu} = \frac{1}{W} f^*(\sigma) = f(\sigma) \tag{1.6}$$

describes the rate at which the regulator changes from one position to another, as a function of the argument σ; the function $f^*(\sigma)$ represents the acting generalized force* developed by the response of the actuator to the value of the argument σ.

---

* V. A. Kotel'nikov, Longitudinal Stability of an Airplane with a Type AVP-12 Autopilot. Trudy LII, No. 2, 1941; A. M. Letov, Concerning the Autopilot Problem, Vestnik MGU, No. 1, 1946.

In the general case the argument σ is given by the following expression

$$\sigma = \sum_{\alpha=1}^{m} p_\alpha \eta_\alpha - r\mu \quad , \tag{1.7}$$

where p_α and r are the constants of the regulator.

We shall consider functions $f(\sigma)$, satisfying the conditions

$$f(\sigma) = 0 \text{ for } |\sigma| \leq \sigma_*, \; \sigma f(\sigma) > 0 \text{ for } |\sigma| > \sigma_* \quad , \tag{1.8}$$

where σ_* is a certain fixed, non-negative number, characterizing the zone in which the regulator is insensitive to the variations in σ. $f(\sigma)$ is continuous for all values $|\sigma| > \sigma_*$; at the point $\sigma = \pm\, \sigma_*$ it is permissible for the continuity of $f(\sigma)$ to be violated. We shall consider such functions to belong to class (A).

In the investigation of inherently unstable regulated systems we shall sometimes talk of such functions $f(\sigma)$, which in addition to the above, satisfy the conditions

$$\sigma_* = 0; \quad \left[\frac{df(\sigma)}{d\sigma}\right]_{\sigma=0} \geq h > 0 \quad ;$$

(1.9)

$$\sigma\varphi(\sigma) > 0 \text{ for } \sigma \neq 0, \text{ where } \varphi(\sigma) = f(\sigma) - h\sigma \quad ,$$

and h is a specified constant.

We shall say of such functions that they form a subclass (A') of the functions $f(\sigma)$ in class (A). Introducing the subclass (A') has as its purpose to distinguish actuators that have sufficient speed of response to the incoming pulse signals σ. Thus, if h is a fixed number, then the third condition of (1.9) signifies that the modulus of the velocity $f(\sigma)$ of the resetting of the regulating device is greater than the value of $h|\sigma|$. If the actuator is characterized by a function of subclass (A'), then the regulator has no insensitivity zone.

For functions of subclass (A') we shall consider also the straight line $H\sigma$, which bounds the curve $f(\sigma)$ from above in such a way that $|H\sigma| \geq |f(\sigma)|$. Consequently, all functions of subclass (A') are represented by curves, which are all located between the lines $y = h\sigma$ and $y = H\sigma$, with $H > h$ (Figure 2).

If, however, the function $f(\sigma)$ nevertheless intersects somehow the line $h\sigma$ at a point with abscissa $\bar{\sigma}$, we shall speak in this case of a regulation range of the system relative to σ, the range being equal to $2|\bar{\sigma}|$.

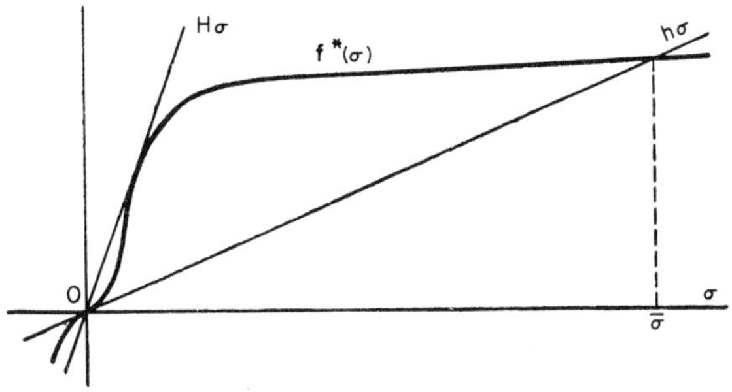

FIGURE 2

Also included among the functions of subclass (A') is the step function $f(\sigma)$, which belongs to the class (A) of the functions:

$$f(\sigma) = \begin{cases} +Q & \text{if} \quad \sigma > 0 , \\ 0 & \text{if} \quad \sigma = 0 , \\ -Q & \text{if} \quad \sigma < 0 . \end{cases} \tag{1.10}$$

For step functions of subclass (A'), we assume $H = \tan(\pi/2)$. In the particular and limiting cases, it may turn out that for certain step functions of the subclass (A') we have $Q = \infty$ and the range $2|\sigma|$ is equal to zero. Then we shall say that an actuator characterized by such a function is ideal, i.e., it can reset the regulating device from one position to another with infinite speed.

The equation of such an ideal actuator is written in the following manner:

$$\sum_{\alpha=1}^{m} p_\alpha \eta_\alpha - r\mu = 0 . \tag{1.11}$$

The class of functions $f(\sigma)$ considered above encompasses the characteristics of a great majority of actuators used in modern technology.

The advisability of singling this class out is dictated by the following considerations:

The determination of the function $f(\sigma)$ in each particular case is carried out experimentally, by recording the velocity of the actuator. Each such experiment is carried out for a definite fixed load. Therefore the function $f(\sigma)$ depends on this load, and the experiment yields, generally speaking, an entire family of such functions. However, under the real operating conditions of the actuator, there is a continuous change of the applied load, which leads to a noticeable distortion of the function $f(\sigma)$, as determined from the static experiment. This distortion cannot be fully calculated, but it does not affect the fact that the function $f(\sigma)$ belongs to the class (A).

In addition, noticeable distortions of the function $f(\sigma)$ may be caused also by other factors, which can be predicted beforehand and eliminated. One of the widespread causes of this kind is change in the level of energy supplied to the regulator by the outside source.

Therefore, in each particular problem it is impossible to fix rigorously the function $f(\sigma)$, and, moreover, it is impossible to carry out its correct linearization by rigorously determining the coefficient of linear approximation. The above argument shows that in such problems it is possible always only to establish the fact that this function belongs either to class (A) or to subclass (A'). We shall see later that this makes it possible, with the aid of the direct Lyapunov method, to obtain the sufficient stability conditions for the control system under consideration.

Naturally, (1.5) is a rather primitive equation for actuators. However, in many cases of practical importance it reflects correctly their fundamental physical features, and therefore it can be used in theoretical investigations. In particular, the use of the even more primitive (1.6), in which $f(\sigma)$ is either a linear or a step function of the argument σ, is the basis of the classical theory of automatic control. (1.5) is a sufficiently broad generalization of the classical equation (1.6), which takes into account many important mechanical properties of modern actuators.

3. COLLECTIVE EQUATIONS OF CONTROL SYSTEMS

Thus, the collective differential equations of the disturbed motion of a control system, containing one regulation organ, are of the form

$$\dot{\eta}_k = \sum_{\alpha=1}^{m} b_{k\alpha}\eta_\alpha + \eta_k\mu \qquad (k = 1, \ldots, m) ,$$

$$V^2\ddot{\mu} + W\dot{\mu} + S\mu = f^*(\sigma) , \tag{1.12}$$

$$\sigma = \sum_{\alpha=1}^{m} p_\alpha\eta_\alpha - r\mu .$$

These equations are defined for all values of variables η_α and μ for which they retain their physical meaning of equations of disturbed motion, and for all values of the variable σ, lying inside the bounded or unbounded interval $2|\bar{\sigma}|$.

4. POSSIBLE STEADY STATE MODES OF CONTROL SYSTEMS

All possible steady state modes of control systems of a given class (1.12), which must be maintained by the regulator, are described by solutions of the systems of algebraic equations

$$\begin{gathered} \sum_{\alpha=1}^{m} b_{k\alpha}\eta_\alpha + n_k\mu = 0 \qquad (k = 1, \ldots, m) , \\ S\mu = f^*(\sigma) , \\ \sigma = \sum_{\alpha=1}^{m} p_\alpha\eta_\alpha - r\mu . \end{gathered} \tag{1.13}$$

In the phase space of the variables η_α and μ, (1.13) determine the so-called singular points M^*, at which our system, describable by (1.12), can generally speaking be in equilibrium. For a control system of the class under consideration, each singular point M^* corresponds to a theoretically feasible steady state process.

Let us consider the auxiliary system of equations

$$\begin{gathered} \sum_{\alpha=1}^{m} b_{k\alpha}\eta_\alpha + n_k\mu = 0 \qquad (k = 1, \ldots, m) , \\ \sum_{\alpha=1}^{m} p_\alpha\eta_\alpha - r\mu = \sigma , \end{gathered} \tag{1.14}$$

and let us assume that the determinant

$$\begin{vmatrix} b_{11} & \cdots & b_{1m} & n_1 \\ \cdot & \cdot & \cdot & \cdot \\ b_{m1} & \cdots & b_{mm} & n_m \\ p_1 & \cdots & p_m & -r \end{vmatrix} \tag{1.15}$$

does not vanish. In this case the system of (1.14) has a non-trivial solution of the form

$$(1.16) \qquad \eta_k = A_k\sigma, \ \mu = B\sigma \qquad (k = 1, \ldots, m) ,$$

where A_k and B are certain numbers, and

$$(1.17) \qquad B = \frac{\begin{vmatrix} b_{11} & \cdots & b_{1m} \\ \cdot & \cdot & \cdot \\ b_{m1} & \cdots & b_{mm} \end{vmatrix}}{\begin{vmatrix} b_{11} & \cdots & b_{1m} & n_1 \\ \cdot & \cdot & \cdot & \cdot \\ b_{m1} & \cdots & b_{mm} & n_m \\ p_1 & \cdots & p_m & -r \end{vmatrix}}$$

In the study of possible solutions of (1.13), it is necessary to distinguish between the following two cases:

Let us assume that the function $f^*(\sigma)$ belongs to the subclass (A'). If the load on the regulating organ is negligibly small and it is possible to set $S = 0$, then the second equation of (1.13) yields $f^*(\sigma) = 0$, i.e., $\sigma^* = 0$, and according to (1.16) we find one singular point M^*

$$(1.18) \qquad \eta_k^* = 0 \quad (k = 1, \ldots, m), \qquad \mu^* = 0 .$$

If, however, the load reaction of the regulating organ must be taken into account, so that the quantity S must be different from 0, then (1.13) can have, generally speaking, several discrete solutions. Actually, according to (1.13) and (1.16) we find

$$(1.19) \qquad SB\sigma = f^*(\sigma) .$$

(1.19) can have various cases of solutions depending on the sign of the quantity SB and on the form of the curve $f^*(\sigma)$.

In fact, let us assume that $SB \leq 0$; in this case (1.19) has the unique solution $\sigma^* = 0$, and the solution of (1.13) is determined by (1.18). If, however, $SB > 0$, then, depending on the shape of the curve $f^*(\sigma)$, (1.19) can have a set of discrete solutions σ_s^* $(s = 1, 2, \ldots)$, whose character is determined by the abscissas of the intersections of the curve $f^*(\sigma)$ with the line $y = SB\sigma$. In accordance with these solutions, we can, using (1.16), find the set of discrete solutions of (1.13):

$$\begin{aligned} \eta^*_{ks} &= A_k\sigma^*_s , \\ \mu^*_s &= B\sigma^*_s \end{aligned} \qquad \begin{pmatrix} k = 1, \ldots, m, \\ s = 1, 2, \ldots \end{pmatrix} . \tag{1.20}$$

Obviously, in an efficiently constructed control system, we should have $B \leq 0$ with $S > 0$.

Let us consider now the second possible solution of (1.13), in which $f^*(\sigma)$ is any function of class (A). It is easy to check, that if $\sigma_* = 0$, in this case, we obtain the same system of solutions (1.18) or (1.20). But if $\sigma_* \neq 0$, then (1.13) can have substantially new and different solutions, which appear in the case when $S = 0$, or when $B = 0$.

Actually, according to the definition of the function $f^*(\sigma)$ of class (A), in this case the (1.19) has a continuum of solutions σ^*, lying in the interval $|\sigma^*| \leq \sigma_*$. Thanks to this, (1.13) will also have a continuum of solutions η^*_k, μ^*, which fill the region

$$|\eta^*_k| \leq \eta_{k*}, \ |\mu^*| \leq \mu_* \qquad (k = 1, \ldots, m) , \tag{1.21}$$

the bounadries of which are determined in accordance with (1.16) as follows:

$$\eta_{k*} = |A_k|\sigma_*, \ \mu_* = |B|\sigma_* \qquad (k = 1, \ldots, m) . \tag{1.22}$$

The continuum of these solutions represents the insensitivity region of the control system, since a change in its coordinates inside this region does not cause any response on the part of the regulator.

If, however, $SB \neq 0$, we obtain the previous system of discrete solutions (1.20) or else the solution (1.18).

Thus, for a control system with one regulating organ, we obtain in any case:

1) either a single solution of the form (1.18),
2) a system of discrete solutions of the form (1.20),
3) a continuum of solutions of the form (1.21).

Hereafter we shall study either the stability of the obvious solution (1.18), the stability of one of the obvious solutions (1.20) of equations of control systems, or the stability of any previously selected solution, belonging to the continuum (1.21). This solution will be denoted

$$\eta_k = \eta^*_k, \ \mu = \mu^* \qquad (k = 1, \ldots, m) . \tag{1.23}$$

It corresponds to a quite definite steady state process in the control system, or, according to Lyapunov, to a fully determined undisturbed motion of the system.

5. NORMAL FORM OF COLLECTIVE EQUATIONS OF CONTROL SYSTEMS

In the future it will be convenient to write the collective equations (1.12) in the normal form.

We introduce a new variable ξ, defined as

(1.24) $$\xi = p\dot{\mu} + q\mu \quad ,$$

in which the constants p and q are to be defined later. Differentiating ξ, we get

(1.25) $$\dot{\xi} = p\ddot{\mu} + q\dot{\mu} \ .$$

Relations (1.24) and (1.25) permit expressing $\dot{\mu}$, $\ddot{\mu}$, in terms of μ, ξ and $\dot{\xi}$; by substituting these into the second equation (1.12) of the actuator we can write the latter as

(1.26) $$\frac{V^2}{p}\dot{\xi} + \frac{1}{p}(W - \frac{q}{p}V^2)\xi + [S - W\frac{q}{p} + V^2(\frac{q}{p})^2]\mu = f^*(\sigma) \ .$$

If the ratio q/p is chosen to be the root of the equation

(1.27) $$V^2(\frac{q}{p})^2 - W(\frac{q}{p}) + S = 0 \quad ,$$

then, by putting

(1.28) $$(\frac{q}{p})_1 = \rho_{m+1}, \quad (\frac{q}{p})_2 = \frac{S}{V^2}\frac{1}{(\frac{q}{p})_1} = \frac{1}{V^2}[W - (\frac{q}{p})_1 V^2] = \rho_{m+2} \quad ,$$

we get instead of one second-order equation two first-order equations of the form

$$\dot{\mu} = -\rho_{m+1}\mu + \frac{1}{p}\xi \quad ,$$

$$\dot{\xi} = -\rho_{m+2}\xi + \frac{p}{V^2}f^*(\sigma) \ .$$

Since if S. W, and V^2 are positive numbers, (1.27) admits of roots with the property

(1.29) $$\operatorname{Re} \rho_{m+1} > 0, \quad \operatorname{Re} \rho_{m+2} > 0 \quad .$$

Finally, putting

$$\mu = \eta_{m+1}, \ n_k = b_{k,m+1}, \ b_{m+1,k} = 0 \quad (k = 1, \ldots, m) \ ,$$

(1.30) $$b_{m+1,m+1} = - \rho_{m+1}, \ h_k = 0, \quad (k = 1, \ldots, m) \ ,$$

$$h_{m+1} = \frac{1}{p} \ , \ \frac{p}{V^2} f^*(0) = f(\sigma), \ m + 1 = n, \ p_{m+1} = - r \ ,$$

we reduce the initial equations for the general case to the normal form

$$\dot{\eta}_k = \sum_{\alpha=1}^{n} b_{k\alpha}\eta_\alpha + n_k\xi \qquad (k = 1, \ldots, n) \ ,$$

(1.31) $$\dot{\xi} = - \rho_{n+1}\xi + f(\sigma) \ ,$$

$$\sigma = \sum_{\alpha=1}^{n} p_\alpha \eta_\alpha \ .$$

We now consider certain important particular cases.

CASE 1. Let us assume $S = 0$. Then (1.27) yields

$$\rho_{m+1} = \frac{W}{V^2} \ ,$$

$$\rho_{m+2} = 0 \ ,$$

and (1.31) become

$$\dot{\eta}_k = \sum_{\alpha=1}^{n} b_{k\alpha}\eta_\alpha + n_k\xi \qquad (k = 1, \ldots, n) \ ,$$

(1.32) $$\dot{\xi} = f(\sigma) \ ,$$

$$\sigma = \sum_{\alpha=1}^{n} p_\alpha \eta_\alpha \ .$$

CASE 2. Let us assume $V^2 = 0$. Then (1.27) yields $\rho_{m+1} = S/W$ and $\rho_{m+2} = \infty$; putting in this case

$$\frac{1}{\rho_{m+2}} \cdot \frac{1}{p} \cdot \frac{p}{V^2} f^*(\sigma) = \frac{1}{W} f^*(\sigma) = f(\sigma) \quad ,$$

we find

$$\dot{\eta}_k = \sum_{\alpha=1}^{m} b_{k\alpha}\eta_\alpha + n_k\mu \qquad (k = 1, \ldots, m) \; ,$$

$$(1.33) \qquad \dot{\mu} = -\rho_{m+1}\mu + f(\sigma) \quad ,$$

$$\sigma = \sum_{\alpha=1}^{m} p_\alpha \eta_\alpha + p_{m+1}\mu \quad .$$

CASE 3. Let us assume $S = V^2 = 0$. Then (1.27) yields $\rho_{m+1} = 0$, $\rho_{m+2} = \infty$, and (1.31) become

$$\dot{\eta}_k = \sum_{\alpha=1}^{m} b_{k\alpha}\eta_\alpha + n_k\mu \qquad (k = 1, \ldots, m) \; ,$$

$$(1.34) \qquad \dot{\mu} = f(\sigma), \qquad \sigma = \sum_{\alpha=1}^{m} p_\alpha \eta_\alpha + p_{m+1}\mu \quad .$$

(1.34) were first studied by A. I. Lur'e. Later he introduced a more general form of normal equations for control systems, namely

$$\dot{\eta}_k = \sum_{\alpha=1}^{n} b_{k\alpha}\eta_\alpha + n_k f(\sigma) \quad ,$$

$$(1.35) \qquad \sigma = \sum_{\alpha=1}^{n} p_\alpha \eta_\alpha \quad ,$$

where n_k are specified constants. This manner of writing the equations offers certain advantages for general analysis and corresponds to physically realized control systems, which differ substantially from those considered above.

6. FIRST FUNDAMENTAL PROBLEM IN THE THEORY OF AUTOMATIC CONTROL

The first and fundamental problem in the theory of automatic control consists of determining all the values of the parameters of the regulator which are required to guarantee stability of the obvious solution (1.23). In the space of the regulator parameters, these values comprise a certain region, which for brevity we shall call region B (the Vyshnegradskiĭ regions).

Hereafter, wherever no special conditions are stated, we speak of stability under all disturbances, at which the equations of motions are valid. With this we shall require stability for all functions $f(\sigma)$ of class (A) or subclass (A'). For sake of brevity we shall call such a stability an absolute stability of the control system.

A. I. Lur'e and V. N. Postnikov were the first who posed and solved the problem of absolute stability of one particular system, which is neutral with respect to one of the coordinates. Following this, Lur'e solved this problem for one class of control systems, widely employed in engineering, which are stable when the regulator is disconnected, and also for systems that are neutral with respect to one coordinate. Later attempts to obtain an analogous solution for a wider class of control systems have disclosed the advisability of solving the problem of absolute stability for all functions $f(\sigma)$, belonging to subclass (A'). The advantages of this approach are that the observed actual variation of the function $f(\sigma)$ occurs for the most part within the angle formed by the lines $y = h$ and $y = H$, where $h > 0$ and $H > h$ are numbers that are fixed in each particular case. In particular, it becomes necessary to restrict the functions $f(\sigma)$ to the subclass (A') [and not to the class (A)] in all cases of control systems that are inherently unstable.

As already mentioned, in the case of absolute stability of the control system, there is guaranteed the physical realizability of the same desirable steady state of the regulated object, a state described by (1.23). By virtue of the absolute stability of this solution, the regulation process converges everywhere, regardless of disturbances to the control system.

The direct method chosen by us to solve this problem makes it possible to obtain, generally speaking, sufficient conditions for stability, given a stability region B', which is within the region B, i.e., $B' \in B$.

In each particular problem it is possible to construct an

infinite set of sufficient conditions and regions B' for absolute stability, whose form is connected with the form of the Lyapunov function that solves the problem.

There may arise the important practical problem of the degree to which the sufficient conditions, obtained by the direct method, differ from the necessary conditions. When this difference is considerable, the solution obtained can be nonconstructive, i.e., it will be very difficult or impossible to realize in practice.

At the present time, it is impossible to give an exhaustive answer to this problem. It is also impossible to guarantee unconditional success of the applied method in the case of any problem in theory of control. However, an analysis of particular cases, and certain general considerations* will show that there are no particular grounds for fearing failure when this method is used.

---

* See footnotes on pp. i and 1.

CHAPTER II: FIRST CANONICAL FORM OF EQUATIONS OF CONTROL SYSTEMS

1. THE LUR'E TRANSFORMATION

To solve the first fundamental problem in the theory of control it is not essential to transform the initial equations into the canonical form. This is due solely to the difficulties of constructing the Lyapunov function for the initial equations, and if the canonical form is used, these difficulties are considerably diminished.

We can propose several forms of canonical equations for control systems. The first to be considered will be the Lur'e form, which is convenient for the investigation of inherently-stable control systems, as well as systems that are neutral with respect to one coordinate.

The canonical variables will be determined with the aid of the equation

$$x_s = \sum_{\alpha=1}^{n} c_\alpha^{(s)} \eta_\alpha + \xi \quad . \tag{2.1}$$

Differentiating relations (2.1) and using (1.31) to eliminate the derivatives, we get:

$$\dot{x}_s = \sum_{\alpha=1}^{n} c_\alpha^{(s)} \left[\sum_{\beta=1}^{n} b_{\alpha\beta}\eta_\beta + n_\alpha \xi \right] + [- \rho_{n+1}\xi + f(\sigma)] \quad .$$

If it is required that the equations in the new variables be in canonical form

$$\dot{x}_s = - \rho_s x_s + f(\sigma) \qquad (s = 1, \ldots, n) \quad , \tag{2.2}$$

then the choice of the transformation constants must be subjected to the relations

$$- \rho_s c_\beta^{(s)} = \sum_{\alpha=1}^{n} b_{\alpha\beta} c_\alpha^{(s)} \qquad (s, \beta = 1, \ldots, n) \quad , \tag{2.3}$$

$$\rho_{n+1} - \rho_s = \sum_{\alpha=1}^{n} n_\alpha C_\alpha^{(s)} \qquad (s = 1, \ldots, n) \quad . \tag{2.4}$$

Here the quantities ρ_s are the transformation parameters, which must be chosen to be the roots of the equation

$$D(\rho) = \begin{vmatrix} b_{11} + \rho & b_{21} & \cdots & b_{n1} \\ \cdot & \cdot & \cdot & \cdot \\ b_{1n} & b_{2n} & \cdots & b_{nn} + \rho \end{vmatrix} = 0 \quad . \tag{2.5}$$

Actually, turning to relations (2.3), we see that only for such values of ρ_s is it possible to construct the transformation (2.1), for in the opposite case the equations yield $C_\alpha^{(s)} = 0$ $(\alpha, s = 1, \ldots, n)$. Let us assume that all roots ρ_s of this equation are simple and have the property

$$\operatorname{Re} \rho_k \geq 0 \qquad (k = 1, \ldots, n) \quad . \tag{2.6}$$

It is assumed that the equality sign, if it applies, pertains to ρ_1. The latter limitation involves the necessity of satisfying the inequalities

$$\Delta_1 > 0, \ldots, \Delta_n \geq 0 \quad , \tag{2.7}$$

where Δ_k $(k = 1, \ldots, n)$ are the Hurwitz determinants for the equation obtained from (2.5) by replacing ρ by $-\rho$. Obviously, the equal sign $(\Delta_n = 0)$ occurs in the sequence (2.7) if $\rho_1 = 0$.

For the first canonical form of the equations of control systems, the characteristic feature is that the coefficients of (2.5) do not contain the parameters of the regulator, while inequalities (2.7) are always satisfied if the control system is inherently stable or neutral with respect to one coordinate. In fact, according to (1.30) we have

$$D(\rho) = (\rho - \rho_n)\,\text{Д}(\rho) \quad . \tag{2.8}$$

from which this statement follows.

If all the roots of (2.5) are simple, then, as is proven in algebra, (2.3) can always be solved, and in this case the transformation (2.1) exists and is not singular. This latter circumstance makes it possible to solve relations (2.1) with respect to η_α. Let us assume that this operation has been performed and that we have found

$$\eta_\alpha = \sum_{k=1}^{n} D_k^{(\alpha)} x_k + G_\alpha \xi \qquad (\alpha = 1, \ldots, n) \quad , \tag{2.9}$$

where $D_k^{(\alpha)}$ and G_α are known constants.

To complete the reduction of the initial equations into the canonical form, it remains for us to express the quantities σ and $\dot{\sigma}$ in terms of new variables. According to (2.9) and (1.31) we get

$$\sigma = \sum_{k=1}^{n} \left[\sum_{\alpha=1}^{n} p_\alpha D_k^{(\alpha)} \right] x_k + \sum_{\alpha=1}^{n} p_\alpha G_\alpha \ \xi \quad .$$

If we now introduce the symbols

$$\gamma_k = \sum_{\alpha=1}^{n} p_\alpha D_k^{(\alpha)} \quad (k = 1, \ldots, n), \qquad \gamma_{n+1} = \sum_{\alpha=1}^{n} G_\alpha p_\alpha \ ,$$

(2.10)

$$- \gamma_k \rho_k = \beta_k, \quad \sum_{k=1}^{n+1} \gamma_k = - r', \quad \xi = x_{n+1} \ ,$$

then the collective canonical equations assume the final form

$$\dot{x}_k = - \rho_k x_k + f(\sigma) \qquad (k = 1, \ldots, n + 1) \ ,$$

(2.11) $$\sigma = \sum_{k=1}^{n+1} \gamma_k x_k \ ,$$

$$\dot{\sigma} = \sum_{k=1}^{n+1} \beta_k x_k - r' f(\sigma) \quad .$$

The last equation in (2.11) is not necessary to make this system complete, but we have included it in view of forthcoming applications.

Let us consider certain known particular cases.

CASE 1. We have $\rho_{n+1} = \beta_{n+1} = 0$, consequently

$$\dot{x}_k = - \rho_k x_k + f(\sigma) \qquad (k = 1, \ldots, n) \ ,$$

(2.12) $$\dot{x}_{n+1} = f(\sigma) \ ,$$

$$\sigma = \sum_{k=1}^{n+1} \gamma_k x_k \ , \quad \dot{\sigma} = \sum_{k=1}^{n} \beta_k x_k - r' f(\sigma) \quad .$$

Thus, the canonical equations can in this case be written in two forms:

(a)

$$\dot{x}_k = -\rho_k x_k + f(\sigma) \qquad (k = 1, \ldots, n) ,$$

$$(2.12a) \qquad \dot{\sigma} = \sum_{k=1}^{n} \beta_k x_k - r' f(\sigma) ,$$

or

(b)

$$\dot{x}_k = -\rho_k x_k + f(\sigma) \qquad (k = 1, \ldots, n) ,$$

$$(2.12b) \qquad \dot{x}_{n+1} = f(\sigma) ,$$

$$\sigma = \sum_{k=1}^{n+1} \gamma_k x_k .$$

CASE 2. We have $\rho_{n+1} = \infty$; in accordance with (1.33), the formulas for the new variables must be written in the following manner:

$$(2.13) \qquad x_s = \sum_{\alpha=1}^{m} C_\alpha^{(s)} \eta_\alpha + \mu \qquad (s = 1, \ldots, m) ,$$

where the coefficients $C_\alpha^{(s)}$ are determined from the relations

$$(2.14) \qquad -\rho_s C_\beta^{(s)} = \sum_{\alpha=1}^{m} C_\alpha^{(s)} b_{\alpha\beta} \qquad (\beta, s = 1, \ldots, m) ,$$

$$(2.15) \qquad \dot{\rho}_{m+1} - \rho_s = \sum_{\alpha=1}^{m} n_\alpha C_\alpha^{(s)} .$$

It is obvious that the transformation parameters ρ_s are the roots of (1.2); in that case, when these roots are simple, (2.14) can be solved with respect to $C_\alpha^{(s)}$ and the transformation (2.13) is not singular. The inverse transformation has the form

$$(2.16) \qquad \eta_\alpha = \sum_{k=1}^{m} D_k^{(\alpha)} x_k + G_\alpha \mu .$$

If we now introduce the notation

$$\gamma_k = \sum_{\alpha=1}^{m} p_\alpha D_k^{(\alpha)} \qquad (k = 1, \ldots, m),$$

(2.17)
$$\sum_{\alpha=1}^{m} p_\alpha G_\alpha + p_{m+1} = \gamma_{m+1}, \qquad \beta_k = -\rho_k \gamma_k,$$

$$\sum_{k=1}^{m+1} \gamma_k = -r', \qquad \mu = x_{m+1},$$

then the initial equations are reduced to the form

$$\dot{x}_k = -\rho_k x_k + f(\sigma) \qquad (k = 1, \ldots, m+1),$$

(2.18)
$$\sigma = \sum_{k=1}^{m+1} \gamma_k x_k, \qquad \dot{\sigma} = \sum_{k=1}^{m+1} \beta_k x_k - r' f(\sigma).$$

CASE 3. We have $\rho_{m+1} = 0$; $\rho_{m+2} = \infty$, and the canonical equations of the problem can be written, as desired:

(a) either in the form

$$\dot{x}_k = -\rho_k x_k + f(\sigma) \qquad (k = 1, \ldots, m),$$

(2.19a)
$$\dot{\sigma} = \sum_{k=1}^{m} \beta_k x_k - r' f(\sigma),$$

(b) or in the form

$$\dot{x}_k = -\rho_k x_k + f(\sigma) \qquad (k = 1, \ldots, m),$$

(2.19b)
$$\dot{x}_{m+1} = f(\sigma),$$

$$\sigma = \sum_{k=1}^{m+1} \gamma_k x_k.$$

Here the quantities γ_k, β_k, and r' are determined from (2.17), in which we must put $\rho_{m+1} = 0$.

2. FORMULAS FOR THE COEFFICIENTS OF THE DIRECT AND INVERSE LUR'E TRANSFORMATION

In spite of the fact that linear transformations have been treated in detail in the mathematical literature, the formulas considered were, apparently, first derived by A. I. Lur'e for the case in which (2.5) has simple roots. It is appropriate to repeat their derivation here,* since they comprise part of the algorithm sought for the solution of the posed problem, and make it possible to express the final results exclusively in terms of the initial data.

Turning to (2.3), in the case of simple roots ρ_s we can always determine the unknown $c_k^{(s)}$ as numbers which are proportional to the cofactors $D_{ik}(\rho_s)$ of the elements of a certain i'th row of the determinant $D(\rho)$ (2.5)

$$(2.20) \qquad c_k^{(s)} = A_i^{(s)} D_{ik}(\rho_s) \qquad (k, s = 1, \ldots, n) .$$

To determine the coefficient of proportionality $A_i^{(s)}$ we obtain, according to relations (2.4), the equation

$$(2.21) \qquad A_i^{(s)} H_i(\rho_s) = \rho_{n+1} - \rho_s \qquad (s = 1, \ldots, n) ,$$

where the factor

$$(2.22) \qquad H_i(\rho_s) = \sum_{k=1}^{n} n_k D_{ik}(\rho_s) \qquad (s = 1, \ldots, n)$$

represents the determinant (2.5), in which the elements of the i'th row are replaced by the numbers n_k $(k = 1, \ldots, n)$. Consequently, we obtain from formulas (2.20) and (2.21)

$$(2.23) \qquad c_k^{(s)} = \frac{\rho_{n+1} - \rho_s}{H_i(\rho_s)} D_{ik}(\rho_s) \qquad (k, s = 1, \ldots, n)$$

Since it is possible, when setting up the formulas for $c_k^{(s)}$, to choose any i'th row of the determinant (2.5) (such that $H_i(\rho_s) \neq 0$), it is possible, generally speaking, to construct n groups of formulas (2.23), which yield the desired canonical transformation.

---

* A. I. Lur'e, On the Canonical Form of the Equations of the Theory of Automatic Control, PMM, Vol. XII, No. 5, 1948.

To find the inverse transformation, we again turn to (2.1), now written as

$$\sum_{k=1}^{n} D_{ik}(\rho_s)\eta_k = \frac{H_i(\rho_s)}{\rho_{n+1} - \rho_s}(x_s - \xi) \qquad (s = 1, \ldots, n) \; . \tag{2.24}$$

Let us consider the determinant

$$\Delta(\rho_1, \ldots, \rho_n) = \begin{vmatrix} D_{i1}(\rho_1) & \ldots & D_{ii}(\rho_1) & \ldots & D_{in}(\rho_1) \\ \cdot & \cdot & \cdot & \cdot & \cdot \\ D_{i1}(\rho_n) & \ldots & D_{ii}(\rho_n) & \ldots & D_{in}(\rho_n) \end{vmatrix} \tag{2.25}$$

and assume that it differs from zero. Let us also consider the system of determinants

$$\Delta_k = \begin{vmatrix} D_{i1}(\rho_1) & \ldots & D_{ik-1}(\rho_1) & \dfrac{H_i(\rho_1)}{\rho_{n+1} - \rho_1}(x_1 - \xi) & \ldots & D_{in}(\rho_1) \\ \cdot & \cdot & \cdot & \cdot & \cdot & \cdot \\ D_{i1}(\rho_n) & \ldots & D_{ik-1}(\rho_n) & \dfrac{H_i(\rho_n)}{\rho_{n+1} - \rho_n}(x_n - \xi) & \ldots & D_{in}(\rho_n) \end{vmatrix} \tag{2.26}$$

which are obtained from the determinant (2.25) by replacing in it the elements of the k'th column by the right sides of (2.24). From Carmer's rule we have

$$\eta_k = \frac{\Delta_k}{\Delta} \qquad (k = 1, \ldots, n) \; . \tag{2.27}$$

Let Δ_{jk} represent the cofactors of the elements of the j'th row and the k'th column of the determinant (2.26). We have:

$$\Delta_k = \sum_{j=1}^{n} \frac{H_i(\rho_j}{\rho_{n+1} - \rho_j}(x_j - \xi)\Delta_{jk} \qquad (k = 1, \ldots, n) \; . \tag{2.28}$$

Let us now prove the validity of the following formulas:

$$\Delta(\rho_1, \ldots, \rho_n) = D'(\rho_j)\Delta_{jk} \; . \tag{2.29}$$

Actually, since the values we have obtained for η_k are solutions of (2.24), the substitution of the above equations will turn them into identities. These identities are of the form

$$\sum_{k=1}^{n} D_{ik}(\rho_s) \left[\frac{1}{\Delta} \sum_{j=1}^{n} \frac{H_i(\rho_j)}{\rho_{n+1} - \rho_j} (x_j - \xi)\Delta_{jk} \right] =$$

$$= \frac{H_i(\rho_s)}{\rho_{n+1} - \rho_s} (x_s - \xi), \qquad (s = 1, \ldots, n) .$$

Comparison of the coefficients of like coordinates of the identities gives the following n^2 relations:

(2.30)
$$\begin{aligned}
&\sum_{k=1}^{n} D_{ik}(\rho_1)\Delta_{1k} = \Delta, \ \ldots, \ \sum_{k=1}^{n} D_{ik}(\rho_1)\Delta_{nk} = 0 , \\
&\sum_{k=1}^{n} D_{ik}(\rho_2)\Delta_{1k} = 0, \ \ldots, \ \sum_{k=1}^{n} D_{ik}(\rho_2)\Delta_{nk} = 0 , \\
&\ldots\ldots\ldots\ldots\ldots\ldots\ldots\ldots \\
&\sum_{k=1}^{n} D_{ik}(\rho_n)\Delta_{1k} = 0, \ \ldots, \ \sum_{k=1}^{n} D_{ik}(\rho_n)\Delta_{nk} = \Delta .
\end{aligned}$$

Let us consider, for example, the first column of these relations. Since, in the case of simple ρ_s we have

$$D'(\rho_s) = \left[\frac{dD}{d\rho}\right]_{\rho=\rho_2} \neq 0 \qquad (s = 1, \ldots, n) ,$$

it is possible to divide the j'th relation of each column by a corresponding quantity $D'(\rho_j)$ in such a way, that after they are added term by term we have

(2.31)
$$\sum_{k=1}^{n} \left[\frac{D_{ik}(\rho_1)}{D'(\rho_1)} + \frac{D_{ik}(\rho_2)}{D'(\rho_2)} + \ldots + \frac{D_{ik}(\rho_n)}{D'(\rho_n)} \right] \Delta_{1k} = \frac{\Delta}{D'(\rho_1)} .$$

If we use the formula*

* See A. I. Lur'e, Certain Nonlinear Problems in the Theory of Automatic Control, Gostekhizdat, 1951.

$$\sum_{j=1}^{n} \frac{D_{ik}(\rho_j)}{D'(\rho_j)} = \delta_{ik} \;, \tag{2.32}$$

then we obtain from (2.31) for any $i = k$

$$\Delta = D'(\rho_1)\Delta_{1k} \;.$$

Analogously, considering each j'th column in (2.30), we obtain formulas (2.29), which we shall use to rewrite the transformation (2.27) in the final form

$$\eta_k = \sum_{j=1}^{n} \frac{H_k(\rho_j)(x_j - \xi)}{(\rho_{n+1} - \rho_j)D'(\rho_j)} \qquad (k = 1, \ldots, n) \;. \tag{2.33}$$

Comparison of formulas (2.9) and (2.33) makes it possible to write down expressions for the coefficients of the inverse transformation. These are

$$D_j^{(k)} = \frac{H_k(\rho_j)}{(\rho_{n+1} - \rho_j)D'(\rho_j)} \;, \quad G_k = -\sum_{j=1}^{n} D_j^{(k)} \qquad (j,\, k = 1, \ldots, n) \;. \tag{2.34}$$

It is now easy to extend formulas (2.34) to include the particular cases considered above. Thus, for example, in order to obtain the transformation coefficients we need in Case 1 it is enough to put $\rho_{n+1} = 0$ in formulas (2.23) and (2.34); in Case 2 these formulas must be written in the following manner:

$$C_k^{(s)} = \frac{\rho_{m+1} - \rho_s}{N_i(\rho_s)} Д_{ik}(\rho_s) \qquad (k,\, s = 1, \ldots, m) \;,$$

(2.36)

$$D_j^{(k)} = \frac{N_k(\rho_j)}{(\rho_{m+1} - \rho_j)\, Д'(\rho_j)} \;, \qquad G_k = -\sum_{f=1}^{m} D_j^{(k)}$$

$$(k,\, j = 1, \ldots, m) \;.$$

Here $N_i(\rho_s)$ must be taken to mean the determinant

$$N_i(\rho_s) = \sum_{k=1}^{m} n_k\, Д_{ik}(\rho_s) \qquad (i,\, s = 1, \ldots, m) \;, \tag{2.37}$$

where Δ_{ik} represent the cofactors of the elements of the i'th row and the k'th column of the determinant $\Delta(\rho)$, while ρ_s are the roots of (1.2). Finally, in Case 3 it is possible to employ the formulas (2.35) and (2.36), but we must put in them $\rho_{m+1} = 0$.

To shorten the calculations it will prove useful hereafter to obtain a more simple formula for the number r'. According to formulas (2.10) and (2.34) we have

$$- r' = \sum_{k=1}^{n} \gamma_k + \gamma_{n+1} = \sum_{\alpha=1}^{n} \left[\sum_{k=1}^{n} D_k^{(\alpha)} + G_\alpha \right] p_\alpha - r = - r \quad .$$

From this we obtain the general relation

$$r' = r \quad .$$

Bearing in mind this result, we shall henceforth write everywhere in the equations of canonical form the value of r in lieu of the quantity r'.

3. CASE OF MULTIPLE ROOTS

Let us assume that (2.5) has a root ρ_1 of multiplicity ℓ, and let us further assume that the nullity of the matrix $B + \rho E$ is also equal to ℓ.

As previously, for various roots $\rho_{\ell+1}, \ldots, \rho_n$, for which the nullity of the matrix $B + \rho E$ is unity, (2.3) can be solved with respect to $C_\alpha^{(s)}$ $(\alpha = 1, \ldots, n)$, $(s = \ell + 1, \ldots, n)$. For the variables $x_{\ell+1}, \ldots, x_n$, it is possible to write the transformation

$$x_s = \sum_{\alpha=1}^{n} C_\alpha^{(s)} \eta_\alpha + \xi \qquad (s = \ell + 1, \ldots, n) \quad , \tag{2.38}$$

whose coefficients are determined by the formulas

$$C_k^{(s)} = \frac{\rho_{n+1} - \rho_s}{H_1(\rho_s)} D_{1k}(\rho_s) \qquad \begin{matrix} k = 1, \ldots, n \; , \\ s = \ell + 1, \ldots, n \end{matrix} \quad . \tag{2.39}$$

Corresponding to transformation (2.38) are the canonical equations

$$\dot{x}_s = - \rho_s x_s + f(\sigma) \qquad (s = \ell + 1, \ldots, n) \quad . \tag{2.40}$$

Since the transformation (2.38) is non-singular, it can be solved with respect to the variables $\eta_{\ell+1}, \ldots, \eta_n$. Let us assume that this solution is of the form

(2.41)
$$\eta_{\ell+\alpha} = \sum_{k=1}^{n-\ell} D_{\ell+k}^{(\ell+\alpha)} x_{\ell+k} + \sum_{k=1}^{\ell} P_k^{(\ell+\alpha)} \eta_k + G_{\ell+\alpha}\xi \qquad (\alpha = 1, \ldots, n - \ell) .$$

Eliminating these variables from the first ℓ equations of (1.31) results in

(2.42)
$$\dot{\eta}_s = \sum_{\alpha=1}^{\ell} b_{s\alpha}^* \eta_\alpha + \sum_{\beta=1}^{n-\ell} b_{s,\ell+\beta}^* x_{\ell+\beta} + n_s^* \xi \qquad (s = 1, \ldots, \ell) ,$$

$$\dot{\xi} = - \rho_{n+1}\xi + f(\sigma) ,$$

$$\sigma = \sum_{\alpha=1}^{\ell} p_\alpha^* \eta_\alpha + \sum_{\beta=1}^{n-\ell} p_{\ell+\beta} x_{\ell+\beta} + R\xi .$$

Here $b_{s\alpha}^*, p_\alpha^*, R$ $(s, \alpha = 1, \ldots, n)$ are constants determined by the transformation itself.

(2.42) can be reduced to the canonical form by two separate methods. Firstly, by means of the transformation

(2.43)
$$\eta_s = \zeta_s + \sum_{\beta=1}^{n} \varepsilon_{s,\ell+\beta} x_{\ell+\beta} + N_s\xi \qquad (s = 1, \ldots, m)$$

it is possible to eliminate from further consideration the group of canonical variables $x_{\ell+\beta}$. Differentiating (2.43) and making use of (2.40) and (2.42), we get

$$\dot{\zeta}_s + \sum_{\beta=1}^{n-\ell} \varepsilon_{s,\ell+\beta} \left[- \rho_{\ell+\beta} x_{\ell+\beta} + f(\sigma) \right] + N_s \left[- \rho_{n+1}\xi + f(\sigma) \right] =$$

$$= \sum_{\alpha=1}^{\ell} b_{s\alpha}^* \left[\zeta_\alpha + \sum_{\beta=1}^{n-\ell} \varepsilon_{\alpha,\ell+\beta} x_{\ell+\beta} + N_\alpha\xi \right] + \sum_{\beta=1}^{n-\ell} b_{s,\ell+\beta}^* x_{\ell+\beta} + n_s^*\xi$$

$$(s = 1, \ldots, \ell) .$$

If the constants $\varepsilon_{s,\ell+\beta}$ and N_s are chosen in accordance with the relations

$$(2.44)\qquad -\rho_{\ell+\beta}\varepsilon_{s,\ell+\beta} = \sum_{\alpha=1}^{\ell} b^*_{s\alpha}\varepsilon_{\alpha,\ell+\beta} + b^*_{s,\ell+\beta} \qquad (\beta = 1, \ldots, n-\ell),$$

$$(2.45)\qquad -\rho_{n+1}N_s = \sum_{\alpha=1}^{\ell} b^*_{s\alpha}N_\alpha + n^*_s \qquad (s = 1, \ldots, \ell),$$

which is always possible for $\rho_k \neq 0$, then the equations for the variables ζ_s are written

$$(2.46)\qquad \dot{\zeta}_s = \sum_{\alpha=1}^{\ell} b^*_{s\alpha}\zeta_\alpha - \left[N_s + \sum_{\beta=1}^{n-\ell} \varepsilon_{s,\ell+\beta}\right] f(\sigma) \quad (s = 1, \ldots, \ell).$$

It is now possible to reduce (2.46) in the usual manner to the canonical form. For this purpose each equation of (2.46) is multiplied by the constant k_s, and after adding term by term we can write the results

$$\frac{d}{dt}[k_1\zeta_1 + \ldots + k_\ell\zeta_\ell] = \sum_{s=1}^{\ell}\sum_{\alpha=1}^{\ell} b^*_{s\alpha}k_s\zeta_\alpha - \sum_{s=1}^{\ell} k_s\left[N_s + \sum_{\beta=1}^{n-\ell}\varepsilon_{s,\ell+\beta}\right] f(\sigma).$$

If the constants $k_s = k_s^{(1)}$ are chosen in accordance with the relations

$$(2.47)\qquad -\bar{\rho}_1 k_\alpha = \sum_{s=1}^{\ell} b^*_{s\alpha}k_\alpha \qquad (\alpha = 1, \ldots, \ell),$$

in which the quantity $\bar{\rho}_1$ is the root of the equation

$$(2.48)\qquad \begin{vmatrix} b^*_{11} + \rho & \cdots & b^*_{\ell 1} \\ \cdot & \cdots & \cdot \\ b^*_{1\ell} & \cdots & b^*_{\ell\ell} + \rho \end{vmatrix} = 0,$$

then, after introducing the notation

$$k_1^{(1)}\zeta_1 + \dots + k_\ell^{(1)}\zeta_\ell = x_1 \quad ,$$

$$(2.49) \qquad -\sum_{s=1}^{\ell} k_s \left[N_s + \sum_{\beta=1}^{n-\ell} \varepsilon_{s,\ell+\beta} \right] = q_1 \quad ,$$

we get

$$\dot{x}_1 = -\overline{\rho_1} x_1 + q_1 f(\sigma) \quad .$$

In view of the invariance of the elementary divisors of the matrix $B + \rho E$ under linear transformations, we have $\bar{\rho}_1 = \rho_1$. Discarding the first of (2.46) and eliminating the variable ζ_1 from the remaining equations, let us rewrite the system in the following manner:

$$\dot{x}_1 = -\rho_1 x_1 + q_1 f(\sigma) \quad ,$$

$$(2.50) \qquad \dot{\zeta}_s = \frac{b_{s1}^*}{k_1^{(1)}} x_1 + \sum_{\alpha=2}^{\ell} b_{s\alpha}^{**}\zeta_\alpha - \left[N_s + \sum_{\beta=1}^{n-\ell} \varepsilon_{s,\ell+\beta} \right] f(\sigma)$$

$$(s = 2, \dots, \ell) \quad .$$

The above arguments can be repeated with respect to (2.50), which contain the variables ζ_α. Again, multiplying each equation by $k_s = k_s^{(2)}$ we obtain after adding term by term

$$\frac{d}{dt}[k_2\zeta_2 + \dots + k_\ell\zeta_\ell] = \sum_{s=2}^{\ell} \frac{k_s b_{s1}^*}{k_1^{(1)}} x_1 + \sum_{s=2}^{\ell}\sum_{\alpha=2}^{\ell} b_{s\alpha}^{**} k_s \zeta_s -$$

$$- \sum_{s=2}^{\ell} k_s \left[N_s + \sum_{\beta=1}^{n-\ell} \varepsilon_{s,\ell+\beta} \right] f(\sigma) \quad .$$

If the constants $k_s = k_s^{(2)}$ are chosen in accordance with the relation

$$(2.51) \qquad -\bar{\rho}_2 k_\alpha = \sum_{s=2}^{\ell} b_{s\alpha}^{**} k_s \qquad (\alpha = 2, \dots, \ell) \quad ,$$

in which $\bar{\rho}_2$ is the root of the equation

$$(2.52)\qquad \begin{vmatrix} b_{22}^{**}+\rho & \cdots & b_{22}^{**} \\ \cdot & \cdots & \cdot \\ b_{2\ell}^{**} & \cdots & b_{\ell\ell}^{**}+\rho \end{vmatrix} = 0 ,$$

then after introducing the symbols

$$k_2^{(2)}\zeta_2 + \cdots + k_\ell^{(2)}\zeta_\ell = x_2 ,$$

$$(2.53)\qquad \sum_{s=2}^{\ell} \frac{b_{s1}^{*}k_s^{(2)}}{k_1^{(1)}} = \varepsilon_1, \qquad -\sum_{s=2}^{\ell} k_s^{(2)}\left[N_s + \sum_{\beta=1}^{n-\ell}\varepsilon_{s\ell+\beta}\right] = q_2$$

and eliminating the variable ζ_2, we get

$$\dot{x}_1 = \rho_1 x_1 + q_1 f(\sigma)$$

$$\dot{x}_2 = \varepsilon_1 x_1 - \overline{\rho_2} x_2 + q_2 f(\sigma) ,$$

$$\cdots\cdots\cdots\cdots\cdots\cdots\cdots$$

$$\dot{\zeta}_s = \frac{b_{s1}^{*}}{k_1^{(2)}} x_1 + \frac{b_{s2}^{**}}{k_2^{(2)}} x_2 + \sum_{\alpha=3}^{\ell} b_{s\alpha}^{***}\zeta_\alpha - \left[N_s + \sum_{\beta=1}^{n-\ell}\varepsilon_{s,\ell+\beta}\right] f(\sigma) .$$

The above process of reduction can be completed in exactly the same manner as was done in the reduction of linear differential equations with constant coefficients to the canonical form. By virtue of the invariance of the elementary divisors in the linear transformations, we have $\bar{\rho}_1 = \bar{\rho}_2 = \cdots = \rho_1$, with, as follows from the definitions in (2.53), the constants $\varepsilon_1, \varepsilon_2, \ldots, \varepsilon_{\ell-1}$ containing arbitrary factors, thanks to which they can assume all values differing from zero. As to the last equation of the system for the variable σ, it can be obtained in the usual manner by eliminating the old variables with the aid of the transformation formulas (2.38), (2.43), (2.49), (2.53), etc.

Thus, in the case of a multiple root we obtain finally,

$$\begin{aligned}
\dot{x}_1 &= -\rho_1 x_1 + q_1 f(\sigma) \ , \\
\dot{x}_2 &= \varepsilon_1 x_1 - \rho_1 x_2 + q_2 f(\sigma) \ , \\
&\cdots\cdots\cdots \\
\dot{x}_\ell &= \varepsilon_{\ell-1} x_{\ell-1} - \rho_1 x_\ell + q_\ell f(\sigma) \ , \\
\dot{x}_{\ell+1} &= -\rho_{\ell+1} x_{\ell+1} + f(\sigma) \ , \\
&\cdots\cdots\cdots \\
\dot{x}_{n+1} &= -\rho_{n+1} x_{n+1} + f(\sigma) \ , \\
\sigma &= \sum_{k=1}^{n+1} \gamma_k x_k \ , \\
\dot{\sigma} &= \sum_{k=1}^{n+1} \beta_k x_k - f(\sigma) \ .
\end{aligned} \tag{2.54}$$

The constants γ_s and β_s are determined during the process of performing the transformation itself; to give general formulas for their determination makes no sense, since these formulas become quite cumbersome.

4. THE FIRST BULGAKOV PROBLEM*

Let us proceed to examine several examples. These will be used to illustrate the application of the general formulas and to develop a step by step calculation procedure.

These examples recur throughout the book. In the first chapter we consider the canonical transformation of the initial equations. Study of the examples will permit a final clarification, in all details, of the method of investigation presented and will provide preliminary skill in its application.

By way of the applications of this theory, let us consider the problem first investigated by B. V. Bulgakov.**

---

* B. V. Bulgakov, Self-Excited Oscillations of Regulated Systems, Dokl. AN SSSR, Vol. XXXVII, No. 9, 1942.

** The solution of (2.55) for a linear $f^*(\sigma)$ was first considered by N. Minorsky ("Directional Stability of Automatically Controlled Bodies," J. Am. Soc. Naval Arch. XXXIV, 1922). It was treated later by W. Oppelt ("Die Flugzeugkurssteurung im Geradeausflug," Luftfahrtforschung, Bd. 14, Nos. 4-5, 1937; "Vergleichende Betrachtung Verschiedener Regelaufgaben hinsichtlich der geeigenten Regelsetzpassigkeit," Luftfahrtforschung, Bd. 16, No. 8, 1939). The possible physical interpretation of (2.55) and its various generalizations, considered in this book, can also be deduced from these papers.

We have

(2.55)
$$T^2\ddot{\psi} + U\dot{\psi} + \mu = 0 \quad ,$$
$$\dot{\mu} = f^*(\sigma), \qquad \sigma = a\psi + E\dot{\psi} + G^2\ddot{\psi} - \frac{1}{\ell}\mu \quad .$$

Here the constant T^2 characterizes the inertia of the regulated object, U is the natural damping, and a, E, G^2, and ℓ are the constants of the regulator.

If we denote

$$b_{22} = -\frac{U}{T^2\sqrt{r}}, \qquad \frac{\ell T^2}{T^2 + \ell G^2} = 1, \qquad r = \frac{1}{T^2}, \qquad n_2 = -\frac{1}{rT^2} = -1 \quad ,$$

(2.56)
$$p_1 = a, \qquad p_2 = \left(E - \frac{UG^2}{T^2}\right)\sqrt{r}, \qquad t = \frac{\tau}{\sqrt{r}} \quad ,$$

$$\frac{1}{1\sqrt{r}} f^*(\sigma) = f(\sigma), \qquad \psi = \eta_1, \qquad \dot{\psi} = \sqrt{r}\eta_2, \qquad \mu = 1\xi \quad ,$$

we can reduce the initial equations to a normal form, which at the same time is dimensionless

(2.57)
$$\dot{\eta}_1 = \eta_2 \quad ,$$
$$\dot{\eta}_2 = b_{22}\eta_2 + \eta_2\xi \quad ,$$
$$\dot{\xi} = f(\sigma), \qquad \sigma = p_1\eta_1 + p_2\eta_2 - \xi \quad .$$

The dot indicates here the derivative with respect to the dimensionless time τ. The steady state of this system is determined by the obvious solution of the equations

(2.58)
$$\eta_2 = 0 \quad ,$$
$$b_{22}\eta_2 + n_2\xi = 0 \quad .$$
$$|p_1\eta_1 + p_2\eta_2 - \xi| \le \sigma_* \quad .$$

It is clear from this that if $f(\sigma)$ is a function of class (A) for which $\sigma_* > 0$, we obtain a continuum of solutions

(2.59)
$$|\eta_1^*| \le \frac{\sigma_*}{p_1}, \qquad \eta_2^* = 0, \qquad \xi^* = 0 \quad ,$$

** (cont.) The formulation of the problem as given by (2.55) gave rise to the three trends in the investigation of nonlinear systems, briefly mentioned in p. 120.

which determines the region of insensitivity of the regulator.

Using (2.57), we find that we deal here with Case 3.

To construct the canonical transformation we set up the determinant $Д(\rho)$ and its derivative. We have

(2.60)
$$Д(\rho) = \begin{vmatrix} \rho & 0 \\ 1 & \rho + b_{22} \end{vmatrix} = \rho(\rho + b_{22})$$
$$Д'(\rho) = 2\rho + b_{22} \quad .$$

(1.2) will have roots $\rho_1 = 0$, $\rho_2 = -b_{22}$. We choose the number $i = 1$. Let us set up a row of minors $Д_{ik}$ and use formulas (2.37) to set up a column of quantities N_α.

We have

(2.61)
$$Д_{11} = \rho + b_{22}, \quad Д_{12} = -1 \quad ,$$
$$N_1 = \begin{vmatrix} 0 & n_2 \\ 1 & \rho + b_{22} \end{vmatrix} = 1, \quad N_2 = \begin{vmatrix} \rho & 0 \\ 0 & n_2 \end{vmatrix} = +\rho n_2 = -\rho \quad .$$

According to fomrulas (2.35) we obtain the coefficients of direct transformation

(2.62) $$C_1^{(s)} = -\rho_s(\rho_s + b_{22}), \quad C_2^{(s)} = \rho_s \qquad (s = 1, 2)$$

and the transformation itself

(2.63)
$$x_1 = \xi \ ,$$
$$x_2 = \rho_2\eta_2 + \xi \quad .$$

Inasmuch as the first row of the transformation is degenerate, the coefficients of the inverse transformation will be sought only for the coordinate η_2. We have

$$D_1^{(2)} = \frac{N_2(\rho_1)}{-\rho_1(2\rho_1+b_{22})} = \frac{\rho_1 n_2}{-\rho_1(2\rho_1+b_{22})} = \frac{n_2}{-(2\rho_1+b_{22})} = -\frac{1}{\rho_2}$$

$$D_2^{(2)} = \frac{N_2(\rho_2)}{-\rho_2(2\rho_2+b_{22})} = \frac{\rho_2 n_2}{-\rho_2^2} = \frac{1}{\rho_2} \qquad ,$$

thanks to which we find

$$\xi = x_1 \ , \qquad \eta_2 = \frac{1}{\rho_2}(x_2 - x_1) \ . \tag{2.64}$$

Differentiating σ, we obtain according to (2.57)

$$\dot{\sigma} = p_1\eta_2 + p_2(b_{22}\eta_2 + n_2\xi) - f(\sigma) \ .$$

If we now employ formulas (2.64) and set

$$\beta_1 = -\frac{p_1}{\rho_2}, \qquad \beta_2 = \frac{p_1 - \eta_2 p_2}{\rho_2} \tag{2.65}$$

we finally obtain

$$\begin{aligned} \dot{x}_1 &= f(\sigma) \ , \\ \dot{x}_2 &= -\rho_2 x_2 + f(\sigma) \ , \\ \dot{\sigma} &= \beta_1 x_1 + \beta_2 x_2 - f(\sigma) \ . \end{aligned} \tag{2.66}$$

Let us note that the expressions for β_1 and β_2 can be obtained from the general formulas directly. In fact, according to (2.10) and (2.36) we have

$$\beta_k = -\rho_k\gamma_k = -\sum_{\alpha=1}^{m} \frac{p_\alpha \rho_k N_\alpha(\rho_k)}{(\rho_{m+1}-\rho_k)\, Д'(\rho_k)} \ .$$

This leads to the already known formulas (2.65):

$$\beta_1 = \frac{p_1 N_1(\rho_1)}{Д'(\rho_1)} + \frac{p_2 N_2(\rho_1)}{Д'(\rho_1)} = -\frac{p_1 n_2}{b_{22}} = -\frac{p_1}{\rho_2}$$

$$\beta_2 = \frac{p_1 N_1(\rho_2)}{Д'(\rho_2)} + \frac{p_2 N_2(\rho_2)}{Д'(\rho_2)} = \frac{(p_1 + b_{22}p_2)n_2}{b_{22}} = \frac{p_1 - \rho_2 p_2}{\rho_2} \ .$$

A possible generalization of the problem (2.55) can be the complete equation of the control device

$$T^2\ddot{\psi} + U\dot{\psi} + \mu = 0 \quad ,$$

(2.67)

$$V^2\ddot{\mu} + W\dot{\mu} + S\mu = f^*(\sigma), \qquad \sigma = a\psi + E\dot{\psi} + G^2\ddot{\psi} - \frac{1}{\ell}\mu \quad .$$

If we retain all the symbols of (2.56), with the exception of the last one, for which we use (1.24), we obtain

(2.68)

$$\begin{aligned}
\dot{\eta}_1 &= \eta_2 \quad , \\
\dot{\eta}_2 &= b_{22}\eta_2 + n_2\eta_3 \quad , \\
\dot{\eta}_3 &= -\rho_3\eta_3 + \frac{1}{\rho}\xi \\
\dot{\xi} &= -\rho_4\xi + \frac{p}{rV^2} f^*(\sigma), \qquad \sigma = p_1\eta_1 + p_2\eta_2 + p_3\eta_3 \quad ,
\end{aligned}$$

where*

(2.69)

$$\eta_3 = \mu, \qquad \rho_3 + \rho_4 = \frac{W}{\sqrt{rV^2}}, \qquad \rho_3\rho_4 = \frac{S}{rV^2} \quad ,$$

$$p_1 = a, \qquad p_2 = \left(E - \frac{UG^2}{T^2}\right)\sqrt{r}, \qquad p_3 = -\frac{1}{\ell} \quad .$$

From (2.68) we see that we deal with the most general specification of initial equations in the form (1.31).

To construct the transformation, let us set up the determinant $D(\rho)$ and its derivative. We have

(2.70)
$$D(\rho) = \begin{vmatrix} \rho & 0 & 0 \\ 1 & \rho + b_{22} & 0 \\ 0 & n_2 & \rho + b_{33} \end{vmatrix} = \rho(\rho + b_{22})(\rho + b_{33}) \quad ,$$

$$D'(\rho) = 3\rho^2 + 2\rho(b_{22} + b_{33}) + b_{22}b_{33} \quad .$$

The sought roots of (2.5) are

* We henceforth assume $p = 1$ in (2.68). This choice of transformation constant for (1.24) does not make the solution less general.

$$\rho_1 = 0, \quad \rho_2 = -b_{22}, \quad \rho_3 = -b_{33} .$$

Assuming $i = 1$, we find the necessary quantities

$$(2.71) \quad \begin{aligned} &D_{11}(\rho) = (\rho + b_{22})(\rho + b_{33}) , \\ &D_{12}(\rho) = -(\rho + b_{33}), \quad D_{13} = n_2 , \\ &H_1 = n_2 = -1, \quad H_2 = -\rho n_2 = \rho, \\ &H_3 = \rho(\rho + b_{22}) . \end{aligned}$$

Now, using formulas (2.32) and (2.71) we write immediately the coefficients of direct transformation

$$(2.72) \quad \begin{aligned} &C_1^{(1)} = \frac{\rho_4 b_{22} b_{33}}{n_2} = -\rho_4\rho_2\rho_3, \quad C_1^{(2)} = C_1^{(3)} = 0 , \\ &C_2^{(1)} = -\frac{\rho_4 b_{33}}{n_2} = -\rho_4\rho_3, \quad C_2^{(2)} = \frac{(\rho_4-\rho_2)(b_{22}-b_{33})}{n_2} = \\ &\qquad = -(\rho_4 - \rho_2)(\rho_3 - \rho_2); \quad C_2^{(3)} = 0 , \\ &C_3^{(1)} = \rho_4; \quad C_3^{(2)} = \rho_4 - \rho_2, \quad C_3^{(3)} = \rho_4 - \rho_3 . \end{aligned}$$

Next, using formulas (2.34) we find the coefficients of the inverse transformation

$$(2.73) \quad \begin{aligned} &D_1^{(1)} = \frac{n_2}{\rho_4 b_{22} b_{33}} = \frac{1}{\rho_2\rho_3\rho_4}, \quad D_1^{(2)} = D_1^{(3)} = 0 , \\ &D_2^{(1)} = \frac{n_2}{(\rho_4-\rho_2)(\rho_2+b_{33})\rho_2} = \frac{1}{(\rho_4-\rho_2)(\rho_3-\rho_2)\rho_2} , \\ &D_2^{(2)} = -\frac{n_2}{(\rho_4-\rho_2)(\rho_2+b_{33})} = -\frac{1}{(\rho_4-\rho_2)(\rho_3-\rho_2)}, \quad D_2^{(3)} = 0 , \\ &D_3^{(1)} = \frac{n_2}{(\rho_4-\rho_3)(\rho_3+b_{22})\rho_3} = \frac{1}{(\rho_4-\rho_3)(\rho_2-\rho_3)\rho_3} , \\ &D_3^{(2)} = -\frac{n_2}{(\rho_4-\rho_3)(\rho_3+b_{22})\rho_3} = \frac{1}{(\rho_4-\rho_3)(\rho_3-\rho_2)\rho_3} , \quad D_3^{(3)} = \frac{1}{\rho_4 - \rho_3} . \end{aligned}$$

To complete the construction of the transformation, let us calculate the quantities γ_1 and β_1, using formulas (2.10), (2.69), and (2.73).

$$\gamma_1 = -\frac{a}{\rho_2\rho_3\rho_4}, \quad \gamma_2 = -\frac{1}{(\rho_4-\rho_2)(\rho_2-\rho_3)}\left[\frac{a}{\rho_2}\cdot\left(E - \frac{UG^2}{T^2}\right)\sqrt{r}\right],$$

$$\gamma_3 = -\frac{1}{\rho_4 - \rho_3}\times\left[\frac{a}{\rho_3(\rho_3-\rho_2)} - \frac{E - \frac{UG^2}{T^2}}{\rho_3 - \rho_2}\sqrt{r} + \frac{T^2 + \ell G^2}{\ell T^2}\right], \tag{2.74}$$

$$\gamma_4 = \frac{1}{\rho_4 - \rho_3}\times\left[\frac{a}{\rho_4(\rho_4-\rho_2)} - \frac{E - \frac{UG^2}{T^2}}{\rho_4 - \rho_2}\sqrt{r} + \frac{T^2 + \ell G^2}{\ell T^2}\right],$$

$$\beta_2 = -\rho_2\gamma_2, \quad \beta_3 = -\rho_3\gamma_3, \quad \beta_4 = -\rho_4\gamma_4, \quad \beta_1 = 0 . \tag{2.75}$$

The canonical equations of the problem considered here will have the following form:

$$\begin{aligned} \dot{x}_1 &= f(\sigma) \\ \dot{x}_2 &= -\rho_2 x_2 + f(\sigma), \\ \dot{x}_3 &= -\rho_3 x_3 + f(\sigma), \\ \dot{x}_4 &= -\rho_4 x_4 + f(\sigma), \end{aligned} \tag{2.76}$$

$$\sigma = \sum_{k=1}^{4} \gamma_k x_k, \quad \dot{\sigma} = \beta_2 x_2 + \beta_3 x_3 + \beta_4 x_4 .$$

Let us consider various particular cases of the problem. Thus, in Case 1, $\rho_4 = 0$ and the formulas of (2.75) yield

$$\beta_2 = \frac{p_1 - \rho_2 p_2}{\rho_2(\rho_3-\rho_2)},$$

$$\beta_3 = \frac{1}{(\rho_3-\rho_2)\rho_3}[-p_1 + \rho_3 p_2 + \rho_3(\rho_3 - \rho_2)p_3], \tag{2.77}$$

$$\beta_4 = -\frac{p_1}{\rho_2\rho_3} .$$

In Case 2 we have $\rho_4 = \infty$ and to form the necessary expressions we must employ formulas (2.36), (2.37), and (2.34). Thus, using formulas (2.37), we have

$$N_1(\rho_s) = n_1 Д_{11} + n_2 Д_{12} = n_2 Д_{12} = 1 \quad ,$$

$$N_2(\rho_s) = n_1 Д_{21} + n_2 Д_{22} = -\rho_s \quad .$$

Consequently, formulas (2.34) yield

$$D_1^{(1)} = -\frac{1}{\rho_3\rho_2} \;, \qquad D_2^{(1)} = \frac{1}{(\rho_3-\rho_2)\rho_2} \;,$$

$$D_1^{(2)} = 0, \qquad D_2^{(2)} = -\frac{\rho_2}{(\rho_3-\rho_2)\rho_2} = -\frac{1}{\rho_3 - \rho_2} \quad .$$

The expressions obtained for $D_k^{(\alpha)}$ make it possible to use formulas (2.17) to set up expressions for the quantities γ_k:

$$\gamma_1 = p_1D_1^{(1)} + p_2D_1^{(2)} = -\frac{p_1}{\rho_2\rho_3} \;,$$

(2.78) $$\gamma_2 = p_1D_2^{(1)} + p_2D_2^{(2)} = \frac{1}{\rho_3 - \rho_2}\left[\frac{p_1}{\rho_2} - p_2\right],$$

$$\gamma_3 = p_3 + \frac{1}{\rho_3 - \rho_2}\left[p_2 - \frac{p_1}{\rho_3}\right]$$

If we now take into account that $\beta_k = -\rho_k\gamma_k$, we get

(2.79) $$\beta_1 = 0; \quad \beta_2 = -\frac{p_1 - \rho_2 p_2}{\rho_3 - \rho_2} \;, \quad \beta_3 = -\rho_3 p_3 + \frac{p_1 - \rho_3 p_2}{\rho_3 - \rho_2} \quad .$$

Thus, in Case 2, the equations of the problem become

(2.80) $$\begin{aligned} \dot{x}_1 &= f(\sigma) \;, \\ \dot{x}_2 &= -\rho_2 x_2 + f(\sigma) \;, \\ \dot{x}_3 &= -\rho_3 x_3 + f(\sigma) \;, \\ \dot{\sigma} &= \beta_2 x_2 + \beta_3 x_3 - f(\sigma) \;. \end{aligned}$$

Case 3 was considered above.

5. BULGAKOV'S SECOND PROBLEM*

For the next example, let us consider an inherently stable control system, which also was first studied by Bulgakov

(2.81)
$$T^2\ddot{\psi} + U\dot{\psi} + k\psi + \mu = 0 \quad ,$$
$$\dot{\mu} = f^*(\sigma), \qquad \sigma = a\psi + E\dot{\psi} + G^2\ddot{\psi} - \frac{1}{\ell}\mu \quad .$$

Here the meaning of all the old symbols is clear from the previous problem; the constant $k > 0$ characterizes the action of the so-called restoring force. Again let us denote

(2.82)
$$\psi = \eta_1, \quad \frac{d\psi}{dt} = \sqrt{r}\,\eta_2, \quad \mu = i\xi, \quad t = \frac{\tau}{\sqrt{r}} \; ,$$
$$p = \frac{U}{T^2}, \quad q = \frac{k}{T^2} \quad r = \frac{1}{T^2}; \quad n_2 = -1 \quad ,$$
$$i = \frac{\ell T^2}{T^2 + \ell G^2}, \quad f(\sigma) = \frac{1}{i\sqrt{r}}\, f^*(\sigma), \quad b_{21} = -\frac{q}{r} \quad ,$$
$$b_{22} = -\frac{p}{\sqrt{r}}; \quad p_1 = a - qG^2; \quad p_2 = (E - pG^2)\sqrt{r} \quad ,$$

and let us reduce (2.81) to normal form in dimensionless variables.

(2.83)
$$\dot{\eta}_1 = \eta_2 \quad ,$$
$$\dot{\eta}_2 = b_{21}\eta_1 + b_{22}\eta_2 + n_2\xi \quad ,$$
$$\dot{\xi} = f(\sigma), \qquad \sigma = p_1\eta_1 + p_2\eta_2 - \xi \quad .$$

The steady state of the control system is determined by the equation

(2.84)
$$\eta_2 = 0 \quad ,$$
$$b_{21}\eta_1 + b_{22}\eta_2 + n_2\xi = 0$$
$$|p_1\eta_1 + p_2\eta_2 - \xi| \le \sigma_* \quad .$$

---

* B. V. Bulgakov. Certain Problems in the Theory of Control with Nonlinear Characteristics, PMM, Vol. X, No. 3, 1946.

For functions of class (A), for which $\sigma_* > 0$, we obtain a continuum of solutions

$$(2.85)\qquad |\eta_1^*| \le \frac{\sigma_*}{|p_1 - b_{21}|} \quad \eta_2^* = 0, \quad \xi^* = b_{21}\eta_1^* ,$$

which characterize the region of insensitivity of the regulator. Here again we are dealing with Case 3.

To construct the transformations of interest to us, we perform calculations that are already known. Let us set up the determinant $Д(\rho)$ and its derivative. We have

$$(2.86)\quad Д(\rho) = \begin{vmatrix} \rho & b_{21} \\ 1 & \rho + b_{22} \end{vmatrix} = \rho(\rho + b_{22}) - b_{21}, \qquad Д'(\rho) = 2\rho + b_{22} .$$

Consequently, we find two roots for (1.2)

$$(2.87)\quad \begin{gathered} \rho_1, \quad \rho_2 = -\frac{b_{22}}{2} \pm \sqrt{\left(\frac{b_{22}}{2}\right)^2 + b_{21}}, \quad \rho_1 + \rho_2 = -b_{22} , \\ \rho_1\rho_2 = -b_{21} . \end{gathered}$$

Let us put $i = 1$ and, replacing the elements of the first and second columns by the quantities 0 and n_2, calculate the necessary quantities N_1 and N_2

$$(2.88)\quad \begin{gathered} N_1 = \begin{vmatrix} 0 & n_2 \\ 1 & \rho + b_{22} \end{vmatrix} = 1, \quad N_2 = \begin{vmatrix} \rho & b_2 \\ 0 & n_2 \end{vmatrix} = n_2\rho = -\rho , \\ Д_{11} = (\rho + b_{22}), \quad Д_{12} = -1 . \end{gathered}$$

According to formulas (2.35), the coefficients of the direct transformation will be of the form

$$(2.89)\quad c_1^{(s)} = \frac{\rho_s(\rho_s + b_{22})}{n_2} = -b_{21}, \quad c_2^{(s)} = -\frac{\rho_s}{n_2} = \rho_2 \qquad (s = 1, 2) .$$

Insertion of the resultant values of ρ yields the canonical transformation

(2.90)
$$x_1 = b_{21}\eta_1 + \rho_1\eta_2 + \xi \quad ,$$
$$x_2 = - b_{21}\eta_1 + \rho_2\eta_2 + \xi \quad .$$

It is obviously non-singular if $\rho_1 \neq \rho_2$. The inverse transformation can be set up by means of formulas (2.36). We find successively

(2.91)
$$D_1^{(1)} = \frac{n_2}{\rho_1(2\rho_1+b_{22})} = \frac{1}{\rho_1(\rho_2-\rho_1)} \quad ,$$
$$D_2^{(1)} = \frac{n_2}{\rho_2(2\rho_1+b_{22})} = - \frac{1}{\rho_2(\rho_2-\rho_1)} \quad ,$$
$$D_1^{(2)} = - \frac{n_2}{2\rho_1 + b_{22}} = - \frac{1}{\rho_2 - \rho_1}, \qquad D_2^{(2)} = - \frac{n_2}{2\rho_2 + b_{22}} = \frac{1}{\rho_2 - \rho_1} \quad ,$$
$$G_1 = \frac{n_2}{\rho_1\rho_2} = - \frac{1}{\rho_1\rho_2}; \qquad G_2 = 0 \quad .$$

Finally, formulas (2.17), (2.82) and (2.91) yield

(2.92)
$$\beta_1 = - \rho_1\gamma_1 = - \rho_1[p_1D_1^{(1)} + p_2D_1^{(2)}] = \frac{1}{\rho_2 - \rho_1}[- p_1 + \rho_1p_2] \quad ,$$
$$\beta_2 = - \rho_2\gamma_2 = - \rho_2[p_1D_2^{(1)} + p_2D_2^{(2)}] =$$
$$= \frac{1}{\rho_2 - \rho_1}[p_1 - \rho_2p_2], \qquad \gamma_3 = - 1 - \frac{p_1}{\rho_1\rho_2}$$

and the canonical equations assume the form

(2.93a)
$$\dot{x}_1 = - \rho_1x_1 + f(\sigma) \quad ,$$
$$\dot{x}_2 = - \rho_2x_2 + f(\sigma) \quad ,$$
$$\sigma = \sum_{k=1}^{3} \gamma_kx_k \quad ,$$

(2.93b)
$$\dot{\sigma} = \beta_1x_1 + \beta_2x_2 - f(\sigma) \quad .$$

Let us now assume that roots ρ_1 and ρ_2 are the same. This case corresponds to specifying the parameters of the regulated object such that

$$p_2 = 4q \quad .$$

According to general theory of transformations, we can write any one of the transformations (2.90), for example, as one, for example, the first canonical equation (2.93). Next, with the aid of the relation

$$(2.94) \qquad b_{21}\eta_1 = n_2 = \left(x_1 + \frac{\rho_1}{n_1}\eta_2 - \xi \right) = \rho_1\eta_2 - x_1 + \xi$$

we eliminate the variable η_1 in the second equation of (2.83), to get

$$\dot{\eta}_2 = \rho_1\eta_2 - x_1 + \xi + b_{22}\eta_2 - \xi = - x_1 - \rho_1\eta_2 \quad .$$

Finally, let us put

$$(2.95) \qquad \eta_2 = C_1x_1 + C_2x_2 \quad ,$$

where C_1 and C_2 are constants, so far arbitrary. The elementary transformation of this relation gives us a second canonical equation

$$\dot{x}_2 = \varepsilon x_1 - \rho_1x_2 + q_2f(\sigma) \quad ,$$

where

$$(2.96) \qquad \varepsilon_1 = - \frac{1}{C_2} , \qquad q_2 = - \frac{C_1}{C_2} \quad ,$$

Obviously the numbers ε_1 and q_2 are arbitrary and can be chosen to be as small as convenient. To obtain the third equation, we have

$$\dot{\sigma} = p_1\eta_2 + p_2(b_{21}\eta_1 + b_{22}\eta_2 - \xi) - f(\sigma) \quad .$$

If we employ relations (2.94) and (2.95) to eliminate the variables η_1 and η_2, then after putting

$$(2.97) \qquad \begin{aligned} \beta_1 &= - p_2 + C_1(p_1 + b_{22}p_2 + \rho_1p_1) \quad , \\ \beta_2 &= C_2(p_1 + b_{22}p_2 + \rho_1p_2), \; q_1 = 1 \end{aligned}$$

we finally obtain

$$
(2.98) \qquad \begin{aligned}
\dot{x}_1 &= -\rho_1 x_1 + q_1 f(\sigma) \quad , \\
\dot{x}_2 &= \varepsilon_1 x_1 - \rho_1 x_2 + q_2 f(\sigma) \quad , \\
\dot{\sigma} &= \beta_1 x_1 + \beta_2 x_2 - f(\sigma) \quad .
\end{aligned}
$$

6. ON THE THEORY OF THE ISODROME REGULATOR*

Let us now consider the second Bulgakov problem in the case of the so-called isodrome regulator.

We have the initial equations

$$
(2.99) \qquad \begin{aligned}
& T^2\ddot{\psi} + U\dot{\psi} + k\psi + \mu = \kappa \quad , \\
& \dot{\mu} = f^*(\sigma) \quad , \\
& \sigma = a\psi + E\dot{\psi} + G^2\ddot{\psi} + \frac{1}{N}\int_0^t \psi \, dt - \frac{1}{\ell}\mu \quad .
\end{aligned}
$$

Here all the old symbols retain their previous meanings, and the constant κ characterizes the disturbing action of the external force on the regulated object; the constant N characterizes the isodrome nature of the regulator.

If we denote

$$
(2.100) \qquad \int_0^t \psi \, dt = \varphi \quad ,
$$

then the initial equation can be written in the following manner

$$
(2.101) \qquad \begin{aligned}
& T^2\ddot{\psi} + U\dot{\psi} + k\psi + \mu = \kappa \quad , \\
& \dot{\varphi} = \psi, \qquad \dot{\mu} = f(\sigma) \quad , \\
& \sigma = (a - qG^2)\psi + \frac{1}{N}\varphi + (E - pG^2)\dot{\psi} - \frac{1}{i}\mu + \frac{G^2}{T^2}\kappa \quad .
\end{aligned}
$$

The obvious solution of these equations

$$
(2.102) \qquad \begin{aligned}
& \psi^* = 0, \qquad \dot{\psi}^* = 0, \qquad \mu^* = \kappa, \qquad |\sigma|^* \leq \sigma_* \quad , \\
& \varphi^* = \frac{N}{i}\kappa - \frac{NG^2}{T^2}\kappa
\end{aligned}
$$

* A. M. Letov, On the Theory of the Isodrome Regulator, PMM, Vol. XII, No. 4, 1948.
An isodrome regulator maintains constant output velocity independent of the load.

characterizes the undisturbed steady state that must be maintained by a properly constructed regulator in the control system.

Formulas (2.102) show that although a constant disturbance acts on the regulated object, nevertheless in the steady state it is compensated for by a corresponding deviation μ^* (error) of the regulator so that the regulated object itself is not subject to deviation (a regulator having the above property is called an isodrome regulator).

The above is attained (even if $G^2 = 0$) because the function σ includes a term

$$\frac{1}{N} \int_0^t \psi \, dt \ ,$$

which causes a regulator with positive action to acquire isodrome properties. If the regulator does not have positive action $(\ell = \infty)$, one can put $N = \infty$, and the regulator with a control function

(2.103) $$\sigma = a\psi + E\dot{\psi} + G^2\ddot{\psi}$$

will also have isodrome properties. Denoting

$$\psi = \eta_1, \quad \frac{d\psi}{dt} = \sqrt{r}\,\eta_2, \quad \varphi - \varphi^* = Np_1\eta_3, \quad \mu - \mu^* = i\xi, \quad t = \frac{\tau}{\sqrt{r}} \ ,$$

(2.104) $$i = \frac{\ell T^2}{T^2 + \ell G^2}, \quad r = \frac{i}{T^2}, \quad n_2 = -\frac{i}{rT^2}, \quad b_{21} = -\frac{k}{rT^2} \ , \quad b_{22} = -\frac{U}{T^2\sqrt{r}}$$

$$p_1 = a - \frac{kG^2}{T^2} \ , \quad p_2 = \left(E - \frac{UG^2}{T^2} \right) \sqrt{r}, \quad p_1 = p_3 \ ,$$

$$b_{31} = \frac{1}{\sqrt{r}\ p_1 N} \ , \quad f(\sigma) = \frac{1}{i\sqrt{r}} f^*(\sigma) \ ,$$

we reduce (2.101) to normal form in dimensionless variables

(2.105) $$\begin{aligned} \dot{\eta}_1 &= \eta_2 \ , \\ \dot{\eta}_2 &= b_{21}\eta_1 + b_{22}\eta_2 + n_2\xi \ , \\ \dot{\eta}_3 &= b_{31}\eta_1 \ , \\ \dot{\xi} &= f(\sigma), \quad \sigma = p_1\eta_1 + p_2\eta_2 + p_3\eta_3 - \xi \ . \end{aligned}$$

The steady state of the control system is characterized by the solution (2.102) or by the solution

$$\eta_1^* = \eta_2^* = \xi^* = 0, \qquad |\eta_3^*| \le \frac{\sigma_*}{|p_3|} . \tag{2.106}$$

Let us note that with respect to the coordinates η_1, η_2, and ξ, this solution does not depend on whether the function $f(\sigma)$ belongs to class (A) or subclass (A'). The isodrome properties (2.99) of this regulator at $N \neq \infty$ are explained by the fact that no continuum of solutions, determined by the insensitivity zone $2\sigma_*$, exists for the coordinates η_1, η_2, and ξ. It is easy to verify that in the limiting case, when $N = \ell = \infty$, the regulator remains isodrome, but the effect of the insensitivity of the actuator, $2\sigma_*$, is fully retained.

Proceeding to the construction of the canonical transformation and noting that (2.104) pertains to Case 3, let us set up the determinant (1.2) and its derivative:

$$(2.107)\qquad \begin{aligned} Д(\rho) &= \begin{vmatrix} \rho & b_{21} & b_{31} \\ 1 & \rho + b_{22} & 0 \\ 0 & 0 & \rho \end{vmatrix} = \rho[\rho(\rho + b_{22}) - b_{21}] , \\ Д'(\rho) &= 3\rho^2 + 2\rho b_{22} - b_{21} . \end{aligned}$$

The roots ρ_1 and ρ_2 of the equation $Д(\rho) = 0$ have their previous values (2.87), while $\rho_3 = 0$.

Taking, as before, $i = 1$, we form the three minors

$$Д_{11} = \rho(\rho + b_{22}), \qquad Д_{12} = -\rho, \qquad Д_{13} = 0 . \tag{2.108}$$

In our case $n_1 = n_3 = 0$, $n_2 = -1$, and formulas (2.37) yield

$$(2.109)\qquad \begin{aligned} N_1 &= \begin{vmatrix} 0 & n_2 & 0 \\ 1 & \rho + b_{22} & 0 \\ 0 & 0 & \rho \end{vmatrix} = -n_2\rho, \qquad N_2 = \begin{vmatrix} \rho & b_{21} & b_{31} \\ 0 & n_2 & 0 \\ 0 & 0 & \rho \end{vmatrix} = n_2\rho^2 , \\ N_3 &= \begin{vmatrix} \rho & b_{21} & b_{31} \\ 1 & \rho + b_{22} & 0 \\ 0 & n_2 & 0 \end{vmatrix} = n_2 b_{31} . \end{aligned}$$

The coefficients of direct transformation are found with formulas (2.35). Since $\rho_3 = 0$, we have

$$C_1^{(s)} = -\rho_s(\rho_s + b_{22}), \quad C_2^{(s)} = \rho_s, \quad C_3^{(s)} = 0 \ ,$$

For the given values of ρ these formulas yield

$$\begin{aligned} &C_1^{(1)} = \rho_1\rho_2, \quad C_1^{(2)} = \rho_1\rho_2, \quad C_1^{(3)} = 0 \ , \\ &C_2^{(1)} = \rho_1 \ , \quad C_2^{(2)} = \rho_2, \quad C_2^{(3)} = 0 \ , \\ &C_3^{(1)} = C_3^{(2)} = C_3^{(3)} = 0 \ , \end{aligned} \tag{2.110}$$

Thanks to which the sought transformation becomes

$$\begin{aligned} x_1 &= \rho_1(\rho_2\eta_1 + \eta_2) + \xi \ , \\ x_2 &= \rho_2(\rho_1\eta_1 + \eta_2) + \xi \ , \\ x_3 &= \xi \ . \end{aligned} \tag{2.111}$$

Since the third row is degenerate, the coefficients of the inverse transformation will be sought only for the coordinates η_1 and η_2. From formulas (2.36) and (2.109) we get

$$D_1^{(1)} = \frac{1}{\rho_1(\rho_2-\rho_1)}, \quad D_1^{(2)} = -\frac{1}{\rho_2-\rho_1}, \quad D_1^{(3)} = -\frac{b_{31}}{\rho_1^2(\rho_2-\rho)_1} \ ,$$

$$D_2^{(1)} = -\frac{1}{\rho_2(\rho_2-\rho_1)}, \quad D_2^{(2)} = \frac{1}{\rho_2-\rho_1}, \quad D_2^{(3)} = \frac{b_{31}}{\rho_2^2(\rho_2-\rho_1)} \ ,$$

$$D_3^{(1)} = -\frac{1}{\rho_1\rho_2}, \quad D_2^{(2)} = 0, \quad D_3^{(3)} = \infty; \quad G_1 = -\frac{1}{\rho_1\rho_2}, \quad G_2 = 0 \ .$$

Thus, the inverse transformation is of the form

$$\begin{aligned} \eta_1 &= \frac{1}{\rho_1\rho_2(\rho_2-\rho_1)}\,[\rho_2 x_1 - \rho_1 x_2 - (\rho_2 - \rho_1)x_3] \ , \\ \eta_2 &= \frac{1}{\rho_2-\rho_1}\,(x_2 - x_1) \ , \\ \xi &= x_3 \ . \end{aligned} \tag{2.112}$$

Finally, using formulas (2.112) and introducing

(2.113)
$$\beta_1 = \frac{p_2b_{21} + p_3b_{31}}{\rho_1(\rho_2-\rho_1)} - \frac{p_1 + p_2b_{22}}{\rho_2 - \rho_1} = \frac{1}{\rho_2 - \rho_1}\left[- p_1 + \rho_1p_2 + \frac{b_{31}p_3}{\rho_1}\right] ,$$
$$\beta_2 = \frac{p_2b_{21} + p_3b_{31}}{\rho_2(\rho_2-\rho_1)} + \frac{p_1 + p_2b_{22}}{\rho_2 - \rho_1} = \frac{1}{\rho_2 - \rho_1}\left[p_1 - \rho_2p_2 - \frac{b_{31}p_3}{\rho_2}\right] ,$$
$$\beta_3 = - p_2 - \frac{b_{21}p_2 + p_3b_{31}}{\rho_1\rho_2} = - \frac{b_{31}p_3}{\rho_1\rho_2} ,$$

we get

(2.114)
$$\dot{x}_1 = - \rho_1x_1 + f(\sigma) ,$$
$$\dot{x}_2 = - \rho_2x_2 + f(\sigma) ,$$
$$\dot{x}_3 = f(\sigma) ,$$

$$\sigma = \sum_{k=1}^{3} \gamma_kx_k, \quad \dot{\sigma} = \beta_1x_1 + \beta_2x_2 + \beta_3x_3 - f(\sigma) .$$

Let us now assume that the parameters of the regulated object are such that $U = kT^2$. In this case $\rho_1 = \rho_2$ and (2.11) will determine only two rows of the transformation.

(2.115)
$$x_1 = \rho_1\rho_2\eta_1 + \rho_1\eta_2 + \xi ,$$
$$x_3 = \xi .$$

Corresponding to them are two canonical equations

$$\dot{x}_1 = - \rho_1x_1 + f(\sigma)$$
$$\dot{x}_3 = f(\sigma) .$$

From the first equation of (2.115) we get

$$\eta_1 = \frac{1}{\rho_1\rho_2} [x_1 - \rho_1\eta_2 - \xi] .$$

Eliminating the variable η_1 from the second equation of (2.105) we get

(2.116) $$\dot{\eta}_2 = \frac{b_{21}}{\rho_1\rho_2}[x_1 - \rho_1\eta_2 - \xi] + b_{22}\eta_2 + n_2\xi = - x_1 - \rho_1\eta_2 \ .$$

Finally, let us put

$$\eta_2 = C_1x_1 + C_2x_2 \ ,$$

where C_1 and C_2 are arbitrarily chosen numbers. Substitution of this equation into (2.116) yields

$$C_1[- \rho_1x_1 + f(\sigma)] + C_2\dot{x}_2 = - x_1 - \rho_1[C_1x_1 + C_2x_2] \ ,$$

and after elementary transformations we get

$$\dot{x}_2 = - \rho_1x_2 + \varepsilon_1x_1 + q_2f(\sigma) \ .$$

Here ε_1 and q_2 denote the two arbitrary numbers

(2.117) $$\varepsilon_1 = - \frac{1}{C_2}, \quad q_2 = - \frac{C_1}{C_2} \ ,$$

which can be chosen as small as convenient.

To form the last equation we have

$$\dot{\sigma} = p_1\eta_2 + p_2(b_{21}\eta_1 + b_{22}\eta_2 + n_2\xi) + p_3b_{31}\eta_1 - f(\sigma) =$$

$$= \frac{n_2}{b_{21}}(p_2b_{21} + p_3b_{31})(x_1 - \rho_1\eta_2 - \xi) + (p_2b_{22} + p_1)\eta_2 + \\ + n_2p_2\xi - f(\sigma)$$

or, changing over to the canonical variables

$$\dot{\sigma} = \left[C_1(p_1 + b_{22}p_2) + \frac{n_2}{b_{21}}\left(1 + \frac{\rho_1}{n_2} C_1 \right)(b_{21}p_2 + b_{31}p_3)\right] x_1 + \\ + \left[C_2(p_1 + b_{22}p_2) + \frac{n_2}{b_{21}}\frac{\rho_1}{n_2} C_2(b_{21}p_2 + b_{31}p_3)\right] x_2 + \\ + \left[n_2p_2 - \frac{n_2}{b_{21}}(p_2b_{21} + p_3b_{31})\right] \eta_3 - f(\sigma) \ .$$

Finally, denoting for brevity

$$\beta_1 = C_1(p_1 + b_{22}p_2) - \frac{1}{b_{21}}(1 - \rho_1 C_1)(b_{21}p_2 + b_{31}p_3) \ ,$$

(2.118)

$$\beta_2 = C_2(p_1 + b_{22}p_2) + \frac{\rho_1}{b_{21}} C_2(b_{21}p_2 + b_{31}p_3), \quad \beta_3 + \frac{b_{31}p_3}{b_{21}}, \quad q_1 = 1,$$

we obtain as the end result

(2.119)

$$\begin{aligned} \dot{x}_1 &= -\rho_1 x_1 + q_1 f(\sigma) \ , \\ \dot{x}_2 &= \varepsilon_1 x_1 - \rho_1 x_2 + q_2 f(\sigma) \ , \\ \dot{x}_3 &= f(\sigma) \ , \\ \dot{\sigma} &= \beta_1 x_1 + \beta_2 x_2 + \beta_3 x_3 - f(\sigma) \ . \end{aligned}$$

7. REGULATION OF THE STEADY STATE OF A SYSTEM SUBJECTED TO THE ACTION OF CONSTANT DISTURBING FORCES*

Let us consider a mechanical system, describable by the equation

(2.120)

$$\begin{aligned} T^2\ddot{\psi} + U\dot{\psi} + k\beta + \mu &= \chi \ , \\ \dot{\psi} - \dot{\beta} - n\beta &= -n\varepsilon \ , \\ \dot{\mu} + \kappa\mu &= f^*(\sigma) \ , \\ \sigma &= a\psi + E\dot{\psi} + G^2\ddot{\psi} - \frac{1}{\ell}\mu \ . \end{aligned}$$

Here ψ and β are the coordinates; T^2, U, k and n are the constants of the regulated object; μ is the coordinate of the regulator; a, E, G^2, ℓ, and κ are the constants of the regulator; χ and ε are constants that characterize the external continuously acting disturbing forces.

All the constants are positives with the exception of k, which can be any real number. The old constants retain here the same values as in the Bulgakov problem; the new constants n and κ characterize new applied forces.

The force $n\beta$ appears here as the result of the new degree of freedom that is added to the Bulgakov problem, while the force $\kappa\mu$ is the

* A. M. Letov. Regulation of the Stationary State of a System Subjected to the Action of Constant Disturbing Forces, PMM, Vol. XII, No. 2, 1948.

result of a load acting on the regulator. At $n = \kappa = 0$ (2.120) becomes (2.81).

The desired steady state of the control system, which must be maintained by the regulator, is characterized by the coordinates

$$(2.121) \quad \psi^{**} = 0, \ \dot{\psi}^{**} = 0, \ \beta^{*} = \varepsilon .$$

In actuality, however, in the presence of constantly-applied disturbing forces, the steady state of the control system will be described by the obvious solutions of (2.120), i.e.,

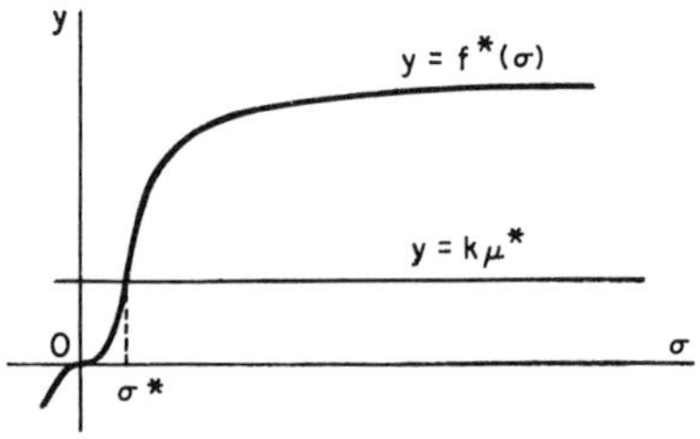

FIGURE 3

$$(2.122) \qquad a\psi^{*} = \frac{1}{\ell}\mu^{*} + \sigma^{*}, \quad \dot{\psi}^{*} = 0, \quad \mu^{*} = \chi - k\varepsilon; \quad \beta^{*} = \varepsilon ,$$

where the constant σ^{*} is the real root of the equation

$$(2.123) \qquad \kappa\mu^{*} = f^{*}(\sigma)$$

(Figure 3). From the physical nature of the relation between the force developed by the regulator and the force $\kappa\mu^{*}$, it follows that such a root must exist for any sensibly constructed regulator, for the latter must always be able to cope with the external force acting on it. In the particular case of a hydraulic motor, for which we can assume $\kappa = 0$, (2.123) yields $|\sigma^{*}| \leq \sigma_{*}$.

Obviously, the actual steady state (2.122) becomes the desired steady state (2.121) when the speed $\dot{\mu}$ with which the regulator organ changes its setting is described by a function $f^{*}(\sigma)$ of subclass (A') and is independent of the load acting on it $(\kappa = 0)$, and the regulator does not include a proportional feedback element.

Let us put

$$\psi - \psi^{*} = \eta_1, \quad \frac{d\dot{\psi}}{dt} = \sqrt{r}\eta_2, \quad \beta - \beta^{*} = \eta_3, \quad \mu - \mu^{*} = i\xi ,$$

$$(2.124) \qquad \sigma - \sigma^{*} = \zeta, \quad t = \frac{\tau}{\sqrt{r}}, \quad p = \frac{U}{T^2} ,$$

$$q = \frac{k}{T^2}, \quad i = \frac{\ell T^2}{T^2 + \ell G^2}, \quad r = \frac{1}{T^2}, \quad b_{22} = -\frac{p}{\sqrt{r}} ,$$

(2.124)
(continued)

$$b_{23} = -\frac{q}{r}, \quad b_{33} = -\frac{n}{\sqrt{r}}, \quad p_1 = a, \quad n_2 = -1,$$

$$p_2 = (E - pG^2)\sqrt{r}, \quad p_3 = -qG^2, \quad p_4 = -1,$$

$$\chi^* = \frac{\chi}{\sqrt{r}}, \quad \Phi(\zeta) = \frac{1}{i\sqrt{r}}\,[f(\sigma) - \chi\mu^*],$$

and let us reduce the original equations to normal form in dimensionless coordinates

(2.125)

$$\begin{aligned} \dot{\eta}_1 &= \eta_2, \\ \dot{\eta}_2 &= b_{22}\eta_2 + b_{23}\eta_3 + n_2\xi, \\ \dot{\eta}_3 &= \eta_2 + b_{33}\eta_3, \\ \dot{\xi} &= -\kappa^*\xi + \Phi(\zeta), \quad \zeta = p_1\eta_1 + p_2\eta_2 + p_3\eta_3 - \xi. \end{aligned}$$

Here the function $\Phi(\zeta)$ is a function of class (A), or more likely of subclass (A'), and we deal with Case 3.

Proceeding to the construction of the canonical transformations, let us examine the determinant $Д(\rho)$ and its derivative

(2.126)

$$Д(\rho) = \begin{vmatrix} \rho & 0 & 0 \\ 1 & \rho + b_{22} & 1 \\ 0 & b_{23} & \rho + b_{33} \end{vmatrix} = \rho[(\rho + b_{22})(\rho + b_{33}) - b_{23}],$$

$$Д'(\rho) = 3\rho^2 + 2\rho(b_{22} + b_{33}) + b_{22}b_{33} - b_{23}.$$

Setting this determinant equal to zero, we get the roots of (1.2):

$$\rho_1 = 0, \quad \rho_2 + \rho_3 = \frac{p + n}{\sqrt{r}}, \quad \rho_2\rho_3 = \frac{q + pn}{r}.$$

As in the preceding examples, we find from (2.37) with $i = 1$:

(2.127)

$$Д_{11}(\rho) = (\rho + b_{22})(\rho + b_{33}) - b_{23}, \quad Д_{12} = -(\rho + b_{33}), \quad Д_{13} = b_{23},$$

and also

(2.128)

$$N_1 = \rho + b_{33}, \quad N_2 = -\rho(\rho + b_{33}), \quad N_3 = \rho.$$

If we employ formulas (2.35), we get

$$C_1^{(s)} = \frac{\rho_4 - \rho_s}{\rho_2 + b_{33}} \, \Delta_{11}(\rho_s), \quad C_2^{(s)} = \rho_s - \rho_4, \quad C_3^{(s)} = \frac{\rho_4 - \rho_s}{\rho_s + b_{33}} \, b_{23} \, .$$

For known values of the quantities ρ_s these formulas yield the following coefficients of direct transformation:

$$C_1^{(1)} = - \frac{\rho_4(pn+q)}{n\sqrt{r}}, \quad C_1^{(2)} = C_1^{(3)} = 0 \; ,$$

$$C_2^{(1)} = - \rho_4, \quad C_2^{(2)} = \rho_2 - \rho_4, \quad C_2^{(3)} = \rho_3 - \rho_4 \; ,$$

$$C_3^{(1)} = \frac{\rho_4 q}{n\sqrt{r}}, \quad C_3^{(2)} = \frac{\rho_2 - \rho_4}{r\rho_2 - n\sqrt{r}} \, q, \quad C_3^{(3)} = \frac{\rho_3 - \rho_4}{r\rho_3 - n\sqrt{r}} \, q \, .$$

Thus, the sought transformation is of the form

$$(2.129) \qquad \begin{aligned} x_1 &= - \frac{\rho_4(pn+q)}{n\sqrt{r}} \eta_1 - \rho_4\eta_2 + \frac{\rho_4 q}{n\sqrt{r}} \eta_3 + \xi \; , \\ x_2 &= (\rho_2 - \rho_4)\eta_2 + \frac{(\rho_2 - \rho_4)q}{r\rho_2 - n\sqrt{r}} \eta_3 + \xi \; , \\ x_3 &= (\rho_3 - \rho_4)\eta_2 + \frac{(\rho_3 - \rho_4)q}{r\rho_3 - n\sqrt{r}} \eta_3 + \xi \, . \end{aligned}$$

This transformation is nonsingular if the quantities ρ_2, ρ_3, and ρ_4 differ from each other and are not equal to zero.

To construct the inverse transformation we employ

$$(2.130) \quad \Delta'(\rho_1) = \rho_2\rho_3, \quad \Delta'(\rho_2) = \rho_2(\rho_2 - \rho_3), \quad \Delta'(\rho_3) = \rho_3(\rho_3 - \rho_2) \, .$$

It now remains for us to employ formulas (2.36), (2.127), and (2.130) to calculate the coefficients of the inverse transformation. We get

$$D_1^{(1)} = \frac{b_{33}}{\rho_2\rho_3\rho_4}, \quad D_2^{(1)} = \frac{\rho_2 + b_{33}}{\rho_2(\rho_4-\rho_2)(\rho_2-\rho_3)} ,$$

$$D_3^{(1)} = \frac{\rho_3 + b_{33}}{\rho_3(\rho_4-\rho_3)(\rho_3-\rho_2)} ,$$

$$D_1^{(2)} = 0, \quad D_2^{(2)} = \frac{\rho_2 + b_{33}}{(\rho_4-\rho_2)(\rho_3-\rho_2)} ,$$

(2.131) $$D_3^{(2)} = - \frac{\rho_3 + b_{33}}{(\rho_4-\rho_3)(\rho_3-\rho_2)} ,$$

$$D_1^{(3)} = 0, \quad D_2^{(3)} = \frac{1}{(\rho_4-\rho_2)(\rho_2-\rho_3)}, \quad D_3^{(3)} = \frac{1}{(\rho_4-\rho_3)(\rho_3-\rho_2)} ,$$

$$G_1 = - \frac{\rho_4 + b_{33}}{\rho_4(\rho_4-\rho_2)(\rho_4-\rho_3)} ,$$

$$G_2 = \frac{\rho_4 + b_{33}}{(\rho_4-\rho_3)(\rho_4-\rho_2)} , \quad G_3 = - \frac{1}{(\rho_4-\rho_2)(\rho_4-\rho_3)} .$$

To obtain the canonical form of the initial equations, it remains to calculate the numbers γ_k and β_k. Using (2.17) we get

$$\gamma_1 = p_1D_1^{(1)} + p_2D_1^{(2)} + p_3D_1^{(3)} = \frac{p_1b_{33}}{\rho_2\rho_3\rho_4} ,$$

$$\gamma_2 = p_1D_2^{(1)} + p_2D_2^{(2)} + p_3D_2^{(3)} =$$

$$= \frac{1}{(\rho_4-\rho_2)(\rho_2-\rho_3)} \left[\frac{p_1(\rho_2+b_{33})}{\rho_2} - p_2(\rho_2 + b_{33}) + p_3 \right] ,$$

(2.132)

$$\gamma_3 = p_1D_3^{(1)} + p_2D_3^{(2)} + p_3D_3^{(3)} =$$

$$= \frac{1}{(\rho_4-\rho_3)(\rho_3-\rho_2)} \left[\frac{p_1(\rho_3+b_{33})}{\rho_3} - p_2(\rho_3 + b_{33}) + p_3 \right] ,$$

$$\gamma_4 = - 1 - \frac{1}{(\rho_4-\rho_2(\rho_4-\rho_3)} \left[\frac{p_1(\rho_4+b_{33})}{\rho_4} - p_2(\rho_4+b_{33}) + p_3 \right] ,$$

$$\beta_1 = 0, \quad \beta_2 = - \rho_2\gamma_2, \quad \beta_3 = - \beta_3\gamma_3, \quad \beta_4 = - \rho_4\gamma_4 .$$

Consequently, the canonical equations assume the following final form:

$$(2.133)\qquad \begin{aligned} \dot{x}_1 &= \Phi(\zeta) \;, \\ \dot{x}_2 &= -\rho_2 x_2 + \Phi(\zeta) \;, \\ x_3 &= -\rho_3 x_3 + \Phi(\zeta) \;, \\ \dot{x}_4 &= -\rho_4 x_4 + \Phi(\zeta) \;, \\ \dot{\zeta} &= \beta_2 x_2 + \beta_3 x_3 + \beta_4 x_4 - \Phi(\zeta) \;, \\ \zeta &= \sum_{k=1}^{4} \gamma_k x_k \;. \end{aligned}$$

8. THE FIRST BULGAKOV PROBLEM IN THE CASE OF NON-IDEAL SENSING DEVICES

Let us assume that the sensing device that determines the quantity $\alpha = E\dot{\psi} + G^2\ddot{\psi}$ is not ideal, so that its intrinsic parameters have a substantial influence on the equilibrium stability of the system.

We assume that the coordinate α of the sensing device satisfies a second order linear differential equation. We then have

$$(2.134)\qquad \begin{aligned} T^2\ddot{\psi} &= U\dot{\psi} + \mu = 0 \;, \\ \dot{\mu} &= f^*(\sigma), \qquad \sigma = a\psi + \tilde{E}\alpha - \frac{1}{\ell}\mu \;, \\ T_s^2\ddot{\alpha} + H\dot{\alpha} + \alpha &= M\dot{\psi} + N^2\ddot{\psi} \;, \end{aligned}$$

where T_s and H are the parameters of the measuring device; if $T_s = H = 0$, then, using the designations

$$(2.135)\qquad \tilde{E}M = E, \qquad \tilde{E}N^2 = G^2 \;,$$

we obtain the problem already considered.

Before we proceed to any construction whatever, let us reduce the equations to the normal and dimensionless form, putting

$$\psi = \eta_1, \quad \frac{d\psi}{dt} = \sqrt{r^*}\eta_2, \quad \mu = \ell\xi, \quad t = \frac{\tau}{\sqrt{r^*}},$$

$$r = \frac{i}{T^2}, \quad i = \frac{\ell T^2}{T^2 + \ell G^2},$$

$$b_{22} = -\frac{p}{\sqrt{r^*}}, \quad b_{42} = \frac{MT^2 - UN^2}{T^2T_s^2\sqrt{r^*}}, \quad b_{43} = -\frac{1}{r^*T_s^2},$$

(2.136)

$$b_{44} = -\frac{H}{T_s^2\sqrt{r^*}}, \quad r^* = -n_2 r, \quad r^* = \frac{\ell}{T\sqrt{T^2 + \ell G^2}},$$

$$n_2 = -\frac{\ell}{T^2 r^*}, \quad n_4 = -\frac{N^2\ell}{T^2T_s^2 r^*},$$

$$\frac{1}{i\sqrt{r^*}} f^*(\sigma) = f(\sigma), \quad \alpha = \eta_3, \quad \dot{\alpha} = \eta_4 .$$

Using the dot as before to denote differentiation with respect to the dimensionless quantity τ, we get

(2.137)

$$\begin{aligned}
\dot{\eta}_1 &= \eta_2 , \\
\dot{\eta}_2 &= b_{22}\eta_2 + n_2\xi , \\
\dot{\eta}_3 &= \eta_4 , \\
\dot{\eta}_4 &= b_{42}\eta_2 + b_{43}\eta_3 + b_{44}\eta_4 + n_4\xi , \\
\dot{\xi} &= f(\sigma), \quad \sigma = a\eta_1 + \tilde{E}\eta_3 - \xi , \\
p_1 &= a, \quad p_2 = p_4 = 0, \quad p_3 = \tilde{E} .
\end{aligned}$$

Before we proceed to the construction of the canonical transformations, we first calculate the determinant $Д(\rho)$ and its derivative. We have

(2.138)

$$Д(\rho) = \begin{vmatrix} \rho & 0 & 0 & 0 \\ 1 & \rho + b_{22} & 0 & b_{42} \\ 0 & 0 & \rho & b_{43} \\ 0 & 0 & 1 & \rho + b_{44} \end{vmatrix},$$

$$Д'(\rho) = 4\rho^3 + 3\rho^2(b_{22} + b_{44}) + 2\rho[b_{22}b_{44} - b_{43}] - b_{22}b_{43} .$$

From this we obtain the roots of (1.2)

$$\rho_1 = 0,\ \rho_2 = -b_{22},\ \rho_3 + \rho_4 = -b_{44},\ \rho_3\rho_4 = -b_{43} . \tag{2.139}$$

Next, putting $i = 4$, we obtain the minors $Д_{4k}$ of the determinant (2.138):

$$Д_{41} = 0,\ Д_{42} = \begin{vmatrix} \rho & 0 & 0 \\ 1 & 0 & b_{42} \\ 0 & \rho & b_{43} \end{vmatrix} = -\rho^2 b_{42} ,$$

$$Д_{43} = -\begin{vmatrix} \rho & 0 & 0 \\ 1 & \rho + b_{22} & b_{42} \\ 0 & 0 & b_{43} \end{vmatrix} = -\rho(\rho + b_{22})b_{43},\ Д_{44} = \rho^2(\rho + b_{22}) .$$

Using formulas (2.37) we obtain the quantities

$$N_1 = \begin{vmatrix} 0 & n_2 & 0 & n_4 \\ 1 & \rho + b_{22} & 0 & b_{42} \\ 0 & 0 & \rho & b_{43} \\ 0 & 0 & 1 & \rho + b_{44} \end{vmatrix} = [\rho(\rho + b_{44}) - b_{43}](-n_2) ,$$

$$N_2 = \begin{vmatrix} \rho & 0 & 0 & 0 \\ 0 & n_2 & 0 & n_4 \\ 0 & 0 & \rho & b_{43} \\ 0 & 0 & 1 & \rho + b_{44} \end{vmatrix} = n_2\rho[\rho(\rho + b_{44}) - b_{43}] ,$$

$$N_3 = \begin{vmatrix} \rho & 0 & 0 & 0 \\ 1 & \rho + b_{22} & 0 & b_{42} \\ 0 & n_2 & 0 & n_4 \\ 0 & 0 & 1 & \rho + b_{44} \end{vmatrix} = -\rho[n_4(\rho + b_{22}) - n_2 b_{42}] ,$$

$$N_4 = \begin{vmatrix} \rho & 0 & 0 & 0 \\ 1 & \rho + b_{22} & 0 & b_{42} \\ 0 & 0 & \rho & b_{43} \\ 0 & n_2 & 0 & n_4 \end{vmatrix} = \rho^2[n_4(\rho + b_{22}) - n_2 b_{42}] .$$

Consequently, in accordance with formulas (2.35), the sought transformation becomes

$$x_1 = \xi ,$$

$$x_2 = \rho_2\eta_2 + \xi ,$$

$$x_3 = \frac{\rho_3 b_{42}}{n_4(\rho_3+b_{22}) - b_{42}n_2}\eta_2 + \frac{b_{43}(\rho_3+b_{22})}{n_4(\rho_3+b_{22}) - b_{42}n_2}\eta_3 - \frac{\rho_3(\rho_3+b_{22})}{n_4(\rho_3+b_{22}) - b_{42}n_2}\eta_4 + \xi ,$$

$$x_4 = \frac{\rho_4 b_{42}}{n_4(\rho_4+b_{22}) - b_{42}n_2}\eta_2 + \frac{b_{43}(\rho_4+b_{22})}{n_4(\rho_4+b_{22}) - b_{42}n_2}\eta_3 - \frac{\rho_4(\rho_4+b_{22})}{n_4(\rho_4+b_{22}) - b_{42}}\eta_4 + \xi .$$

To calculate the coefficients of the inverse transformation, and also to calculate the quantity β_k, we find:

$$Д'(\rho_1) = - b_{22}b_{43}, \qquad Д'(\rho_3) = - \rho_3(\rho_4 - \rho_3)(\rho_3 + b_{22}) ,$$

$$Д'(\rho_2) = \rho_2[\rho_2(\rho_2 + b_{44}) - b_{43}], \qquad Д'(\rho_4) = \rho_4(\rho_4 - \rho_3)(\rho_4 + b_{22}) .$$

Since $p_2 = p_4 = 0$, we need only the quantities

$$D_k^{(1)} = \frac{N_1(\rho_k)}{\rho_k\, Д'(\rho_k)} , \quad D_k^{(3)} = - \frac{N_3(\rho_k)}{\rho_k\, Д'(\rho_k)} , \quad (k = 1, \ldots, 4) , \tag{2.140}$$

and then we employ (2.10) and (2.36) to find

$$\begin{aligned}
\beta_1 &= \frac{p_1N_1(\rho_1) + p_3N_3(\rho_1)}{D'(\rho_1)} = - \frac{p_1}{\rho_2}(- n_2) , \\
\beta_2 &= \frac{p_1N_1(\rho_2) + p_3N_3(\rho_2)}{Д'(\rho_2)} = \left[\frac{p_1}{\rho_2} - \frac{b_{42}p_3}{\rho_2(\rho_2+b_{44}) - b_{43}} \right] (- n_2) , \\
\beta_3 &= \frac{n_4(\rho_3+b_{22}) - n_2b_{42}}{(\rho_3+b_{22})(\rho_4-\rho_3)} p_s , \\
\beta_4 &= - \frac{n_4(\rho_4+b_{22}) - n_2b_{42}}{(\rho_4+b_{22})(\rho_4-\rho_3)} p_3 .
\end{aligned} \tag{2.141}$$

The canonical equations of the problem become

$$\begin{aligned}
\dot{x}_1 &= f(\sigma) \ , \\
\dot{x}_2 &= -\rho_2 x_2 + f(\sigma) \ , \\
\dot{x}_3 &= -\rho_3 x_3 + f(\sigma) \ , \\
\dot{x}_4 &= -\rho_4 x_4 + f(\sigma) \ , \\
\dot{\sigma} &= \beta_1 x_1 + \beta_2 x_2 + \beta_3 x_3 + \beta_4 x_4 - f(\sigma) \ .
\end{aligned} \tag{2.142}$$

9. BULGAKOV'S SECOND PROBLEM IN THE CASE OF NON-IDEAL SENSING DEVICES

In this case we take into account the restoring force $k\psi$ $(k > 0)$ of the object of regulation, and therefore the initial equations have the following form:

$$\begin{aligned}
&T^2\ddot{\psi} + U\dot{\psi} + k\psi + \mu = 0 \ , \\
&\dot{\mu} = f^*(\sigma), \qquad \sigma = a\psi + \tilde{E}\alpha - \frac{1}{\ell}\mu \ , \\
&T_s^2\ddot{\alpha} = H\dot{\alpha} + \alpha = M\dot{\psi} + N^2\ddot{\psi} \ .
\end{aligned} \tag{2.143}$$

Retaining the previous notation of (2.136), and also putting

$$q = \frac{k}{T^2}, \qquad b_{21} = -\frac{q}{r^*}, \qquad b_{41} = -\frac{qN^2}{r^* T_s^2} \ , \tag{2.144}$$

we write the initial equations in normal form

$$\begin{aligned}
\dot{\eta}_1 &= \eta_2 \ , \\
\dot{\eta}_2 &= b_{21}\eta_1 + b_{22}\eta_2 + n_2\xi \ , \\
\dot{\eta}_3 &= \eta_4 \ , \\
\dot{\eta}_4 &= b_{41}\eta_1 + b_{42}\eta_2 + b_{43}\eta_3 + b_{44}\eta_4 + n_4\xi \ , \\
\dot{\xi} &= f(\sigma), \qquad \sigma = p_1\eta_1 + p_2\eta_3 - \xi \ .
\end{aligned} \tag{2.145}$$

In order to obtain the equations in the canonical form, we must carry out the necessary calculations.

We set up the determinant $Д(\rho)$ and its derivatives. We obtain

(2.146)
$$Д(\rho) = \begin{vmatrix} \rho & b_{21} & 0 & b_{41} \\ 1 & \rho + b_{22} & 0 & b_{42} \\ 0 & 0 & \rho & b_{43} \\ 0 & 0 & 1 & \rho + b_{44} \end{vmatrix} = \Delta_1(\rho)\Delta_3(\rho) ,$$

$$Д'(\rho) = (2\rho + b_{22})\Delta_3(\rho) - (2\rho + b_{44})\Delta_1(\rho) .$$

Here we denote

(2.147) $$\Delta_1(\rho) = \rho(\rho + b_{22}) - b_{21}, \quad \Delta_3(\rho) = \rho(\rho + b_{44}) - b_{43} .$$

The roots of (1.2) are determined by the equalities

(2.148)
$$\rho_1 + \rho_2 = - b_{22}, \quad \rho_1\rho_2 = - b_{21} ,$$
$$\rho_3 + \rho_4 = - b_{44}, \quad \rho_3\rho_4 = - b_{43} .$$

When $k > 0$ we find that ρ_1 and ρ_2 satisfy as before the condition $\mathrm{Re}\rho_k > 0$ $(k = 1, 2)$.

Next, putting $i = 4$, we calculate the minors $Д_{ik}$ of the determinant (2.146). We find

(2.149)
$$Д_{41} = - \rho b_{21}b_{42} + \rho(\rho + b_{22})b_{41}, \quad Д_{42} = - \rho^2 b_{42} + \rho b_{41} ,$$
$$Д_{43} = - b_{43}\Delta_1(\rho), \quad Д_{44} = \rho\Delta_1(\rho) .$$

We obtain from (2.37):

(2.150)
$$N_1(\rho) = - n_2\Delta_3(\rho) ,$$
$$N_2(\rho) = n_2\rho\Delta_3(\rho) ,$$
$$N_3(\rho) = - n_4\Delta_1(\rho) + n_2(\rho b_{42} - b_{41}) ,$$
$$N_4(\rho) = n_4\rho\Delta_1(\rho) - n_2\rho(\rho b_{42} - b_{41}) .$$

The canonical transformation of this problem is as follows:

$$x_1 = \frac{Д_{41}(\rho_1)}{n_2(\rho_1 b_{42} - b_{41})}\eta_1 + \frac{\rho_1}{(-n_2)}\eta_2 + \xi \quad ,$$

$$x_2 = \frac{Д_{41}(\rho_2)}{n_2(\rho_2 b_{42} - b_{41})}\eta_1 + \frac{\rho_2}{(-n_2)}\eta_2 + \xi \quad ,$$

(2.151)
$$x_3 = \frac{\rho_3 Д_{41}(\rho_3)}{N_4(\rho_3)}\eta_1 - \frac{\rho_3 Д_{42}(\rho_3)}{N_4(\rho_3)}\eta_2 - \frac{\rho_3 b_{43}\Delta_1(\rho_3)}{N_4(\rho_3)}\eta_3 - \frac{\rho_3^2\Delta_1(\rho_3)}{N_4(\rho_3)}\eta_4 + \xi \quad ,$$

$$x_4 = -\frac{\rho_4 Д_{41}(\rho_4)}{N_4(\rho_4)}\eta_1 - \frac{\rho_4 Д_{42}(\rho_4)}{N_4(\rho_4)}\eta_2 - \frac{\rho_4 b_{43}\Delta_1(\rho_4)}{N_4(\rho_4)}\eta_3 - \frac{\rho_4^2\Delta_1(\rho_4)}{N_4(\rho_4)}\eta_4 + \xi \quad .$$

When $k = 0$ this transformation coincides with the canonical transformation derived in the preceding problem.

To write the canonical equations, we still have to calculate the coefficients of the inverse transformation and the quantity β_k. Since $Д(\rho)$ is a fourth-degree polynomial, then

(2.152)
$$Д'(\rho_1) = (\rho_1 - \rho_2)(\rho_1 - \rho_3)(\rho_1 - \rho_4) \quad ,$$
$$Д'(\rho_2) = (\rho_2 - \rho_1)(\rho_2 - \rho_3)(\rho_2 - \rho_4) \quad ,$$
$$Д'(\rho_3) = (\rho_3 - \rho_1)(\rho_3 - \rho_2)(\rho_3 - \rho_4) \quad ,$$
$$Д'(\rho_4) = (\rho_4 - \rho_1)(\rho_4 - \rho_2)(\rho_4 - \rho_3) \quad .$$

Inserting quantities (2.150) and (2.152) into formulas (2.10) and (2.36) we get

(2.153)
$$\beta_1 = -\frac{n_2}{\rho_1 - \rho_2}\left[p_1 + \frac{b_{41}p_3 - \rho_1 b_{42}p_3}{(\rho_1-\rho_3)(\rho_1-\rho_4)}\right] ,$$

$$\beta_2 = \frac{n_2}{\rho_1 - \rho_2}\left[p_1 + \frac{b_{41}p_3 - \rho_2 b_{42}p_3}{(\rho_2-\rho_3)(\rho_2-\rho_4)}\right] ,$$

$$\beta_3 = -\frac{n_4(\rho_3-\rho_1)(\rho_3-\rho_2) + n_2 b_{41} - n_2 b_{42}\rho_3}{(\rho_3-\rho_1)(\rho_3-\rho_2)(\rho_3-\rho_4)}\,p_3 \quad ,$$

(2.153) (continued)
$$\beta_4 = - \frac{n_4(\rho_4-\rho_1)(\rho_4-\rho_2) + n_2 b_{41} - n_2 b_{42}\rho_4}{(\rho_4-\rho_1)(\rho_4-\rho_2)(\rho_4-\rho_3)} p_3 \quad .$$

The canonical equations of the problem become

(2.154)
$$\begin{aligned}
\dot{x}_1 &= - \rho_1 x_1 + f(\sigma) \quad , \\
\dot{x}_2 &= - \rho_2 x_2 + f(\sigma) \quad , \\
\dot{x}_3 &= - \rho_3 x_3 + f(\sigma) \quad , \\
\dot{x}_4 &= - \rho_4 x_4 + f(\sigma) \quad , \\
\dot{\sigma} &= \beta_1 x_1 + \beta_2 x_2 + \beta_3 x_3 + \beta_4 x_4 - Rf(\sigma) \quad .
\end{aligned}$$

When $k = 0$ we get $\rho_1 = 0$ and $\rho_2 = - b_{22}$, and by proper choice of the multiplier R, (2.154) are transformed into (2.142). In fact, it is seen from (2.153) that when $k = 0$ all the quantities β_k coincide with the values obtained from formulas (2.141).

10. INDIRECT CONTROL OF MACHINES

Let us consider two other particular problems in automatic control. These will be used to verify that the preceding arguments in connection with the contruction of the canonical transformation cannot be used and require a certain generalization.

Let us first consider the classical scheme of indirect control of a machine in the presence of an isodrome.*

The initial equations have in this case the following form:

(1) Equation of a machine with self-equalization

$$T_a\dot{\varphi} + n\varphi = - \xi \quad .$$

(2) Equation of a sensing element having mass and viscous friction,

$$T_r^2\ddot{\eta} + T_k\dot{\eta} + \delta\eta = \varphi \quad ,$$

(3) Equation of the isodrome element and of the servo motor,

* A. I. Lur'e, Nekotorye nelineĭnye zadachi teorii avtomaticheskovo regulirovaniya (Certain Nonlinear Problems in the Theory of Automatic Control), Gostekhizdat, 1951.

$$T_1\dot{\zeta} = \zeta = \beta T_1\dot{\xi} + \gamma\xi ,$$

$$\dot{\xi} = f^*(\sigma), \quad \sigma = \eta - \zeta .$$

Here T_a, n, T_r, T_k, δ, β, T_1, and γ are positive constants. Introducing now the symbols

$$t = T_a\tau, \quad \eta = \eta_1, \quad T_a\frac{d}{dt} = \eta_2, \quad \varphi = \eta_3, \quad \zeta = \eta_4, \quad \xi = \eta_5 ,$$

$$-\frac{T_aT_k}{T_r^2} = b_{22}, \quad -\frac{\delta T_a^2}{T_r^2} = b_{21}, \quad \frac{T_a^2}{T_r^2} = b_{23}, \quad T_af^*(\sigma) = f(\sigma) ,$$

(2.155)
$$h_4 = \beta, \quad -n = b_{33}, \quad -1 = b_{35}, \quad -\frac{T_a}{T_1} = b_{44} ,$$

$$\frac{\gamma T_a}{T_1} = b_{45}, \quad h_5 = 1 ,$$

we reduce the above equation to the normal form in dimensionless variables

(2.156)
$$\begin{aligned}
\dot{\eta}_1 &= \eta_2 , \\
\dot{\eta}_2 &= b_{21}\eta_1 + b_{22}\eta_2 + b_{23}\eta_3 , \\
\dot{\eta}_3 &= b_{33}\eta_3 + b_{35}\eta_5 , \\
\dot{\eta}_4 &= b_{44}\eta_4 + b_{45}\eta_5 + h_4f(\sigma) , \\
\dot{\eta}_5 &= h_5f(\sigma) , \\
\sigma &= \eta_1 - \eta_4 .
\end{aligned}$$

A more complicated example is the problem of indirect control of a machine in the presence of an isodrome that has a "mass".*

In this case the equation of the isodrome is written as

$$\alpha T_1^2\ddot{\zeta} + T_1\dot{\zeta} + \zeta = \beta T_1\dot{\xi} + \gamma\xi ,$$

where α is a positive constant, characterizing the "mass". Retaining here the notation of (2.155), and introducing the new symbols

(2.157) $$T_a\frac{d\eta_4}{dt} = \eta_6, \quad \frac{\beta T_a}{aT_1} = h_6, \quad b_{64} = -\frac{T_a^2}{aT_1^2}, \quad b_{65} = \frac{\gamma T_a^2}{aT_1^2}, \quad b_{66} = -\frac{T_a}{aT_1} ,$$

* V. L. Lossievskiĭ, Avtomaticheskie regulyatory (Automatic Regulators), NKAP, Oborongiz, 1944.

we get a normal system of equations in dimensionless variables

$$\begin{aligned}
\dot{\eta}_1 &= \eta_2 \quad , \\
\dot{\eta}_2 &= b_{21}\eta_1 + b_{22}\eta_2 + b_{23}\eta_3 \quad , \\
\dot{\eta}_3 &= b_{33}\eta_3 + b_{35}\eta_5 \quad , \\
\dot{\eta}_4 &= \eta_6 \quad , \\
\dot{\eta}_5 &= f(\sigma) \quad , \\
\dot{\eta}_6 &= b_{64}\eta_4 + b_{65}\eta_5 + b_{66}\eta_6 + h_6 f(\sigma) \quad , \\
\sigma &= \eta_1 - \eta_4 \quad .
\end{aligned} \tag{2.158}$$

Each of these problems is characterized by the presence of the function $f(\sigma)$ in the right sides of (2.156) and (2.158).

As can be seen from the formulas that determine h_4 and h_6, we can obtain a problem of the type already considered only if the constant α of the isodrome vanishes.

Let us consider still another particular problem in automatic control. We have the following initial equations:

$$T^2\ddot{\psi} + U\dot{\psi} + k\psi + \mu = 0 \quad ,$$

$$\dot{\mu} = f(\sigma) \quad ,$$

$$T_1\dot{\sigma} = \sigma = \zeta \quad ,$$

$$\zeta = a\psi + E\dot{\psi} + G^2\ddot{\psi} - \frac{1}{\ell}\mu \quad .$$

Obviously, when $T_1 = 0$ we obtain (2.81) of the second Bulgakov problem for the case of ideal sensing elements. If, however, $T_1 \neq 0$, there is a certain time delay of the relay that controls the actuator,* and this is taken into account by the third equation.

In the latter case the characteristic of the relay is assumed to be linear, and its gain is assumed to be unity.

Retaining the notation of (2.82) for this problem, we can write the initial equations as follows:

$$\dot{\eta}_1 = \eta_2, \quad \dot{\eta}_2 = b_{21}\eta_1 + b_{22}\eta_1 + n_2\xi, \quad \dot{\xi} = f(\sigma) \quad ,$$

$$T_1\sqrt{r}\,\dot{\sigma} + \sigma = \chi\zeta, \quad \chi\zeta = p_1\eta_1 + p_2\eta_2 - \xi \quad .$$

---

* A. M. Letov. Concerning the Autopilot Problem, Vestnik MGU, No. 1, 1946.

Let us introduce a new variable η_3, defined by the equality

$$\sigma = \zeta + s\eta_3 \quad ,$$

where s is a constant, so far undetermined, which vanishes when $T_1 = 0$. Let us differentiate σ and find with the aid of (2.156) an equation for the new variable

$$sT_1 \sqrt{r}\, \dot{\eta}_3 + s\eta_3 + T_1 \sqrt{r}[p_1\eta_2 + p_2(b_{21}\eta_1 + b_{22}\eta_2 - \xi) - f(\sigma)] = 0 \cdot$$

After an elementary transformation we have

$$\begin{aligned}
\eta_1 &= \eta_2 \quad , \\
\dot{\eta}_2 &= b_{21}\eta_1 + b_{22}\eta_2 + n_2\xi \quad , \\
\dot{\eta}_3 &= -\frac{p_2 b_{21}}{s}\eta_1 - \frac{p_1 + b_{22}p_2}{s}\eta_2 - \frac{1}{T_1\sqrt{r}}\eta_3 + \frac{p_2}{s}\xi + \frac{1}{s}f(\sigma) \quad , \\
\dot{\xi} &= f(\sigma) \quad , \\
\sigma &= p_1\eta_1 + p_2\eta_2 + s\eta_s - \xi \quad .
\end{aligned}$$

Introducing the additional symbols

$$\xi = \eta_4, \quad -\frac{p_2 b_{21}}{s} = b_{31}, \quad -\frac{p_1 + b_{22}p_2}{s} = b_{32} \quad ,$$

$$-\frac{1}{T_1\sqrt{r}} = b_{33} \quad ,$$

(2.158a)

$$-\frac{n_2 p_2}{2} = b_{34}, \quad \frac{1}{s} = h_3, \quad 1 = h_4 \quad ,$$

$$n_2 = b_{24}, \quad p_3 = s, \quad p_4 = -1 \quad ,$$

we can write the original equations in a final form

(2.159)

$$\begin{aligned}
\dot{\eta}_1 &= \eta_2 \quad , \\
\dot{\eta}_2 &= b_{21}\eta_1 + b_{22}\eta_2 + b_{24}\eta_4 \quad , \\
\dot{\eta}_3 &= b_{31}\eta_1 + b_{32}\eta_2 + b_{33}\eta_3 + b_{34}\eta_4 + h_3 f(\sigma) \quad , \\
\dot{\eta}_4 &= h_4 f(\sigma) \quad , \\
\sigma &= p_1\eta_1 + p_2\eta_2 + p_3\eta_3 + p_4\eta_4 \quad .
\end{aligned}$$

Thus, it can be concluded that this control system is described by (2.156), (2.158), and (2.159), which belong to the following general class of equations

$$(2.160)\qquad \dot{\eta}_k = \sum_{\alpha=1}^{n} b_{k\alpha}\eta_\alpha + h_k f(\sigma) \qquad (k = 1, \ldots, n) \; , \qquad \sigma = \sum_{\alpha=1}^{n} p_\alpha \eta_\alpha \; .$$

Equations of this class were first considered by A. I. Lur'e.*

We could multiply the number of examples of control systems with isodromes, whose original equations belong to the class of equations (2.160). This is why an examination of the class of equations is determined not only by considerations of mathematical elegance,** which is attained here by introducing the symmetrical and compact form (2.160), but also by the nature of the matter.

11. SECOND LUR'E TRANSFORMATION

Before we reduce (2.160) to the canonical form we shall explain here the construction of the second canonical Lur'e transformation as given in the monograph cited above.

Let the equations

$$(2.161)\qquad x_s = \sum_{\alpha=1}^{n} c_\alpha^{(s)} \eta_\alpha \qquad (s = 1, \ldots, n)$$

determine a system of canonical variables.

Differentiating (2.161) and using (2.160), we obtain

$$\dot{x}_s = \sum_{\alpha=1}^{n} c_\alpha^{(s)} \left[\sum_{\beta=1}^{n} b_{\alpha\beta}\eta_\beta + h_\alpha f(\sigma) \right] \qquad (s = 1, \ldots, n) \; .$$

Since we wish to transform these equations to the canonical form

---

* A. I. Lur'e. Nekotorye nelineinye zadachi teorii avtomaticheskovo regulirovaniya (Certain Nonlinear Problems in the Theory of Automatic Control), Gostekhizdat, 1951.

** In connection with the above, the author acknowledges the error in his estimate of the role and significance of these general equations in the theory of automatic control, given in his review of the book by A. I. Lur'e (A. M. Letov, Review of the book by A. I. Lur'e, "Certain Nonlinear Problems in the Theory of Automatic Control", Avtomatika i Telemekhanika, Vol. XIII, No. 5, 1952).

$$\dot{x}_s = -\rho_s x_s + f(\sigma) \qquad (s = 1, \ldots, n) \quad , \tag{2.162}$$

we determine the constants $c_\alpha^{(s)}$ of the transformation (2.161) with the aid of the system of equations

$$-\rho_s c_\beta^{(s)} = \sum_{\alpha=1}^{n} c_\alpha^{(s)} b_{\alpha\beta} \, , \tag{2.163}$$

$$1 = \sum_{\alpha=1}^{n} c_\alpha^{(s)} h_\alpha \quad . \tag{2.164}$$

For these relations to be solvable with respect to the unknowns $c_\alpha^{(s)}$, it is enough that the parameters of transformation ρ_s be simple roots of the equation

$$D(\rho) = 0 \quad . \tag{2.165}$$

Here the determinant $D(\rho)$ is of the form*

$$D(\rho) = \begin{vmatrix} b_{11} + \rho & b_{12} & \cdots & b_{1n} \\ b_{21} & b_{22} + \rho & \cdots & b_{2n} \\ \cdots & \cdots & \cdots & \cdots \\ b_{n1} & b_{n2} & & b_{nn} + \rho \end{vmatrix} . \tag{2.166}$$

Let $D_{ik}(\rho)$ represent the cofactor of the element located in the i'th row and the k'th column. Repeating the arguments of Chapter II, Section 2, we construct the canonical transformation

$$x_s = \frac{1}{H_m(\rho_s)} \sum_{\alpha=1}^{n} D_{\alpha m}(\rho_s)\eta_\alpha \qquad (s = 1, \ldots, n) \quad . \tag{2.167}$$

Here $H_m(\rho_s)$ is obtained by replacing the elements of the m'th column of the determinant (2.166) by the numbers $h_1, \ldots, h_n$:

$$H_k(\rho) = \sum_{i=1}^{n} h_i D_{ik}(\rho) \qquad (k = 1, \ldots, n) \quad , \tag{2.168}$$

* Note that (2.166) is the transposed determinant (2.5).

and the index m is chosen from the condition that $H_m(\rho_s) \neq 0$ for all ρ_s.

The transformation inverse to the one given in (2.167) is determined by the system of equations

(2.169)
$$\eta_k = \sum_{\alpha=1}^{n} \frac{H_k(\rho_\alpha)}{D'(\rho_\alpha)} x_\alpha .$$

The above is verified by direct substitution of the quantities from (2.167) into (2.169):

$$\eta_k = \sum_{\alpha=1}^{n} \frac{H_k(\rho_\alpha)}{D'(\rho_\alpha)} \cdot \frac{1}{H_m(\rho_\alpha)} \sum_{\beta=1}^{n} D_{\beta m}(\rho_\alpha)\eta_\beta =$$

$$= \sum_{\beta=1}^{n} \left[\sum_{\alpha=1}^{n} \frac{H_k(\rho_\alpha)D_{\beta m}(\rho_\alpha)}{D'(\rho_\alpha)H_m(\rho_\alpha)} \right] \eta_\beta .$$

Continuing the calculations, we replace $H_k(\rho_\alpha)$ by its value determined from (2.168). We obtain

$$\eta_k = \sum_{\beta=1}^{n} \left[\sum_{\alpha=1}^{n} \sum_{i=1}^{n} \frac{h_i D_{ik}(\rho_\alpha)D_{\beta m}(\rho_\alpha)}{D'(\rho_\alpha)H_m(\rho_\alpha)} \right] \eta_\beta .$$

Using the well-known relation between the minors of the determinant (2.166)

$$D_{ik}(\rho)D_{\beta m}(\rho) = D_{im}(\rho)D_{\beta k}(\rho)$$

and using again (2.168), we obtain

$$\eta_k = \sum_{\beta=1}^{n} \left[\sum_{\alpha=1}^{n} \frac{D_{\beta k}(\rho_\alpha)}{D'(\rho_\alpha)H_m(\rho_\alpha)} \sum_{i=1}^{n} h_i D_{im}(\rho_\alpha) \right] \eta_\beta =$$

$$= \sum_{\beta=1}^{n} \left[\sum_{\alpha=1}^{n} \frac{D_{\beta k}(\rho_\alpha)}{D'(\rho_\alpha)} \right] \eta_\beta .$$

If we now employ formula (2.32), we can finally write

$$\eta_k = \sum_{\beta=1}^{n} \delta_{k\beta}\eta_\beta = \eta_k \quad ,$$

which was to be proved.

To obtain the canonical equations of the problem with the aid of transformation (2.167), let us first introduce one auxiliary formula.

Let m be the index of some column of the determinant (2.166). If $m = \beta$, then obviously

$$D(\rho) = \sum_{\alpha=1}^{n} (b_{\alpha\beta} + \rho\delta_{\alpha\beta})D_{\alpha\beta}(\rho) \quad .$$

If, however, $m \neq \beta$, we obtain the identity

$$0 = \sum_{\alpha=1}^{n} (b_{\alpha\beta} + \rho\delta_{\alpha\beta})D_{\alpha m}(\rho) \quad .$$

If in each of these formulas we insert instead of ρ any root of (2.165), these formulas become identical. Thus, for any m we have

$$\sum_{\alpha=1}^{n} b_{\alpha\beta}D_{\alpha m}(\rho_s) = -\rho_s D_{\beta m}(\rho_s) \qquad (\beta,\ s = 1,\ \ldots,\ n) \quad . \tag{2.170}$$

We now differentiate the canonical variables (2.167). Using the original (2.160), we calculate

$$\dot{x}_s = \frac{1}{H_m(\rho_s)} \sum_{\alpha=1}^{n} D_{\alpha m}(\rho_s)\eta_\alpha =$$

$$= \frac{1}{H_m(\rho_s)} \sum_{\alpha=1}^{n} D_{\alpha m}(\rho_s) \left[\sum_{\beta=1}^{n} b_{\alpha\beta}\eta_\beta + h_\alpha f(\sigma) \right] .$$

The expressions obtained can be simplified immediately by calling attention to the definition (2.168) of the quantities $H_m(\rho_s)$. We find:

$$\dot{x}_s = \frac{1}{H_m(\rho_s)} \sum_{\beta=1}^{n} \left[\sum_{\alpha=1}^{n} D_{\alpha m}(\rho_s) b_{\alpha\beta} \right] \eta_\beta + f(\sigma) \tag{2.171}$$

$$(s = 1, \ldots, n) \; .$$

but, on the other hand, thanks to relations (2.170), the sum of the first terms in (2.171) can be transformed into

$$\frac{1}{H_m(\rho_s)} \sum_{\beta=1}^{n} \left[\sum_{\alpha=1}^{n} D_{\alpha m}(\rho_s) b_{\alpha\beta} \right] \eta_\beta = \frac{1}{H_m(\rho_s)} \sum_{\beta=1}^{n} (-\rho_s) D_{\beta m} \eta_\beta \; ,$$

which, by definition of the canonical variables, leads us to the canonical equations (2.162).

It now remains to obtain expressions for σ and $\dot{\sigma}$. The latter is readily done by using the formulas of the inverse transformation and (2.160) and (2.162).

Let us use the symbols

$$\gamma_\alpha = - \sum_{s=1}^{n} p_s \frac{H_s(\rho_s)}{D'(\rho_s)} \; ,$$

$$r = - \sum_{\alpha=1}^{n} \gamma_\alpha = - \sum_{\alpha=1}^{n} p_\alpha h_\alpha \; , \tag{2.172}$$

$$\beta_\alpha = - \rho_\alpha \gamma_\alpha \; .$$

The second Lur'e transformation reduces the original equations to the canonical form:

$$\dot{x}_s = - \rho_s x_s + f(\sigma) \qquad (s = 1, \ldots, n) \; ,$$

$$\sigma = \sum_{\alpha=1}^{n} \gamma_\alpha x_\alpha \; , \tag{2.173}$$

$$\dot{\sigma} = \sum_{\alpha=1}^{n} \beta_\alpha x_\alpha - r f(\sigma) \; .$$

In the presence of one zero root of (2.165), for example $\rho_n = 0$, the quantity β_n will also vanish, and the term $\rho_n x_n$ and $\beta_n x_n$ will drop out of the equations for $\dot{x}_n$ and $\dot{\sigma}$ in the system (2.173).

Let us find now the canonical equations for the problem considered in Section 10. In the first case we compile the determinant (2.166). We have

$$(2.174)\qquad D(\rho) = \begin{vmatrix} \rho & 1 & 0 & 0 & 0 \\ b_{21} & b_{22}+\rho & b_{23} & 0 & 0 \\ 0 & 0 & b_{33}+\rho & 0 & b_{35} \\ 0 & 0 & 0 & b_{44}+\rho & b_{45} \\ 0 & 0 & 0 & 0 & \rho \end{vmatrix} .$$

The roots of (2.165) are determined by the relations

$$(2.175)\qquad \begin{aligned} &\rho_1 + \rho_2 = -b_{22}, \quad \rho_1\rho_2 = -b_{21} , \\ &\rho_3 = -b_{33}, \quad \rho_4 = -b_{44}, \quad \rho_5 = 0 . \end{aligned}$$

Inasmuch as $p_2 = p_3 = p_5 = 0$ and $p_1 = -p_4 = 1$, then, according to formulas (2.172) we must calculate the determinants $H_1(\rho)$ and $H_4(\rho)$ as well as the derivative $D'(\rho)$ in order to form the canonical equations. For the derivative $D'(\rho)$ we obtain

$$(2.176)\qquad \begin{aligned} D'(\rho_1) &= \rho_1(\rho_4 - \rho_1)(\rho_3 - \rho_1)(\rho_2 - \rho_1) , \\ D'(\rho_2) &= \rho_2(\rho_4 - \rho_2)(\rho_3 - \rho_2)(\rho_1 - \rho_2) , \\ D'(\rho_3) &= \rho_3(\rho_4 - \rho_3)(\rho_2 - \rho_3)(\rho_1 - \rho_3) , \\ D'(\rho_4) &= \rho_4(\rho_3 - \rho_4)(\rho_2 - \rho_4)(\rho_1 - \rho_4) , \\ D'(\rho_5) &= -\rho_4\rho_3\rho_2\rho_1 . \end{aligned}$$

According to formula (2.168) the determinants H_1 and H_4 can be rewritten

$$(2.177)\qquad H_1 = \begin{vmatrix} 0 & 1 & 0 & 0 & 0 \\ 0 & b_{22}+\rho & b_{23} & 0 & 0 \\ 0 & 0 & b_{33}+\rho & 0 & b_{35} \\ h_4 & 0 & 0 & b_{44}+\rho & b_{45} \\ h_5 & 0 & 0 & 0 & \rho \end{vmatrix} =$$

$$= -b_{23}b_{35}h_5(\rho + b_{44}) ,$$

(2.177) (continued)

$$H_4 = \begin{vmatrix} \rho & 1 & 0 & 0 & 0 \\ b_{21} & b_{22}+\rho & b_{23} & 0 & 0 \\ 0 & 0 & b_{33}+\rho & 0 & b_{35} \\ 0 & 0 & 0 & h_4 & b_{45} \\ 0 & 0 & 0 & h_5 & \rho \end{vmatrix} =$$

$$= [\rho(\rho + b_{22}) - b_{21}](\rho + b_{33})(h_4\rho - h_5 b_{45}) \quad .$$

To shorten the calculations, we bypass the direct and inverse transformations and proceed immediately to determine the quantities β_k which are of interest to us. From (2.172) we obtain

$$\beta_\alpha = \frac{\rho_\alpha H_1(\rho_\alpha)}{D'(\rho_\alpha)} - \frac{\rho_\alpha H_4(\rho_\alpha)}{D'(\rho_\alpha)} = \frac{\rho_\alpha[H_1(\rho_\alpha)-H_4(\rho_\alpha)]}{D'(\rho_\alpha)} \quad . \tag{2.178}$$

From this, on the basis of (2.176) and (2.177), we have

$$\begin{aligned} \beta_1 &= \frac{b_{23}b_{35}}{(\rho_3-\rho_1)(\rho_2-\rho_1)} \;, \\ \beta_2 &= \frac{b_{23}b_{35}}{(\rho_3-\rho_2)(\rho_1-\rho_2)} \;, \\ \beta_3 &= \frac{b_{23}b_{35}}{(\rho_2-\rho_3)(\rho_1-\rho_3)} \;. \\ \beta_4 &= h_4\rho_4 - h_5 b_{45} \;, \\ \beta_5 &= 0 \;. \end{aligned} \tag{2.179}$$

The resultant values of β_k can, if desired, be expressed in terms of the initial parameters.

To complete the calculations, let us find

$$r = - h_1 p_1 - h_4 p_4 = \beta \;,$$

and determine the canonical equations as

$$\dot{x}_s = -\rho_s x_s + f(\sigma) \qquad (s = 1, \ldots, 5) \ ,$$

(2.180)

$$\dot{\sigma} = \sum_{\alpha=1}^{5} \beta_\alpha x_\alpha - \beta f(\sigma) \ .$$

We proceed analogously in the second case. We form the determinant (2.166)

(2.181)

$$D(\rho) = \begin{vmatrix} \rho & 1 & 0 & 0 & 0 & 0 \\ b_{21} & \rho + b_{22} & b_{23} & 0 & 0 & 0 \\ 0 & 0 & \rho + b_{33} & 0 & b_{35} & 0 \\ 0 & 0 & 0 & \rho & 0 & 1 \\ 0 & 0 & 0 & 0 & \rho & 0 \\ 0 & 0 & 0 & b_{64} & b_{65} & b_{66} + \rho \end{vmatrix} =$$

$$= [\rho(\rho + b_{22}) - b_{21}] \cdot (\rho + b_{33}) \cdot \rho[\rho(\rho + b_{66}) - b_{64}] \ .$$

The roots of (2.165) are determined by

(2.182)

$$\rho_1 + \rho_2 = -b_{22}, \quad \rho_1\rho_2 = -b_{21}, \quad \rho_3 = -b_{33} \ ,$$

$$\rho_4 + \rho_5 = -b_{66}, \quad \rho_4\rho_5 = -b_{64}, \quad \rho_6 = 0 \ .$$

Let us calculate the derivative of $D(\rho)$. According to (2.181) and (2.182) we have

$$D(\rho) = \rho(\rho - \rho_1)(\rho - \rho_2)(\rho - \rho_3)(\rho - \rho_4)(\rho - \rho_5) \ .$$

The values of $D(\rho)$ of interest to us will be

(2.183)

$$D'(\rho_1) = \rho_1(\rho_2 - \rho_1)(\rho_3 - \rho_1)(\rho_4 - \rho_1)(\rho_5 - \rho_1) \ ,$$
$$D'(\rho_2) = \rho_2(\rho_1 - \rho_2)(\rho_3 - \rho_2)(\rho_4 - \rho_2)(\rho_5 - \rho_2) \ ,$$
$$D'(\rho_3) = \rho_3(\rho_1 - \rho_3)(\rho_2 - \rho_3)(\rho_4 - \rho_3)(\rho_5 - \rho_3) \ ,$$
$$D'(\rho_4) = \rho_4(\rho_1 - \rho_4)(\rho_2 - \rho_4)(\rho_3 - \rho_4)(\rho_5 - \rho_4) \ ,$$
$$D'(\rho_5) = \rho_5(\rho_1 - \rho_5)(\rho_2 - \rho_5)(\rho_3 - \rho_5)(\rho_4 - \rho_5) \ .$$

Further, we are interested as before, only in the determinants H_1 and H_4. Using (2.181) and (2.168) we form these quantities

$$H_1 = \begin{vmatrix} 0 & 1 & 0 & 0 & 0 & 0 \\ 0 & b_{22}+\rho & b_{23} & 0 & 0 & 0 \\ 0 & 0 & b_{33}+\rho & 0 & b_{35} & 0 \\ 0 & 0 & 0 & \rho & 0 & 1 \\ h_4 & 0 & 0 & 0 & \rho & 0 \\ h_5 & 0 & 0 & b_{64} & b_{65} & b_{66}+\rho \end{vmatrix} =$$

(2.184) $$= - b_{23}b_{35}h_5(\rho - \rho_4)(\rho - \rho_5) ,$$

$$H_4 = \begin{vmatrix} \rho & 1 & 0 & 0 & 0 & 0 \\ b_{21} & b_{22}+\rho & b_{23} & 0 & 0 & 0 \\ 0 & 0 & b_{33}+\rho & 0 & b_{35} & 0 \\ 0 & 0 & 0 & 0 & 0 & 1 \\ 0 & 0 & 0 & h_5 & \rho & 0 \\ 0 & 0 & 0 & h_6 & b_{65} & b_{66}+\rho \end{vmatrix} =$$

$$= (\rho - \rho_1)(\rho - \rho_2)(\rho - \rho_3)(b_{65}h_5 - \rho h_6) .$$

We determine β_s from the previous formula

(2.185) $$\beta_s = \frac{H_1(\rho_s) - H_4(\rho_s)}{D'(\rho_s)} \rho_s \qquad (s = 1, \dots, 6) ,$$

where $D'(\rho_s)$, $H_4(\rho_s)$, and $H_1(\rho_s)$ are determined from (2.183) and (2.184). Accordingly, we get

(2.186)
$$\beta_1 = \frac{b_{35}b_{23}}{(\rho_2-\rho_1)(\rho_3-\rho_1)} ,$$
$$\beta_2 = \frac{b_{35}b_{23}}{(\rho_1-\rho_2)(\rho_3-\rho_2)} ,$$
$$\beta_3 = \frac{b_{35}b_{23}}{(\rho_1-\rho_3)(\rho_2-\rho_3)} ,$$
$$\beta_4 = \frac{b_{65} - \rho_4 h_6}{\rho_4 - \rho_5} ,$$
$$\beta_5 = \frac{b_{65} - \rho_5 h_6}{\rho_5 - \rho_4} ,$$
$$\beta_6 = 0 .$$

The quantity r vanishes. Consequently we obtain

$$(2.187)\qquad \begin{aligned} \dot{x}_s &= -\rho_s x_s + f(\sigma) \qquad (s = 1, \ldots, 6) , \\ \dot{\sigma} &= \sum_{s=1}^{6} \beta_s x_s . \end{aligned}$$

For the Bulgakov problem, in the case of a relay with time delay, we have

$$(2.188)\qquad D(\rho) = \begin{vmatrix} \rho & 1 & 0 & 0 \\ b_{21} & \rho + b_{22} & 0 & b_{24} \\ b_{31} & b_{32} & \rho + b_{33} & b_{34} \\ 0 & 0 & 0 & \rho \end{vmatrix} .$$

From this we get

$$D(\rho) = \rho(\rho + b_{33})[\rho^2 + \rho b_{22} - b_{21}] ,$$

$$D'(\rho) = + [\rho^2 + b_{22}\rho - b_{21}]\rho + (\rho + (\rho + b_{33})[\rho^2 + b_{22}\rho - b_{21}] + $$
$$+ \rho(\rho + b_{33})(2\rho + b_{22}) .$$

The roots of (2.165) are

$$(2.189)\qquad \rho_1 + \rho_2 = - b_{22}, \quad \rho_1\rho_2 = - b_{21}, \quad \rho_3 = - b_{33}, \quad \rho_4 = 0 .$$

To obtain the equations we need, it is enough to calculate the quantities r and β_s. We have

$$H_1 = \begin{vmatrix} 0 & 1 & 0 & 0 \\ 0 & \rho + b_{22} & 0 & b_{24} \\ h_3 & b_{32} & \rho + b_{33} & b_{34} \\ h_4 & 0 & 0 & \rho \end{vmatrix} = b_{24}h_4(\rho + b_{33}) ,$$

$$(2.190)\qquad H_2 = \begin{vmatrix} \rho & 0 & 0 & 0 \\ b_{21} & 0 & 0 & b_{24} \\ b_{31} & h_3 & \rho + b_{33} & b_{34} \\ 0 & h_4 & 0 & \rho \end{vmatrix} = - \rho b_{24}h_4(b_{33} + \rho) ,$$

$$H_3 = \begin{vmatrix} \rho & 1 & 0 & 0 \\ b_{21} & \rho + b_{22} & 0 & b_{24} \\ b_{31} & b_{32} & h_3 & b_{34} \\ 0 & 0 & h_4 & \rho \end{vmatrix} =$$
$$= (\rho^2 + b_{22}\rho - b_{21})(\rho h_3 - b_{34}h_4) + b_{24}h_4(b_{32}\rho - b_{31}) ,$$

(2.190) (continued)

$$H_4 = \begin{vmatrix} \rho & 1 & 0 & 0 \\ b_{21} & \rho + b_{22} & 0 & 0 \\ b_{31} & b_{32} & \rho + b_{33} & h_3 \\ 0 & 0 & 0 & h_4 \end{vmatrix} = h_4(\rho + b_{33})[\rho^2 + b_{22}\rho - b_{21}] \quad .$$

We first find γ_k. We have:

(2.191)

$$\gamma_1 = - \frac{(p_1-\rho_1 p_2)b_{24}h_4(\rho_1+b_{33}) + h_4 p_3 b_{24}(\rho_1 b_{32}-b_{31})}{\rho_1(\rho_1+b_{33})(\rho_1-\rho_2)} \quad ,$$

$$\gamma_2 = - \frac{(p_1-\rho_2 p_2)b_{24}h_4(\rho_2+b_{33}) + h_4 p_3 b_{24}(\rho_2 b_{32}-b_{31})}{\rho_2(\rho_2+b_{33})(\rho_2-\rho_1)} \quad ,$$

$$\gamma_3 = - \frac{p_s[(\rho_3^2+b_{22}\rho_3-b_{21})(\rho_3 h_3-b_{34}h_4) + h_4 b_{24}(\rho_3 b_{32}-b_{31})]}{\rho_3(\rho_3^2+b_{22}\rho_3-b_{21})} \quad ,$$

$$\gamma_4 = - \frac{h_4[p_1 b_{24}b_{33}-p_3(b_{24}b_{31}-b_{21}b_{34}) - p_4 b_{33}b_{21}]}{b_{33}b_{21}} \quad .$$

Multiplication of each γ_s by ρ_s yields β_s:

(2.192)

$$\beta_1 = \frac{(p_1-\rho_1 p_2)b_{24}h_4(\rho_1+b_{33}) + h_4 p_3 b_{24}(b_{32}\rho_1-b_{31})}{(\rho_1+b_{33})(\rho_1-\rho_2)} \quad ,$$

$$\beta_2 = \frac{(p_1-\rho_2 p_2)b_{24}h_4(\rho_2+b_{33}) + h_4 p_3 b_{24}(\rho_2 b_{32}-b_{31})}{(\rho_2-b_{33})(\rho_2-\rho_1)} \quad ,$$

$$\beta_3 = p_s \frac{[(\rho_3^2+b_{22}\rho_3-b_{21})(\rho_3 h_3-b_{34}h_4) + h_4 b_{24}(\rho_3 b_{32}-b_{31})]}{\rho_3^2 + b_{22}\rho_3 - b_{21}} \quad ,$$

$$\beta_4 = 0 \quad .$$

Finally, we obtain for r

$$r = p_3 h_3 + p_4 h_4 = 1 - 1 = 0 \quad .$$

The canonical equations of the problem have the following form:

$$(2.193)\qquad \begin{aligned} \dot{x}_1 &= -\rho_1 x_1 + f(\sigma) , \\ \dot{x}_2 &= -\rho_2 x_2 + f(\sigma) , \\ \dot{x}_3 &= -\rho_3 x_3 + f(\sigma) , \\ \dot{\sigma} &= \beta_1 x_1 + \beta_2 x_2 + \beta_3 x_3 . \end{aligned}$$

CHAPTER III: SECOND AND THIRD CANONICAL FORMS OF THE EQUATIONS OF CONTROL SYSTEMS

1. SECOND FORM OF CANONICAL TRANSFORMATION*

If a control system is inherently unstable or neutral with respect to several coordinates, then (2.5) has roots for which $\operatorname{Re} \rho_k < 0$ $(k = 1, 2, \ldots)$. In cases that are very difficult to investigate, it is desirable to employ a different form of canonical transformation, one which permits setting up the Lyapunov function for all problems of this kind. Henceforth we shall assume for simplicity that $V^2 = 0$.

Let us consider again (1.33). First, we shall assume the new variable to be the function

$$\sigma = \sum_{\alpha=1}^{m} p_\alpha \eta_\alpha - r\mu \quad . \tag{3.1}$$

We shall assume that $r \neq 0$. If we set

$$\bar{b}_{k\alpha} = b_{k\alpha} + \frac{n_k p_\alpha}{r} \qquad (\alpha, k = 1, \ldots, m) \quad , \tag{3.2}$$

then, after eliminating the old variable μ, the first m equations of (1.33) can be rewritten as

$$\dot{\eta}_k = \sum_{\alpha=1}^{m} \bar{b}_{k\alpha} \eta_\alpha - \frac{n_k}{r} \sigma \qquad (k = 1, \ldots, m) \quad . \tag{3.3}$$

Next, differentiating σ and denoting

---

* A. M. Letov, Inherently Unstable Control Systems, PMM, Vol. XIIII, No. 2, 1950.

(3.4) $$\sum_{\alpha=1}^{m} p_\alpha \bar{b}_{\alpha\beta} + \rho_{m+1} p_\beta = \bar{p}_\beta, \qquad \sum_{\alpha=1}^{m} \frac{p_\alpha n_\alpha}{r} + \rho_{m+1} = \bar{\rho}$$

we obtain jointly with (3.2)

(3.5)
$$\dot{\eta}_k = \sum_{\alpha=1}^{m} \bar{b}_{k\alpha}\eta_\alpha - \frac{n_k}{r}\sigma \qquad (k = 1, \ldots, m) ,$$
$$\dot{\sigma} = \sum_{\alpha=1}^{m} \bar{p}_\alpha \eta_\alpha - \bar{\rho}\sigma - rf(\sigma) .$$

Let us consider the linear transformation for (3.3).

(3.6) $$x_s = \sum_{\alpha=1}^{m} c_\alpha^{(s} \eta_\alpha \qquad (s = 1, \ldots, m) .$$

Differentiating (3.6) we obtain with the aid of (3.3):

$$\dot{x}_s = \sum_{\alpha=1}^{m} c_\alpha^{(s)} \left[\sum_{\beta=1}^{m} \bar{b}_{\alpha\beta}\eta_\beta - \frac{n_\alpha}{r}\sigma \right] \qquad (s = 1, \ldots, m) .$$

In order to recast the equations with the new variables into the canonical form

$$\dot{x}_s = - r_s x_s + \sigma \qquad (s = 1, \ldots, m) ,$$

we choose the transformation constants of (3.6) such as to satisfy the relation

(3.7) $$- r_s c_\beta^{(s)} = \sum_{\alpha=1}^{m} c_\alpha^{(s)} \bar{b}_{\alpha\beta} \qquad (\beta, s = 1, \ldots, m) ,$$

(3.8) $$- r = \sum_{\alpha=1}^{m} c_\alpha^{(s)} n_\alpha .$$

As before, this transformation is possible only if the parameters r_s are chosen to be the roots of the equation

$$(3.9) \qquad \bar{D}(r) = \begin{vmatrix} \bar{b}_{11} + r & \bar{b}_{21} & \cdots & \bar{b}_{m1} \\ \cdot & \cdot & \cdot & \cdot \\ \bar{b}_{1m} & \bar{b}_{2m} & \cdots & \bar{b}_{mm} + r \end{vmatrix} = 0 .$$

Let us assume that the roots of this equation are different and are such that $\operatorname{Re} r_s > 0$ $(s = 1, \ldots, m)$. The physical meaning of this condition is, in effect, an assumption that the system is stable if an ideal regulator is used. Since the coefficients of (3.9) contain the constants of the regulator, they can always be so chosen to satisfy the above condition. Consequently, in this case the constants of the regulator should be chosen in accordance with the inequality

$$(3.10) \qquad \Delta_1 > 0, \ldots, \Delta_m > 0 ,$$

where Δ_k $(k = 1, \ldots, m)$ are the Hurwitz determinants for the expression obtained from (3.9) by replacing r with $-r$.

We must note here the substantial difference between the determinants (2.7) and the determinants (3.10), in that the latter depend on the parameters p_α, $(\alpha = 1, \ldots, m)$ and r.

Therefore, for certain regulated objects, inequalities (2.7) may not hold while inequalities (3.10) can be satisfied in an infinite number of ways. Here one can encounter two cases:

(a) (3.9) has only simple roots.

(b) Some of the roots are multiple.

In the first case the transformation (3.6) is nonsingular and can be solved with respect to the old variables. Let us assume that we have found

$$(3.11) \qquad \eta_k = \sum_{\alpha=1}^{m} \bar{D}_\alpha^{(k)} x_\alpha \qquad (k = 1, \ldots, m).$$

Equations (3.11) permit us to eliminate all the old variables in the last equation of the system (3.5). If we introduce

$$(3.12) \qquad \bar{\beta}_k = \sum_{\alpha=1}^{m} \bar{D}_k^{(\alpha)} \bar{p}_\alpha \qquad (k = 1, \ldots, m) ,$$

then the canonical equations can be changed to the following form

$$\begin{aligned} \dot{x}_k &= -r_k x_k + \sigma \qquad (k = 1, \ldots, m) , \\ \dot{\sigma} &= \sum_{k=1}^{m} \bar{\beta}_k x_k - \rho\sigma - rf(\sigma) . \end{aligned} \tag{3.13}$$

The system (3.13) includes the two last particular cases of the initial equations considered in Chapter II (pages 36 and 37).

To form the canonical equation in the third case it is necessary to put in (3.13) $\rho_{m+1} = 0$. Actually, the formulas (3.4) are then written

$$\sum_{\alpha=1}^{m} p_\alpha \bar{b}_{\alpha\beta} = \bar{p}_\beta, \quad \sum_{\alpha=1}^{m} p_\alpha n_\alpha = r\bar{\rho} \qquad (\beta = 1, \ldots, m) . \tag{3.14}$$

The canonical equations will in this case have the form (3.13).

If (3.9) has a root of multiplicity ℓ, then the canonical form of equations can be obtained by means of the same arguments employed in Section 3 of the preceding chapter. Therefore, omitting the details of these arguments, we confine ourselves merely to giving the final canonical form of the equations. Let us assume that the multiple root is r_1. We then have

$$\begin{aligned} \dot{x}_1 &= -r_1 x_1 + q_1\sigma , \\ \dot{x}_2 &= \varepsilon_1 x_1 - r_1 x_2 + q_2\sigma , \\ &\cdots\cdots\cdots\cdots \\ \dot{x}_\ell &= \varepsilon_{\ell-1} x_{\ell-1} - r_1 x_\ell + q_\ell \sigma , \\ \dot{x}_{\ell+1} &= -r_{\ell+1} x_{\ell+1} + \sigma , \\ &\cdots\cdots\cdots\cdots \\ \dot{x}_m &= -r_m x_m + \sigma , \\ \dot{\sigma} &= \sum_{k=1}^{m} \bar{\beta}_k x_k - \bar{\rho}\sigma - rf(\sigma) , \end{aligned} \tag{3.15}$$

where $\varepsilon_1, \ldots, \varepsilon_{\ell-1}$ are arbitrary numbers, which can be made as small as desired; $q_1, \ldots, q_\ell, \bar{\beta}_1, \ldots, \bar{\beta}_m$, and $\bar{\rho}$ are known constants, determined in the course of carrying out the transformation itself.

2. FORMULAS FOR THE TRANSFORMATION COEFFICIENTS

These formulas can be obtained by the same method employed in Chapter II. However, the derivation can now be considerably shortened. Thus, comparing (3.7) and (3.8) with (2.3) and (2.4), we see that the formulas for the coefficients $C_\alpha^{(s)}$ can be obtained from (2.23) by replacing in the latter ρ_s by r_s and $\rho_{n+1} - \rho_s$ by $-r$; we then obtain

$$C_k^{(s)} = \frac{-r}{\bar{N}_i(r_s)} \bar{D}_{ik}(r_s) \quad . \tag{3.16}$$

In these formulas the numbers $\bar{D}_{ik}(r_s)$ stand for the cofactors of the elements of the i'th row and the k'th column of determinant (3.9), while the numbers $\bar{N}_i(r_s)$ are

$$\bar{N}_i(r_s) = \sum_{k=1}^{m} n_k \bar{D}_{ik}(r_s) \qquad (s = 1, \ldots, m) \quad . \tag{3.17}$$

It is useful to note that $\bar{N}_i(r_s)$ represents the determinant (3.9), in which the elements of the i'th row are replaced by the numbers n_k. In exactly the same manner, the formulas for the coefficients of the inverse transformation are obtained by substituting in (2.34) $\bar{D}'(r_s)$ for $D'(\rho_s)$ and $-\bar{N}_k(r_j)/r$ for $H_k(\rho_j)/(\rho_{n+1} - \rho_j)$. Therefore

$$\bar{D}_\alpha^{(k)} = \frac{\bar{N}_k(r_\alpha)}{r\bar{D}'(r_\alpha)} \qquad (\alpha, k = 1, \ldots, m) \quad , \tag{3.18}$$

where $\bar{D}'(r_\alpha)$ is the derivative of the determinant (3.9) with respect to the variable r.

As already noted above, if inherently unstable systems are investigated by the above method, there is no need to derive special formulas for particular cases.

3. BULGAKOV'S FIRST AND SECOND PROBLEMS

We continue the examination of the above problems for the case in which the constant k becomes negative or vanishes.

Turning again to (2.83), we eliminate ξ from these equations, and choose σ to be the new variable. This leads to

$$\begin{aligned} \dot{\eta}_1 &= \eta_2 \ , \\ \dot{\eta}_2 &= \bar{b}_{21}\eta_1 + \bar{b}_{22}\eta_2 + \sigma \ , \\ \dot{\sigma} &= \bar{p}_1\eta_1 + \bar{p}_2\eta_2 - \rho\sigma - f(\sigma) \ , \end{aligned} \tag{3.19}$$

where

$$\begin{aligned} &\bar{b}_{21} = b_{21} - p_1, \quad \bar{b}_{22} = b_{22} - p_2, \quad \bar{p}_1 = \bar{b}_{21}p_2 \ , \\ &\bar{p}_2 = p_1 + \bar{b}_{22}p_2, \quad \bar{\rho} = - p_2 \ . \end{aligned} \tag{3.20}$$

We next obtain

$$\bar{D}(r) = \begin{vmatrix} r & \bar{b}_{21} \\ 1 & r + \bar{b}_{22} \end{vmatrix} = 0 \ .$$

Obviously the roots of this equation satisfy the relations

$$\begin{aligned} r_1 + r_2 &= - \bar{b}_{22} = \frac{U + \ell E}{\sqrt{\ell(T^2 + \ell G^2)}} \ , \\ r_1 r_2 &= - \bar{b}_{21} = \frac{k + a\ell}{\ell} \ . \end{aligned} \tag{3.21}$$

No matter what the number k, the conditions (3.10) will always be satisfied if the producal $a\ell$ is sufficiently large, that is, if

$$k + a\ell > 0 \ . \tag{3.22}$$

To form the coefficients of the direct and inverse transformations, it is possible to employ (3.16) and (3.18) by putting $n_1 = 0$ and $n_2 = -1$. First of all, putting $r = -1$, we get

$$\bar{N}_1 = 1, \quad \bar{N}_2 = - r_s \qquad (s = 1, 2) \ , \tag{3.23}$$

$$\bar{D}_{11} = r_s + \bar{b}_{22}, \quad \bar{D}_{12} = - 1 \ . \tag{3.24}$$

Substitution of the quantities from (3.22) and (3.24) into (3.16) yields

$$C_1^{(s)} = - (r_s + \bar{b}_{22}) = - \frac{\bar{b}_{21}}{r_s}, \quad C_2^{(s)} = 1 \ ,$$

from which we obtain

$$C_1^{(1)} = r_2, \quad C_1^{(2)} = r_1, \quad C_2^{(1)} = C_2^{(2)} = 1 \quad .$$

Thus, the direct-transformation formulas become

$$(3.25) \qquad \begin{aligned} x_1 &= r_2\eta_1 + \eta_2 \quad , \\ x_2 &= r_1\eta_1 + \eta_2 \quad . \end{aligned}$$

If $r_1 \neq r_2$ the transformation (3.25) is nonsingular, thanks to which we can determine the coefficients of the inverse transformation by using formulas (3.18). In this case we have

$$\bar{D}'(r_1) = 2r_1 + \bar{b}_{22} = r_1 - r_2 \quad ,$$

$$\bar{D}'(r_2) = 2r_2 + \bar{b}_{22} = r_2 - r_1 \quad ,$$

and with the aid of formulas (3.18) and (3.23) we find

$$\bar{D}_1^{(1)} = \frac{1}{r_2 - r_1}, \qquad \bar{D}_2^{(1)} = -\frac{1}{r_2 - r_1} \quad ,$$

$$\bar{D}_1^{(2)} = -\frac{r_1}{r_2 - r_1}, \qquad \bar{D}_2^{(2)} = \frac{r_2}{r_2 - r_1} \quad .$$

To write finally the canonical equations of the problem, it is necessary now to calculate the quantities $\bar{\beta}_k$ with the aid of formulas (3.12). We have

$$(3.26) \qquad \begin{aligned} \bar{\beta}_1 &= \bar{p}_1\bar{D}_1^{(1)} + \bar{p}_2\bar{D}_1^{(2)} = \frac{1}{r_2 - r_1}(\bar{p}_1 - r_1\bar{p}_2) \quad , \\ \bar{\beta}_2 &= \bar{p}_1\bar{D}_2^{(1)} + \bar{p}_2\bar{D}_2^{(2)} = \frac{1}{r_2 - r_1}(-\bar{p}_1 + r_2\bar{p}_2) \quad , \end{aligned}$$

from which we find

$$(3.27) \qquad \begin{aligned} \dot{x}_1 &= -r_1x_1 + \sigma \quad , \\ \dot{x}_2 &= -r_2x_2 + \sigma \quad , \\ \dot{\sigma} &= \bar{\beta}_1x_1 + \bar{\beta}_2x_2 - \bar{\rho}\sigma - f(\sigma) \quad . \end{aligned}$$

Let us now assume that the roots (3.21) are equal. In this case we need retain in the transformation (3.25) only the first row which corresponds to the first equation of (3.27). Returning to the second equation (2.83), we have, after eliminating the variable η_1,

$$\dot{\eta}_2 = \frac{\bar{b}_{21}}{r_1}(x_1 - \eta_2) = -r_2 x_1 + \left(\bar{b}_{22} - \frac{\bar{b}_{21}}{r_1}\right)\eta_2 + \sigma \ .$$

But

$$\bar{b}_{22} - \frac{\bar{b}_{21}}{r_1} = -r_1 - r_2 + r_2 = -r_1 \ ,$$

therefore

$$\dot{\eta}_2 = -r_1 x_1 - r_1 \eta_2 + \sigma \ .$$

If we introduce the new variable x_2, putting

$$\eta_2 = C_2 x_2, \quad \varepsilon_1 = -\frac{r_1}{C_2}, \quad q_2 = \frac{1}{C_2} \ , \tag{3.28}$$

we shall have

$$\dot{x}_2 = \varepsilon_1 x_1 - r_1 x_2 + q_2 \sigma \ .$$

To form the third canonical equation we eliminate η_1 and η_2 from the third equation of (3.19); therefore, after denoting

$$\bar{\beta}_1 = \frac{\bar{p}_1}{r_1}, \quad \bar{\beta}_2 = C_2\left(\bar{p}_2 - \frac{\bar{p}_1}{r_1}\right) \tag{3.29}$$

we can write

$$\begin{aligned} \dot{x}_1 &= -r_1 x_1 + \sigma \ , \\ \dot{x}_2 &= \varepsilon_1 x_1 - r_1 x_2 + q_2 \sigma \ , \\ \dot{\sigma} &= \bar{\beta}_1 x_1 + \bar{\beta}_2 x_2 - \bar{\rho}\sigma - f(\sigma) \ . \end{aligned} \tag{3.30}$$

One remark must be made concerning the above argument. It may appear that this method of reducing the original equations to the canonical form cannot be used if the regulator does not contain proportional feedback (i.e., when $\ell = \infty$), for in this case the function σ does not contain the variable μ. However, this example shows readily that (3.19) can indeed be obtained in a direct manner.

We thus have

(3.31)
$$T^2 \frac{d^2\psi}{dt^2} + U \frac{d\psi}{dt} + k\psi + \mu = 0 \quad ,$$
$$\frac{d\mu}{dt} = f^*(\sigma), \qquad \sigma = a\psi + E \frac{d\psi}{dt} + G^2 \frac{d^2\psi}{dt^2} \quad .$$

Let us consider the relation

(3.32)
$$\sigma = a\psi + E \frac{d\psi}{dt} + G^2 \frac{d^2\psi}{dt^2}$$

not only as an equation that defines a new variable, but also as an equation that relates the variables σ and ψ. Since this equation is identically satisfied when the corresponding values of σ and ψ are inserted in it, it can be differentiated. We have

(3.33)
$$\frac{d\sigma}{dt} = a \frac{d\psi}{dt} + E \frac{d^2\psi}{dt^2} + G^2 \frac{d^3\psi}{dt^3} \quad .$$

After differentiating (3.32) and the first equation of (3.31) we get

(3.34)
$$\frac{d^2\psi}{dt^2} = \frac{1}{G^2} \left[\sigma - a\psi - E \frac{d\psi}{dt} \right] ,$$
$$\frac{d^3\psi}{dt^3} = \frac{1}{T^2} \left[- U \frac{d^2\psi}{dt^2} - k \frac{d\psi}{dt} - f^*(\sigma) \right] .$$

If we now change to new variables

(3.35)
$$\psi = \eta_1, \quad \frac{d\psi}{dt} = \frac{1}{G} \eta_2, \quad t = G\tau \quad ,$$

we can use formulas (3.34) and (3.35) to reduce (3.21) to the following form

(3.36)
$$\dot{\eta}_1 = \eta_2 \quad , .$$
$$\dot{\eta}_2 = - a\eta_1 - \sqrt{r}\, E\eta_2 + \sigma \quad ,$$
$$\dot{\sigma} = - \frac{a}{G} (E - pG^2)\eta_1 + \left[a - pG^2 - \frac{E}{G^2} (E - pG^2) \right] \eta_2 + \frac{1}{G} (E - pG^2)\sigma - \frac{G^3}{T^2} f^*(\sigma) \quad .$$

Comparison shows that (3.36) are identical with (3.19) if $\ell = \infty$.

4. ON THE THEORY OF THE ISODROME REGULATOR

As the second example of an application of the above theory, let us consider the problem of Section 6 for $k < 0$ (with all the other constants retaining their previous values). Obviously, in this case one of the roots ρ_1 or ρ_2 of (2.107) will be negative and the canonical equation (2.114) cannot be used in the manner desired.

Using the above method, we can eliminate the variable ξ from (2.105) and get

$$\begin{aligned} \dot{\eta}_1 &= \eta_2 \ , \\ \dot{\eta}_2 &= \bar{b}_{21}\eta_1 + \bar{b}_{22}\eta_2 + \bar{b}_{23}\eta_3 + \sigma \ , \\ \dot{\eta}_3 &= \bar{b}_{31}\eta_1 \ , \\ \dot{\sigma} &= \bar{p}_1\eta_1 + \bar{p}_2\eta_2 + \bar{p}_3\eta_3 - \bar{\rho}\sigma - f(\sigma) \ , \end{aligned} \tag{3.37}$$

where

$$\bar{b}_{21} = -\frac{k + a\ell}{\ell}, \quad \bar{b}_{22} = -\frac{U + \ell E}{\sqrt{\ell(T^2+\ell G^2)}} \ ,$$

$$\bar{b}_{31} = \frac{1}{N(a-qG^2)}\sqrt{\frac{T^2 + \ell G^2}{\ell}} \ , \tag{3.38}$$

$$\bar{p}_1 = \bar{b}_{21}p_2 + \bar{b}_{31}p_3, \quad \bar{p}_2 = p_1 + \bar{b}_{22}p_2, \quad \bar{p}_3 = \bar{b}_{23}p_2 \ ,$$

$$\rho = -p_2 \ ,$$

$$r = \frac{\ell}{T^2 + \ell G^2}, \quad \bar{b}_{23} = -p_3, \quad f(\sigma) = \frac{1}{\sqrt{r}} f^*(\sigma) \ .$$

Let us write down the determinant

$$\bar{D}(r) = \begin{vmatrix} r & \bar{b}_{21} & \bar{b}_{31} \\ 1 & r + \bar{b}_{22} & 0 \\ 0 & \bar{b}_{23} & r \end{vmatrix} . \tag{3.39}$$

Expanding this determinant by rows, we get

$$r^3 - \frac{U + \ell E}{\sqrt{\ell(T^2+\ell G^2)}} r^2 + \frac{k + a\ell}{\ell} r - \frac{1}{N}\sqrt{\frac{T^2 + \ell G^2}{\ell}} = 0 \ . \tag{3.40}$$

If the parameters of the control system are made to satisfy the relations

$$(3.41) \qquad \frac{(U+\ell E)(k+a\ell)}{\ell} - \frac{T^2 + \ell G^2}{N} > 0, \qquad \frac{k + a\ell}{\ell} > 0 \quad ,$$

then the roots of (3.40) will have the necessary properties. Obviously, for a given $k < 0$, both inequalities of (3.41) can be satisfied by choosing sufficiently large values of $a\ell$, $E\ell$, and N. Since $n_1 = 0$, $n_2 = -1$, and $n_3 = 0$, we find

$$(3.42) \qquad \bar{N}_1 = \begin{vmatrix} 0 & -1 & 0 \\ 1 & r_s + \bar{b}_{22} & 0 \\ 0 & \bar{b}_{23} & r_s \end{vmatrix} = r_s, \qquad \bar{N}_2 = \begin{vmatrix} r_s & \bar{b}_{21} & \bar{b}_{31} \\ 0 & -1 & 0 \\ 0 & \bar{b}_{23} & r_s \end{vmatrix} = -r_s^2 ,$$

$$\bar{N}_3 = \begin{vmatrix} r_s & \bar{b}_{21} & \bar{b}_{31} \\ 1 & r_s + \bar{b}_{22} & 0 \\ 0 & -1 & 0 \end{vmatrix} = -\bar{b}_{31} \quad .$$

Let us put, as before, $i = 1$. We then get

$$(3.43) \qquad \bar{D}_{11} = r_s(r_s + \bar{b}_{22}), \quad \bar{D}_{12} = -r_s, \quad \bar{D}_{13} = \bar{b}_{23} \qquad (s = 1, 2, 3) \quad .$$

Substituting the corresponding values from (3.42) and (3.43) into (3.16) we get

$$(3.44) \qquad c_1^{(s)} = -(r_r + \bar{b}_{22}), \quad c_2^{(s)} = 1, \quad c_3^{(s)} = -\frac{\bar{b}_{23}}{r_s} \qquad (s = 1, 2, 3).$$

In accordance with formulas (3.44) we obtain the direct transformation

$$(3.45) \qquad \begin{aligned} x_1 &= -(r_1 + \bar{b}_{22})\eta_1 + \eta_2 - \frac{\bar{b}_{23}}{r_1}\eta_3 , \\ x_2 &= -(r_2 + \bar{b}_{22})\eta_1 + \eta_2 - \frac{\bar{b}_{23}}{r_2}\eta_3 , \\ x_3 &= -(r_3 + \bar{b}_{22})\eta_1 + \eta_2 - \frac{\bar{b}_{23}}{r_3}\eta_3 . \end{aligned}$$

Going on to calculate the coefficients of the inverse transformation, let us represent the determinant (3.39) in the following form:

$$\bar{D}(r) = (r - r_1)(r - r_2)(r - r_3) \quad .$$

From this we get

$$\bar{D}'(r_1) = (r_1 - r_2)(r_1 - r_3) \quad ,$$

$$\bar{D}'(r_2) = (r_2 - r_1)(r_2 - r_3) \quad ,$$

$$\bar{D}'(r_3) = (r_3 - r_1)(r_3 - r_2) \quad ,$$

and on the basis of formulas (3.18) and (3.42) we will have

$$D_1^{(1)} = \frac{r_1}{(r_2-r_1)(r_1-r_2)} \, , \quad D_2^{(1)} = \frac{r_2}{(r_3-r_2)(r_2-r_1)} \quad ,$$

$$D_3^{(1)} = \frac{r_3}{(r_1-r_3)(r_3-r_2)} \quad ,$$

$$D_1^{(2)} = -\frac{r_1}{(r_2-r_1)(r_1-r_3)} \, , D_2^{(2)} = -\frac{r_2}{(r_3-r_2)(r_2-r_1)} \quad ,$$

(3.46)

$$D_3^{(2)} = -\frac{r_3}{(r_1-r_3)(r_3-r_2)} \quad ,$$

$$D_1^{(3)} = -\frac{\bar{b}_{31}}{(r_2-r_1)(r_1-r_3)} \, , D_2^{(3)} = -\frac{\bar{b}_{31}}{(r_3-r_2)(r_2-r_1)} \quad ,$$

$$D_3^{(3)} = -\frac{\bar{b}_{31}}{(r_1-r_3)(r_3-r_2)} \quad .$$

To write the canonical equations, it remains for us to calculate the constants $\bar{\beta}_k$. From formulas (3.13) and (3.46) we obtain

$$\bar{\beta}_1 = \frac{1}{(r_2-r_1)(r_1-r_3)} \, [\bar{p}_1 r_1 - \bar{p}_2 r_1 - \bar{p}_3 \bar{b}_{31}] \quad ,$$

(3.47)

$$\bar{\beta}_2 = \frac{1}{(r_3-r_2)(r_2-r_1)} \, [\bar{p}_1 r_2 - \bar{p}_2 r_2 - \bar{p}_3 \bar{b}_{31}] \quad ,$$

$$\bar{\beta}_3 = \frac{1}{(r_1-r_3)(r_3-r_2)} \, [\bar{p}_1 r_3 - \bar{p}_2 r_3 - \bar{p}_3 \bar{b}_{31}] \quad .$$

Thus, the canonical equations of the problem will have the following form:

$$\begin{aligned} \dot{x}_1 &= -r_1x_1 + \sigma , \\ \dot{x}_2 &= -r_2x_2 + \sigma , \\ \dot{x}_3 &= -r_3x_3 + \sigma , \\ \dot{\sigma} &= \bar{\beta}_1x_1 + \bar{\beta}_2x_2 + \bar{\beta}_3x_3 - \rho\sigma - f(\sigma) . \end{aligned} \tag{3.48}$$

5. THIRD FORM OF THE CANONICAL TRANSFORMATION*

The two transformation forms discussed above take care of most cases in which we are interested in reducing the initial equations to the canonical form, and they permit, in principle, construction of the Lyapunov function for a given class of control systems. However, certain computational difficulties, connected with this construction in the case of systems with a large degree of freedom, and also other considerations connected with the quality problem, cause us to seek other forms of canonical transformations.

We shall consider in this section a transformation that leads to a criterion for system stability and permits solving the problem of system quality without explicit derivation of the Lyapunov function, thus ridding us of all limitations superimposed on the initial equations in connection with these constructions. In particular, we shall waive here the requirement that the function $f(\sigma)$ be analytic.

Let us first consider the initial equations in the form (3.5). Let

$$R^2 = F^2 + S^2\sigma^2, \quad F^2 = \sum_{\alpha,\beta=1}^{m} a_{\alpha\beta}\eta_\alpha\eta_\beta \tag{3.49}$$

represent a real, positive definite quadratic form, with $a_{\alpha\beta} = a_{\beta\alpha}$. Consequently, the real numbers $a_{\alpha\beta}$ satisfy the Sylvester inequalities

$$a_{11} > 0, \quad \begin{vmatrix} a_{11} & a_{12} \\ a_{21} & a_{22} \end{vmatrix} > 0, \ \ldots, \ \Delta = \begin{vmatrix} a_{11} & \cdots & a_{1m} \\ \cdot & \cdot\ \cdot\ \cdot & \cdot \\ a_{m1} & \cdots & a_{mm} \end{vmatrix} > 0 , \tag{3.50}$$

---

* A. M. Letov, Contribution to the Theory of the Quality of Nonlinear Control Systems, Avtomatika i Telemekhanika, Vol. XIV, No. 5, 1953.

and $S > 0$.

Let us agree to assign to the expression

$$R = +\sqrt{\sum_{\alpha,\beta=1}^{m} a_{\alpha\beta}\eta_\alpha\eta_\beta + S^2\sigma^2} \tag{3.51}$$

the positive sign only. The remaining variables will be determined from the equations

$$\frac{\sum_{k=1}^{m} a_{ik}\eta_k}{\sqrt{a_{ii}R}} = \zeta_i \quad (i = 1, \ldots, m), \quad \frac{S\sigma}{R} = \zeta \ . \tag{3.52}$$

Geometrically speaking, the new variables ζ_k $(k = 1, \ldots, m)$ and ζ represent the direction cosines of the radius vector $\bar{R}$ of the representative point M of the system in the space of the variables η_k and σ, while the quantity R represents the distance from this point to the origin.

The transformations (3.51) and (3.52) were used by V. S. Vedrov* for the case $a_{ik} = \delta_{ik}$ and $S = 1$ to study stability in the Lyapunov sense in critical cases.

Transformation (3.52), by virtue of the assumption (3.50), is mutually single-valued and can be solved with respect to the old variables. In fact, we have

$$\sum_{k=1}^{m} a_{ik}\eta_k = R\sqrt{a_{ii}}\zeta_i \qquad (i = 1, \ldots, m) \ . \tag{3.53}$$

Let us consider the determinants

$$\Delta_r = R \begin{vmatrix} a_{11} & \cdots & a_{1r-1} & \sqrt{a_{11}}\zeta_1 & a_{1r+1} & \cdots & a_{1m} \\ \cdots & \cdots & \cdots & \cdots & \cdots & \cdots & \cdots \\ a_{m1} & \cdots & a_{mr-1} & \sqrt{a_{mm}}\zeta_m & a_{mr+1} & \cdots & a_{mm} \end{vmatrix},$$

$$(r = 1, \ldots, m) \ .$$

and let $\Delta_{s,r}$ stand for the cofactors of the element in the s'th row and the r'th column of this determinant. We obtain

* V. S. Vedrov, On the Stability of Motion, Trudy TsAGI, No. 327, 1937.

(3.54) $$\eta_r = \frac{R}{\Delta} \sum_{s=1}^{m} \Delta_{sr} \sqrt{a_{ss}}\zeta_s \qquad (r = 1, \ldots, m).$$

Since we want to obtain equations in the new variables, we differentiate (3.49) and employ (3.5) to get

$$2R\dot{R} = \sum_{\alpha,\beta=1}^{m} a_{\alpha\beta}(\eta_\alpha\dot{\eta}_\beta + \dot{\eta}_\alpha\eta_\beta) + 2S^2\dot{\sigma}\sigma \quad .$$

But an elementary transformation yields

$$\sum_{\alpha,\beta=1}^{m} a_{\alpha\beta}(\eta_\alpha\dot{\eta}_\beta + \dot{\eta}_\alpha\eta_\beta) = \sum_{\alpha,\beta=1}^{m} a_{\alpha\beta}\left[\sum_{k=1}^{m} \bar{b}_{\alpha k}\eta_\beta\eta_k + \sum_{k=1}^{m} \bar{b}_{\beta k}\eta_\alpha\eta_k\right] =$$

$$= \sum_{\alpha=1}^{m}\sum_{k=1}^{m}\left[\sum_{\beta=1}^{m} a_{\alpha\beta}\bar{b}_{\beta k}\right]\eta_\alpha\eta_k + \sum_{\beta=1}^{m}\sum_{k=1}^{m}\left[\sum_{\alpha=1}^{m} a_{\alpha\beta}\bar{b}_{\alpha k}\right]\eta_\beta\eta_k =$$

$$= \sum_{\alpha=1}^{m}\sum_{k=1}^{m}\left[\sum_{\beta=1}^{m} a_{\alpha\beta}\bar{b}_{\beta k} + \sum_{\beta=1}^{m} a_{\beta\alpha}\bar{b}_{\beta k}\right]\eta_\alpha\eta_k \quad .$$

Interchanging, for convenience, the summation indices β and k, we get

$$2R\dot{R} = 2\sum_{\alpha,\beta=1}^{m}\left[\sum_{k=1}^{m} a_{\alpha k}\bar{b}_{k\beta}\right]\eta_\alpha\eta_\beta +$$

$$+ 2\sum_{\alpha=1}^{m}\left[-\sum_{k=1}^{m} a_{k\alpha}\frac{n_k}{r} + S^2\bar{p}_\alpha\right]\eta_\alpha\sigma - 2S^2\bar{\rho}\sigma^2 - 2S^2 r\sigma f(\sigma) \ .$$

If we eliminate by means of (3.54) the old variables from this relation and if we denote

$$W = \sum_{s,r=1}^{m} B_{sr}\zeta_s\zeta_r + 2\sum_{s=1}^{m} Q_s\zeta_s\zeta + (\rho + h)\zeta^2 ,$$

(3.55)
$$B_{sr} = -\frac{\sqrt{a_{rr}a_{ss}}}{\Delta_n^2}\sum_{\alpha,\beta=1}^{m}\left[\sum_{k=1}^{m} a_{k\beta}\bar{b}_{k\alpha}\right]\Delta_{\alpha s}\Delta_{\beta r} ,$$

$$Q_s = -\frac{\sqrt{a_{ss}}}{2\Delta_n S}\left[S^2\sum_{\alpha=1}^{m}\bar{p}_\alpha\Delta_{\alpha s} + \sum_{\alpha,\beta=1}^{m} a_{\alpha\beta}\bar{n}_\alpha\Delta_{\beta s}\right] ,$$

we obtain the first equation

(3.56)
$$\dot{R} = -WR - Sr\zeta\varphi(R\zeta) .$$

It is important to note that the transformation employed here retains the known properties of the function $f(\sigma)$, which mark it as belonging to class (A) or subclass (A') of the function $f(\sigma)$.

The equations for the remaining variables can be readily obtained by differentiating (3.52). Omitting detailed derivations, we write the collective equations of the system directly in the final form:

(3.57)
$$\begin{aligned}
\dot{R} &= -WR - Sr\zeta\varphi(R\zeta) ,\\
\dot{\zeta}_1 &= \frac{1}{\sqrt{a_{11}}}\left[W + \frac{S}{R}\zeta\varphi(R\zeta)\right]\zeta_1 - \\
&\quad - \frac{1}{\Delta\sqrt{a_{11}}}\sum_{s,\alpha=1}^{m}\left[\sum_{k=1}^{m} a_{1k}\Delta_{sk}\bar{b}_{k\alpha}\sqrt{a_{ss}}\right]\zeta_s = \frac{\sum_{k=1}^{m} a_{1k}n_k}{rS\sqrt{a_{11}}}\zeta ,\\
\dot{\zeta} &= \frac{S}{\Delta}\left[\sum_{s,r=1}^{m}\Delta_{sr}\bar{\beta}_r\sqrt{a_{ss}}\right]\zeta_s - \frac{\bar{\rho}}{S}\zeta - \frac{S}{R}f(R\zeta) .
\end{aligned}$$

The first integral of (3.57) will be determined in Chapter VIII. The integral represents a certain relation between the direction cosines and is obtained from (3.51) by substituting therein the quantities (3.54). After performing the substitution, we obtain*

---

* It is important to note, for what follows, that the left half of (3.58) is positive-definite.

$$\sum_{s,r=1}^{m} A_{rs}\zeta_r\zeta_s + \zeta^2 = 1 \quad . \tag{3.58}$$

We denote here

$$A_{rs} = \frac{1}{\Delta^2}\sqrt{a_{rr}a_{ss}} \sum_{\alpha,\beta=1}^{m} a_{\alpha\beta}\Delta_{\alpha s}\Delta_{\beta r} \qquad (r,\ s = 1,\ \ldots,\ m) \quad . \tag{3.59}$$

The third form of the canonical transformation can be obtained for any other specification of the initial equations. For example, if we assume the initial equations to be in the form (3.13), we can put

$$R^2 = \sum_{\alpha,\beta=1}^{m} a_{\alpha\beta}x_\alpha x_\beta + S^2\sigma^2 \quad ,$$

$$\frac{\sum_{k=1}^{m} a_{ik}x_k}{R\sqrt{a_{ii}}} = \zeta_i \quad (i = 1,\ \ldots,\ m), \quad \frac{S\sigma}{R} = \zeta \quad . \tag{3.60}$$

Performing the calculations described above, we obtain

$$\dot{R} = -\,WR - \frac{S\zeta}{R}\,\varphi(R\zeta) \quad ,$$

$$\dot{\zeta}_i = -\,\frac{1}{\Delta\sqrt{a_{ii}}}\sum_{s=1}^{m}\left[\sum_{k=1}^{m} a_{ik}\Delta_{sk}r_k\sqrt{a_{ss}}\right]\zeta_s + \frac{\sum_{k=1}^{m} a_{ik}}{S\sqrt{a_{ii}}}\,\zeta + \frac{1}{\sqrt{a_{ii}}}\left[W + \frac{S\zeta}{R^2}\,\varphi(R\zeta)\right]\zeta_i \quad , \tag{3.61}$$

$$\dot{\zeta} = \frac{S}{\Delta}\left(\sum_{s,r=1}^{m} \Delta_{sr}\bar{\beta}_r\sqrt{a_{ss}}\right)\zeta_s - \frac{\bar{\rho}}{S}\,\zeta - \frac{S}{R}\,f(R\zeta) \quad ,$$

where W denotes the form

$$W = \frac{1}{\Delta^2} \sum_{r,s=1}^{m} \left[\sum_{\alpha,\beta=1}^{m} a_{\alpha\beta} \frac{r_\alpha + r_\beta}{2} \Delta_{\alpha s} \Delta_{\beta r} \sqrt{a_{ss} a_{rr}} \right] \zeta_s \zeta_r -$$

(3.62)

$$- \sum_{s=1}^{m} \left[\sum_{\alpha=1}^{m} \left(S^2 \bar{\beta}_\alpha + \sum_{\beta=1}^{m} a_{\alpha\beta} \right) \frac{\Delta_{s\alpha} \sqrt{a_{ss}}}{\Delta S} \right] \zeta_s \zeta + (\bar{\rho} + h) \zeta^2 \quad .$$

This form will certainly be real if all the r_k are real. If, however, r_k includes at least one pair of complex conjugate numbers the transformation can be retained if the initial equations, corresponding to this pair of roots, is reduced to real variables by means of the known customary techniques.

Let

$$x_r = \frac{R}{\Delta} \sum_{s=1}^{m} \Delta_{sr} \sqrt{a_{ss}} \zeta_s \qquad (r = 1, \ldots, m) \quad ,$$

(3.63)

$$\sigma = \frac{R}{S} \zeta$$

be the solutions of (3.60). The first integral of (3.61) is

(3.64)
$$\sum_{s,r=1}^{m} A_{rs} \zeta_r \zeta_s + \zeta^2 = 1 \quad ,$$

where

(3.65)
$$A_{rs} = \frac{\sqrt{a_{rr} a_{ss}}}{\Delta^2} \sum_{\alpha,\beta=1}^{m} a_{\alpha\beta} \Delta_{\alpha s} \Delta_{\beta r} \quad .$$

(3.58) [(3.64)] defines a closed surface of second order in the space of the variables $\zeta_1, \ldots, \zeta_m, \zeta$.

CHAPTER IV: STABILITY OF CONTROL SYSTEMS

1. STATEMENT OF THE PROBLEM

The fundamental difficulty in the construction of the Lyapunov function for each problem lies in the absence of any general indications and principles by which it is possible to select from among the infinite set of sign-definite V-functions determined by Lyapunov with the aid of partial differential equations.

The isolated examples found in the literature give us only a certain amount of practice in such a construction, which practice, unfortunately, has not yet been acquired by a large circle of investigators.

We shall consider in this chapter a general method of constructing the Lyapunov function for inherently-stable control systems and for control systems that are neutral with respect to one coordinate.

As already mentioned, the first to initiate work in this direction were A. I. Lur'e and V. N. Postnikov.

Subsequent general development of the method proposed was carried out by A. I. Lur'e for a given class of control systems.

In particular, he developed an algorithm, with which it becomes possible to solve the problem of stability of the control system studied in this chapter [for any function $f(\sigma)$ of class (A)].

Thus, let us assume a control system whose disturbed motion is described by the following canonical equations

$$\dot{x}_k = -\rho_k x_k + f(\sigma) \qquad (k = 1, \ldots, n+1) ,$$

(4.1)
$$\sigma = \sum_{k=1}^{n+1} \gamma_k x_k ,$$

$$\dot{\sigma} = \sum_{k=1}^{n+1} \beta_k x_k - rf(\sigma) ,$$

where ρ_k, γ_k, β_k, and r are known numbers.

(4.1), as well as the initial equations, have either one obvious solution, or else a continuum of such solutions comprising the insensitive region of the control system. We shall designate the obvious solutions by the symbol

$$x_k = x_k^* \qquad (k = 1, \ldots, n + 1) \quad . \tag{4.2}$$

The solutions (1.23) and (4.2) have a one-to-one correspondence which can always be established with the aid of the canonical-transformation formulas. The stability of solution (4.2) will be accompanied by the physical realization of this steady-state of the control system, described by this solution.

We shall assume that among all the n roots of (2.5) are included s real roots $(\rho_1, \ldots, \rho_s)$ and $\frac{1}{2}(n - s)$ pairs of complex conjugate roots: $(\rho_{s+1}, \ldots, \rho_n)$. Accordingly, the constants $\gamma_1, \ldots, \gamma_s$, $\beta_1, \ldots, \beta_s$ and the variables $x_1, \ldots, x_s$ will be real, while the constants $\gamma_{s+1}, \ldots, \gamma_n$, $\beta_{s+1}, \ldots, \beta_n$ and the variables $x_{s+1}, \ldots, x_n$ will form complex conjugate pairs. All the roots ρ_k will be assumed different and satisfying the condition

$$\operatorname{Re}\rho_k > 0 \qquad (k = 1, \ldots, n) \quad , \tag{4.3}$$

but admitting the possibility of one of the real roots vanishing (thus signifying the neutrality of the system relative to one of its coordinates

The problem consists of determining values of the regulator parameters for which absolute stability of solution (4.2) is assured.

2. TWO QUADRATIC FORMS

Let us consider the quadratic form

$$F(a_1x_1, \ldots, a_{n+1}x_{n+1}) = \sum_{i=1}^{n+1}\sum_{k=1}^{n+1} \frac{a_k a_i}{\rho_k + \rho_i} x_k x_i \quad , \tag{4.4}$$

where $a_1, \ldots, a_s$ are arbitrary real numbers, and $a_{s+1}, \ldots, a_{n+1}$ are arbitrary complex conjugate sets of pairs.

We shall show that this form is a positive definite function of the variables x_k. In fact, we have the obvious equality

$$\frac{1}{\rho_k + \rho_i} = \int_0^\infty e^{-(\rho_i+\rho_k)\tau}\, d\tau \quad ,$$

thanks to which the form (4.4) can be represented as follows:

$$F = \sum_{i=1}^{n+1} \sum_{k=1}^{n+1} \left[a_k a_i \int_0^\infty e^{-(\rho_i+\rho_k)\tau} d\tau \right] x_k x_i =$$

(4.5)

$$= \int_0^\infty \left[\sum_{k=1}^{n+1} a_k x_k e^{-\rho_k \tau} \right]^2 d\tau \quad .$$

The integrand in (4.5) is the square of a real number, for the complex terms enter in it as conjugate pairs, and the integrand can vanish for all τ only if the x_k $(k = 1, \ldots, n + 1)$ vanish. Consequently, the function F assumes only positive real values and vanishes only at the origin, i.e., it is sign-definite* and positive everywhere in the variables $x_1, \ldots, x_{n+1}$.

Let us next consider the quadratic form

$$\Phi(x_1, \ldots, x_{n+1}) = \frac{1}{2} (A_1 x_1^2 + \ldots + A_s x_s^2) +$$

(4.6)

$$+ C_1 x_{s+1} x_{s+2} + \ldots + C_{n-s} x_n x_{n+1} \quad ,$$

in which $A_1, \ldots, A_s, C_1, C_3, \ldots, C_{n-s}$ are positive real numbers.

It is clear that Φ assumes only real positive values, since the product of any pair of complex conjugate numbers of the form $x_{s+1} x_{s+2}$ equals the square of the modulus of the factor. Since Φ vanishes only at the origin, this function is positive definite.

Finally, let us remark that no matter what the function $f(\sigma)$ of class (A) might be, the integral

(4.7) $$\int_0^\sigma f(\sigma) d\sigma$$

---

* This is correct if the ρ_k are all different. Otherwise we have, for example, for $n + 1 = 2$ and $\rho_1 = \rho_2$,

$$\frac{a_1^2 x_1^2}{2\rho_1} + \frac{2a_1 a_2 x_1 x_2}{\rho_1 + \rho_2} + \frac{a_2^2 x_2^2}{2\rho_2} = \frac{1}{2\rho_1} [a_1 x_1 + a_2 x_2]^2$$

and the form $F(a_1 x_1, a_2 x_2)$ is sign-definite.

is positive for all values $|\sigma| > \sigma^*$; for $|\sigma| < \sigma^*$ this integral vanishes. Exactly the same property is possessed by the integral

$$\int_0^\sigma \varphi(\sigma)d\sigma \quad , \tag{4.7a}$$

where $\varphi(\sigma)$ is a function defined by relations (1.9). It is important to note here still another important difference between these integrals.

For a specified actuator with normal characteristics, the integral (4.7) has the property

$$\lim_{\sigma \to \infty} \int_0^\sigma f(\sigma)d\sigma = \infty$$

while the integral (4.7a) can be bounded. This difference must be borne in mind when constructing the Lyapunov function and when conclusions are drawn concerning the stability of the system in the entire space of the variables.

3. THE LUR'E THEOREM

In proceeding to study the stability of solution (4.2), let us consider the function

$$V = \Phi + F + \int_0^\sigma f(\sigma)d\sigma \quad , \tag{4.8}$$

which is sign-definite and positive everywhere. The construction of all possible sufficiency criteria for the stability of the solution under investigation involves the calculation of its total derivative with respect to the variable t. According to (4.1), we get

$$\begin{aligned} \frac{dV}{dt} = \sum_{k=1}^{s} A_k x_k[- \rho_k x_k + f(\sigma)] + C_1 x_{s+2}[- \rho_{s+1} x_{s+1} + f(\sigma)] + \\ + C_1 x_{s+1}[- \rho_{s+2} x_{s+2} + f(\sigma)] + \dots \\ \dots + \sum_{i=1}^{n+1} \sum_{k=1}^{n+1} \frac{a_i a_k}{\rho_k + \rho_i} \{x_k[- \rho_i x_i + f(\sigma)] + \\ + x_i[- \rho_k x_k + f(\sigma)]\} + f(\sigma) \left[\sum_{k=1}^{n+1} \beta_k x_k - r f(\sigma) \right] . \end{aligned} \tag{4.9}$$

Taking into account that

$$\sum_{i=1}^{n+1} \sum_{k=1}^{n+1} a_k a_i x_k x_i = \sum_{k=1}^{n+1} a_k x_k \sum_{i=1}^{n+1} a_i x_i = \left(\sum_{k=1}^{n+1} a_k x_k \right)^2 ,$$

(4.10)

$$\sum_{k=1}^{n+1} \sum_{i=1}^{n+1} \frac{a_k a_i}{\rho_k + \rho_i} (x_k + x_i) = 2 \sum_{i=1}^{n+1} \sum_{k=1}^{n+1} \frac{a_k a_i}{\rho_k + \rho_i} x_k ,$$

and adding and subtracting the expression

$$(4.11) \qquad 2 \sqrt{r} f(\sigma) \sum_{k=1}^{n+1} a_k x_k ,$$

from the right half of (4.9) we can reduce the derivative of the function (4.8) to

$$\frac{dV}{dt} = - \sum_{k=1}^{s} \rho_k A_k x_k^2 - C_1 (\rho_{s+1} + \rho_{s+2}) x_{s+1} x_{s+2} - \cdots$$

$$\cdots - C_{n-s} (\rho_n + \rho_{n+1}) x_n x_{n+1} - \left[\sum_{k=1}^{n+1} a_k x_k \right]^2 -$$

$$- [\sqrt{r} f(\sigma)]^2 - 2 \sqrt{r} f(\sigma) \sum_{k=1}^{n+1} a_k x_k +$$

$$+ f(\sigma) \sum_{k=1}^{s} \left[A_k + \beta_k + 2 \sqrt{r} a_k + 2 a_k \sum_{i=1}^{n+1} \frac{a_i}{\rho_k + \rho_i} \right] x_k +$$

$$+ f(\sigma) \sum_{\alpha=1}^{n+1-s} \left[C_\alpha + \beta_{s+\alpha} + 2 \sqrt{r} a_{s+\alpha} + 2 a_{s+\alpha} \sum_{i=1}^{n+1} \frac{a_i}{\rho_k + \rho_i} \right] x_{s+\alpha} .$$

For convenience in writing the sums in the last brackets of this expression, we use

$$C_1 = C_2, \quad C_3 = C_4, \quad \ldots, \quad C_{n-s} = C_{n-s+1} .$$

We shall obtain without fail a sign-definite derivative DV/dt, opposite in sign to V, if we specify that the following relations be satisfied

$$A_k + \beta_k + 2\sqrt{r}a_k + 2a_k \sum_{i=1}^{n+1} \frac{a_i}{\rho_k + \rho_i} = 0$$

$$(k = 1, \ldots, s) \quad ,$$

(4.12)

$$C_\alpha + \beta_{s+\alpha} + 2\sqrt{r}a_{s+\alpha} + 2a_{s+\alpha} \sum_{i=1}^{n+1} \frac{a_i}{\rho_k + \rho_i} = 0$$

$$\alpha = 1, \ldots, n - s + 1) \quad .$$

This derivative is of the form

$$\frac{dV}{dt} = - \sum_{k=1}^{s} \rho_k A_k x_k^2 -$$

(4.13)

$$- C_1(\rho_{s+1} + \rho_{s+2})x_{s+1}x_{s+2} - \cdots - C_{n-s}(\rho_n + \rho_{n+1})x_n x_{n+1} -$$

$$- \left[\sum_{k=1}^{n+1} a_k x_k + \sqrt{r} f(\sigma) \right]^2 \quad .$$

Thus, if relations (4.12) are satisfied, it is possible to construct for this problem a sign-definite Lyapunov function (4.8), having a sign-definite derivative (4.13) opposite in sign to V. In this case it is possible to guarantee absolute stability of solution (4.2) of (4.1); this stability will be asymptotic.

Relations (4.12) contain the following arbitrary constants: positive -- $A_1, \ldots, A_s, C_1$ and $C_3, \ldots, C_{n-s+1}$; real -- $a_1, \ldots, a_s$; and complex-conjugate pairs $a_{s+1}, \ldots, a_{n+1}$. If the first group of numbers is specified somehow, then relations (4.12) can be considered as equations for the constants $a_1, \ldots, a_{n+1}$. Henceforth we shall be interested only in the criterion for the solvability of these equations, and not in the numbers $a_1, \ldots, a_{n+1}$ themselves.

Let us assume that this criterion can be expressed in the form of a certain number of inequalities

(4.14)

$$F_k(\beta_1, \ldots, \beta_{n+1}, A_1, \ldots, A_s, C_1, \ldots, C_{n-s}) > 0$$

$$(k = 1, 2, \ldots) \quad .$$

Then the choice of the regulated constants, satisfying inequalities (4.14), will guarantee the asymptotic stability of solution (4.2).

Let us consider the particular case in which $A_1 = 0, \ldots, A_s = 0$ and $C_1 = 0, \ldots, C_{n-s} = 0$ (and Φ does not enter into the Lyapunov function). (4.12) have in this case the following form

$$\beta_k + 2\sqrt{r}a_k + 2a_k \sum_{i=1}^{n+1} \frac{a_i}{\rho_k + \rho_i} = 0 \tag{4.15}$$

$$k = 1, \ldots, n+1) \quad .$$

The criterion for the solvability of these equations can be written, as in the preceding case, by means of the same number of inequalities of the same type

$$F_k(\beta_1, \ldots, \beta_{n+1}, 0, \ldots, 0) > 0 \qquad (k = 1, 2, \ldots) \quad . \tag{4.16}$$

If the regulated constants satisfy these inequalities, there exists a sign-definite function

$$V = F + \int_0^\sigma f(\sigma)d\sigma \quad , \tag{4.17}$$

which is positive everywhere and whose derivatives

$$\frac{dV}{dt} = -\left[\sum_{i=1}^{n+1} a_k x_k + \sqrt{r}f(\sigma)\right]^2 \quad , \tag{4.18}$$

calculated according to (4.1) is sign-definite of a sign opposite that of V. This is why the satisfaction of inequalities (4.16) guarantees the stability of solution (4.2).

If, however, the constants $A_1, \ldots, C_{n-s}$ are positive numbers, but as small as desired, the results obtained on the basis of criterion (4.16) will be as close as desired to those obtained by means of the criterion for asymptotic stability (4.14).

Bearing this circumstance in mind, we shall consider henceforth the Lyapunov function (4.17) everywhere, and we shall speak of the absolute asymptotic stability of solution (4.2), guaranteed by the satisfaction of inequalities (4.16).

We have thus obtained the following theorem, first proved by Lur'e: If the regulated constants are such that the system of quadratic equations (4.15), in which $\rho_1, \ldots, \rho_s$, and $\beta_1, \ldots, \beta_s$ are real and $\rho_{s+1}, \ldots, \rho_{n+1}$ and $\beta_{s+1}, \ldots, \beta_{n+1}$ are complex conjugate pairs, has a solution that contains real roots $a_1, \ldots, a_s$ and complex pairs of roots $a_{s+1}, \ldots, a_{n+1}$, then the steady state of the system has absolute asymptotic stability.

Let us analyze certain particular cases. Thus, in Case 1, using (2.12a), we obtain a stability criterion in the form (4.16), which contains only the quantities $\beta_1, \ldots, \beta_n$. Analogously, to obtain the stability criterion in Case 2, it is necessary to put in (4.16) $n = m$. Finally, in Case 3 the stability criterion is obtained by analysis of the solvability of (4.15), the number of which is m.

To conclude this section, we examine an important subcase of Case 3, in which we assume that the control system is neutral relative to any one coordinate. Let us assume that this subcase is characterized by the equation $\rho_1 = 0$. Then, according to (2.17) and (2.36) we obtain

(4.19)
$$\beta_1 = -\rho_1 \sum_{\alpha=1}^{m} p_\alpha D_1^{(\alpha)} = - \sum_{\alpha=1}^{m} p_\alpha \frac{N_\alpha(\rho_1)\rho_1}{(-\rho_1)Д'(\rho_1)} = \sum_{\alpha=1}^{m} p_\alpha \frac{N(0)}{Д'(0)} ,$$

and the canonical equations of the problem have the form

(4.20)
$$\begin{aligned} \dot{x}_1 &= f(\sigma) , \\ \dot{x}_k &= -\rho_k x_k + f(\sigma) \qquad (k = 2, \ldots, m) , \\ \dot{\sigma} &= \beta_1 x_1 + \sum_{k=2}^{m} \beta_k x_k - rf(\sigma) . \end{aligned}$$

The Lyapunov function can be taken here in the following form

(4.21)
$$V = \frac{1}{2} A x_1^2 + \sum_{k=2}^{m} \sum_{i=2}^{m} \frac{a_k a_i}{\rho_k + \rho_i} x_k x_i + \int_0^\sigma f(\sigma) d\sigma + \Phi ,$$

where A is a positive number, and the function Φ no longer contains the term $\frac{1}{2} A x_1^2$. If all operations are performed in the calculation of the total derivative of this function, we obtain instead of conditions (4.15) the following conditions

$$A + \beta_1 = 0 \quad ,$$

$$A_k + \beta_k + 2\sqrt{r}a_k + 2a_k \sum_{i=2}^{m} \frac{a_i}{\rho_k + \rho_i} = 0$$

(4.22) $$(k = 2, \ldots, s) \quad ,$$

$$C_\alpha + \beta_{s+\alpha} + 2\sqrt{r}a_{s+\alpha} + 2a_{s+\alpha} \sum_{i=1}^{m} \frac{a_i}{\rho_k + \rho_i} = 0$$

$$(\alpha = 1, \ldots, m - s) \quad .$$

The first shows that β_1 should be negative; the remaining conditions of (4.22) should be considered from the same points of view as conditions (4.15).

4. THE FIRST BULGAKOV PROBLEM

Let us see now what the Lur'e theorem produces with respect to the first Bulgakov problem. Since $\rho_1 = 0$, the stability criterion of the system becomes

$$A + \beta_1 = 0 \quad ,$$

$$\beta_2 + 2a_2 + \frac{a_2^2}{\rho_2} = 0 \quad .$$

According to formulas (2.65), the first condition can always be satisfied, for

$$\beta_1 = -\frac{\rho_1}{\rho_2} < 0 \quad ;$$

from the second condition we get

$$a_2 = -\rho_2 \pm \sqrt{\rho_2^2 - \rho_2\beta_2} \quad .$$

But $\rho_2 = -b_{22}$ is a real number; consequently, a_2 is certainly going to be real if the parameters of the system satisfy the inequality

(4.23) $$\beta_2 < \rho_2 \quad .$$

According to (2.65) and (2.56) we get

$$\frac{1}{\ell}\left(\frac{U}{T}\right)^2 > a - \frac{U}{T^2}\left(E - \frac{U}{T^2}G^2\right) .$$

An elementary transformation of this inequality yields

$$\frac{U(U+\ell E)}{\ell T^2} > a \quad . \tag{4.24}$$

Inequality (4.24) is the only condition for absolute stability of the control system.

As can be seen, it is independent of $G^2 > 0$ and, for a specified regulator, limits merely the choice of the constants E, a, and ℓ. If the regulator does not contain proportional feedback (i.e., if $\ell = \infty$), then inequality (4.21) becomes simpler and assumes the following form

$$UE > aT^2 \quad . \tag{4.25}$$

The solution obtained illustrates the above method of investigation and serves as an example of the application of the Lur'e theorem. As stated in Chapter II, the first solution of the problem (2.55) for $G^2 = 0$ is that of Bulgakov. In developing his method for solving the problem, thereby making it possible to obtain the necessary and sufficient conditions for stability, Bulgakov arrived at the inequality (4.24). Another solution of this problem was obtained by A. A. Andronov and N. N. Bautin.* The case they considered had one substantial distinction: they assume the function $f(\sigma)$ to be equal to sign (σ):

$$f(\sigma) = \begin{cases} +1 & \text{for} \quad \sigma > 0 \ , \\ 0 & \text{for} \quad \sigma = 0 \ , \\ -1 & \text{for} \quad \sigma < 0 \ . \end{cases}$$

This distinction influenced substantially the choice of method for solving the problem, but it did not affect the result, which agreed with (4.24).**

Finally, the third solution of the problem belongs to Lur'e and Postnikov. That solution is quite equivalent to the statement of the problem of absolute stability, discussed in this study. The comparison will show that the previous predicted fear of obtaining non-constructive solutions of the problem is not well founded.

---

* A. A. Andronov and N. N. Bautin, Motion of Neutral Stand, Equipped with an Autopilot, and the Theory of Point-By-Point Transformation of Surfaces, DAN SSSR, Vol. XIII, No. 9, 1942.

** A. M. Letov, Contribution to the Autopilot Problem. Vestnik MGU, No. 1, 1946.

Let us now study the criterion (4.22), written in the form (4.12). We have

$$a_2 = -\rho_2 \pm \sqrt{\rho_2^2 - \rho_2(\beta_2 + A_2)} \ .$$

The condition that a_2 be real is represented by the inequality

$$\rho_2 > \beta_2 + A_2 \ ,$$

the elementary transformation of which leads to

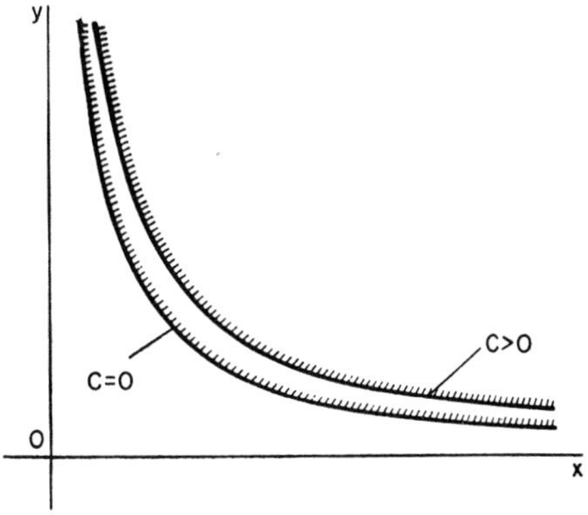

FIGURE 4

(4.26) $$\frac{U(U+\ell E)}{\ell T^2} > a + Ca \ ,$$

where C is an arbitrary positive number. Denoting

(4.27) $$x = \frac{U}{T\sqrt{a\ell}}, \qquad y = \frac{U + \ell E}{T\sqrt{a\ell}} \ ,$$

we plot a family of hyperbolas $xy = 1 + C$ (Figure 4), assigning various values to C. For $C = 0$ we obtain the limit of the exact boundary of the absolute-stability region $(xy = 1)$; consequently, we shall have asymptotic stability of the system everywhere within this region, for C can be taken as small as desired.

This particular example illustrates clearly the significance of the constant A_2, which enters into the stability criterion formulated by the Lur'e method.

The mechanical interpretation of the result obtained is as follows: The quantity U characterizes the coefficient of natural damping of the system, while ℓE represents the coefficient of articicial damping. Then, for a specified $\sqrt{a\ell}T$, the smaller the quantity U the greater must be the coefficient of artificial damping. This interpretation, and also the construction of the diagram (Figure 4) for $C = 0$, was originated by Bulgakov.

5. BULGAKOV'S SECOND PROBLEM

According to the Lur'e theorem, we have the initial equations for the determination of the coefficients of the Lyapunov function in this problem:

$$\frac{a_1^2}{\rho_1} + 2a_1 + \frac{2a_1a_2}{\rho_1 + \rho_2} + \beta_1 = 0 \ ,$$

(4.28)

$$\frac{a_2^2}{\rho_2} + 2a_2 + \frac{2a_1a_2}{\rho_1 + \rho_2} + \beta_2 = 0 \ .$$

Here ρ_1, ρ_2, β_1, and β_2 are determined by formulas (2.87) and (2.92). Let us put

$$a_1 = \rho_1(z_1 - 1), \quad a_2 = \rho_2(z_2 - 1)$$

and let us write (4.28) in new variables

$$\rho_1(z_1^2 - 1) + \frac{2\rho_1\rho_2}{\rho_1 + \rho_2}(z_1 - 1)(z_2 - 1) + \beta_1 = 0 \ ,$$

(4.29)

$$\rho_2(z_2^2 - 1) + \frac{2\rho_1\rho_2}{\rho_1 + \rho_2}(z_1 - 1)(z_2 - 1) + \beta_2 = 0 \ .$$

Dividing the first equation by ρ_1 and the second by ρ_2, and adding, we get

(4.30) $$(z_1 + z_2 - 1)^2 = r^2 \ ,$$

where the real quantity r is

(4.31) $$r^2 = 1 - \frac{\beta_1}{\rho_1} - \frac{\beta_2}{\rho_2} \ .$$

On the other hand, term by term subtraction of the equations (4.29) yields

(4.32) $$\rho_1 z_1^2 - \rho_2 z_2^2 = \rho_1 - \rho_2 - \beta_1 + \beta_2 \ .$$

Obviously, for arbitrary values of z_1 and z_2, the first stability condition assumes the form

(4.33) $$r^2 = 1 - \frac{\beta_1}{\rho_1} - \frac{\beta_2}{\rho_2} = 1 + \gamma_1 + \gamma_2 = 1 + \frac{rp_1}{q} = \frac{T^2(k+a\ell)}{k(T^2+\ell G^2)} > 0$$

and, no matter what the value of $q > 0$, this condition is always satisfied. Thus, (4.30) yields

(4.34) $$z_1 + z_2 = 1 \pm \Gamma \quad .$$

The matter therefore lies in the compatibility of (4.32) and (4.34). Eliminating the unknown z_2 from these equations leads to

(4.35) $$(\rho_2 - \rho_1)z_1^2 - 2\rho_2(1 \pm \Gamma)z_1 + \rho_2(1 \pm \Gamma)^2 + \rho_1 - \rho_2 - \beta_1 + \beta_2 = 0 \ .$$

Let us assume that ρ_1 and ρ_2 are real numbers; then, taking (4.31) into account, we obtain the following condition for the reality of the roots of

(4.36) $$D^2 = \rho_1^2 + \rho_2^2 - \rho_1\beta_1 - \rho_2\beta_2 \pm 2\rho_1\rho_2\Gamma > 0 \quad ,$$

and it is enough to take Γ in this inequality with either the plus or the minus sign. This inequality, in conjunction with inequality (4.33), yields the sought stability criterion.

If, however, ρ_1 and ρ_2 are complex numbers, then inequality (4.36) still retains its meaning as a stability criterion. Actually, let us denote

$$\rho_1 = r + is, \quad \beta_1 = a + ib, \quad z_1 = \mu + i\nu \quad ,$$

and then (4.32) and (4.34) become

$$2\mu = 1 \pm \Gamma, \quad 2s(\mu^2 + \nu^2) + 4r\mu\nu - 2(s - b) = 0 \quad .$$

Eliminating μ from these equations we obtain

$$2s\nu^2 - 2r(1 \pm \Gamma)\nu + 2(s - b) - \frac{s}{2}(1 \pm \Gamma)^2 = 0 \quad .$$

The condition for the reality of the solutions of this equation reduces to the satisfaction of the inequality

$$(r^2 + s^2)(1 \pm \Gamma)^2 - 4s(s - b) > 0$$

or, taking (4.31) into account, of the inequality

$$2[r^2 - s^2 - ar + sb \pm (r^2 + s^2)\Gamma] > 0 \quad .$$

It is easy to verify that this inequality is a consequence of inequality (4.36) when the latter is written in real numbers.

Thus, inequalities (4.32) and (4.36) are the only conditions for the absolute stability of the control system.

Using (2.87) and (2.92) we obtain

$$\begin{aligned} \rho_1^2 + \rho_2^2 &= (\rho_1 + \rho_2)^2 - 2\rho_1\rho_2 = \frac{p^2 - 2q}{r} , \\ \rho_1\beta_1 + \rho_2\beta_2 &= p_1 + b_{22}p_2 , \end{aligned} \tag{4.37}$$

and the stability criterion assumes the following form

$$D^2 = \frac{p^2 - 2q}{r} - p_1 + \frac{pp_2}{\sqrt{r}} \pm 2\frac{q}{r}\sqrt{1 + \frac{p_1 r}{q}} > 0 .$$

If we now employ formulas (2.82), we obtain finally

$$\frac{U}{T^2}\frac{U + \ell E}{T^2 + \ell G^2} \pm 2\sqrt{\frac{k(k+a\ell)}{T^2(T^2+\ell G^2)}} > \frac{2k + a\ell + \frac{k\ell G^2}{T^2}}{T^2 + \ell G^2} . \tag{4.38}$$

This inequality can be suitably interpreted. For this purpose, let us replace it by the corresponding equation, using the notation

$$s^2 = k + a\ell, \quad \alpha = \frac{k}{s^2}\left(1 + \frac{\ell G^2}{T^2}\right), \quad \beta = \frac{U + \ell E}{sT}, \quad M = \frac{U}{sT} \tag{4.39}$$

and let us get rid of the square root in the resulting equation. We then obtain the following equation for the boundary of the stability region:

$$(\beta M - \alpha - 1)^2 = 4\alpha . \tag{4.40}$$

Finally, denoting

$$\beta M = \eta, \quad \alpha = \xi$$

and transforming (4.40), we get

$$(\eta - \xi - 1)^2 = 4\xi .$$

In the variables ξ and η, the boundary line of the sought stability region will be the parabola

$$(\xi + \eta - 1)^2 = 4\xi\eta . \tag{4.41}$$

This was first plotted by Bulgakov (Figure 5) for the case $G^2 = 0$. The inner points of this parabola correspond to the absolute stability of the system. The parabola is determined from the original inequality (4.38).

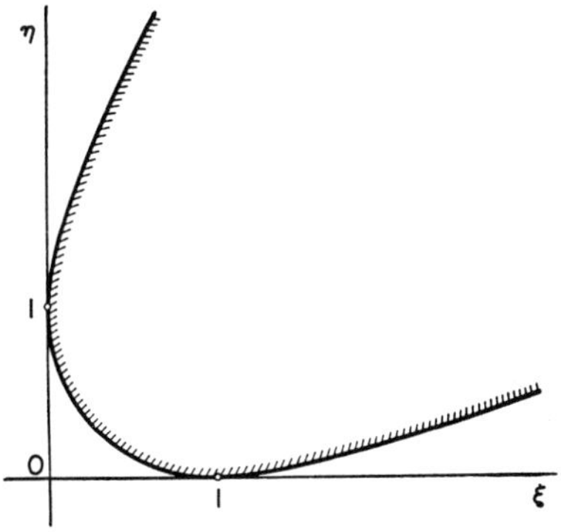

FIGURE 5

Let us note that the solution (4.41), like the solution (4.24), was obtained by Bulgakov with a method that yields the necessary stability conditions. Thus, we again verify the possibility of obtaining constructive solutions from the Lur'e theorem.

6. PROBLEM OF THE ISODROME REGULATOR

This problem is described by (2.114). Turning to these equations, we see that the stability conditions will in this case be of the form (4.22). The first condition becomes $\beta_3 < 0$, or

$$-\frac{b_{31}p_3}{\rho_1\rho_2} = -\frac{\quad}{\sqrt{rN}} < 0 \quad ,$$

and is satisfied for all $N > 0$.

Proceeding to an investigation of equations of the (4.28) type, we obtain, as in the preceding case, two inequalities, (4.33) and (4.36). Using formulas (2.115), (2.113), and (2.104), we recase these inequalities in the following form

$$k(k + a\ell) > \frac{U\ell}{N} \; ,$$

$$\frac{p^2 - 2q}{r} - p_1 + \frac{p}{\sqrt{r}}\, p_2 \pm 2\,\frac{q}{r}\sqrt{1 + \frac{p_1}{\rho_1\rho_2} - \frac{pr^2}{rNq^2}} > 0 \quad .$$

If we now employ formulas (2.104), we get

$$k(k + a\ell) > \frac{U\ell}{N} \quad ,$$

$$\frac{U}{T^2} \cdot \frac{U + \ell E}{T^2 + \ell G^2} \pm 2\sqrt{\frac{k(k+a\ell)}{T^2(T^2+\ell G^2)} - \frac{U\ell}{T^2 N(T^2+\ell G^2)}} > \tag{4.42}$$

$$> \frac{2k + a\ell + \dfrac{k\ell G^2}{T^2}}{T^2 + \ell G^2} \quad .$$

To clarify certain features of isodrome control, let us transform these inequalities to coordinates suitable for comparison with the Bulgakov plot. For this purpose let us denote

$$\rho = \frac{\ell T}{Ns^3}\left(1 + \frac{\ell G^2}{T^2}\right) \tag{4.43}$$

and make use of formulas (4.39). Then, after elementary transformations of inequalities (4.42), we obtain

$$(\beta M - \alpha - 1) > \mp 2\sqrt{\alpha - \rho M} \quad ,$$

$$\alpha > \rho M \quad .$$

Finally, introducing the new symbols

$$\alpha - \rho M = \xi - \rho M = x, \quad \beta M - \rho M = \eta - \rho M = y \quad , \tag{4.44}$$

let us write the latter inequalities in the following final form

$$x > 0, \quad (y - x - 1) > \pm 2\sqrt{x} \quad . \tag{4.45}$$

Obviously, the boundary line of the sought stability region, expressed in the parameters x and y, will be the parabola

$$(x + y - 1)^2 = 4xy \tag{4.46}$$

and the Oy axis. When $\rho = 0$ it coincides with the Bulgakov parabola. The internal points of the parabola correspond to the absolute stability of the system. As in the preceding case, the stability conditions (4.45), can be obtained by the Bulgakov method*, and are therefore exact to the limit.

To estimate the effect of the isodrome coefficient N on the

---

* See footnote on page 59.

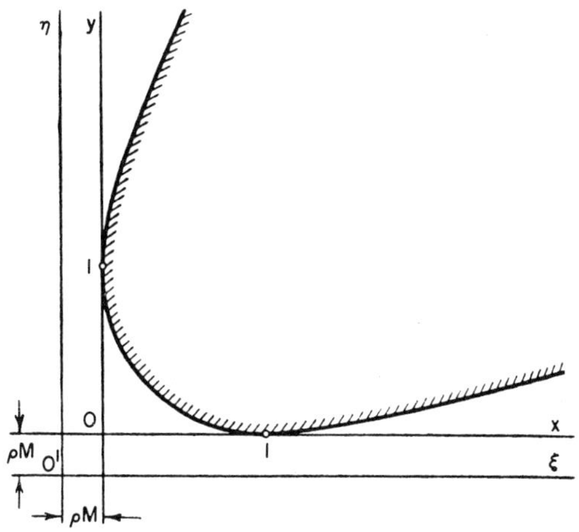

FIGURE 6

configuration of the stability region, we must compare the parabolas (4.41) and (4.46) (Figure 6).

It turns out that if a control system of a given type has too small a restoring force, so that

$$k(k + a\ell) < \frac{U\ell}{N} \quad ,$$

it is impossible to obtain convergence of the control process for any value of the constant E (i.e., for any value of the artificial damping of the system); consequently, it is essential that the first inequality (4.45) be satisfied. On the other hand, for fixed values of α and ρ, E should be greater in the case of isodrome control than in the Bulgakov problem, by an amount proportional to ρ.

The above shows that if an isodrome device, $0 < N < \infty$, is added to an ordinary regulator with proportional feedback $(\ell \neq \infty$, $N = \infty)$, (for the purpose of compensating for the effect of a constant disturbance on the regulated object), it imposes substantially increased requirements on the inherent static stability of the system, as well as on the artificial damping constant E. From this point of view, an isodrome regulator without proportional feedback which we discussed in the preceding problem for $N = \ell = \infty$, is more suitable than an isodrome regulator with proportional feedback.

7. FIRST VARIANT OF THE LUR'E THEOREM

The first variant of this theorem can be obtained directly, without any derivations. Thus, if we calculate the total derivative of V (formula (4.8)) and, when transforming it, do not substitute the term (4.11) into (4.9), we can readily see that instead of (4.15) we obtain

$$(4.47) \qquad \beta_k + 2a_k \sum_{i=1}^{n+1} \frac{a_i}{\rho_k + \rho_i} = 0 \qquad (k = 1, \dots, n + 1)$$

(for the case when $A_k = C_\alpha = 0$). It is clear that the Lur'e theorem remains valid in this case, but instead of (4.15) it will concern (4.47).

To explain the mechanical meaning of this modification, let us turn to the examples.

In the case of the first Bulgakov problem we now have

$$A + \beta_1 = 0 \ ,$$

$$\beta_2 = \frac{a_2^2}{\rho_2} = 0 \ .$$

This is equivalent to requiring that $\beta_2 < 0$. According to formulas (2.56) and (2.65), this inequality gives the following stability criterion:

$$a - \frac{U}{T^2}\left(E - \frac{UG^2}{T^2}\right) < 0 \ ,$$

i.e.,

$$(4.48) \qquad \frac{UE}{T^2} > a + \left(\frac{UG}{T^2}\right)^2 \ .$$

Comparing inequality (4.48) with inequalities (4.24) and (4.25), we see that this modification gives a less constructive (more difficult) solution of the problem.

In the case of the second Bulgakov problem, we shall now have

$$(4.49) \qquad \beta_1 = \frac{a_1^2}{\rho_1} + \frac{2a_1a_2}{\rho_1 + \rho_2} = 0, \qquad \beta_2 + \frac{a_2^2}{\rho_2} + \frac{2a_1a_2}{\rho_1 + \rho_2} = 0 \ .$$

Dividing the first equation by ρ_1 and the second by ρ_2, we get

(4.50) $$\left(\frac{a_1}{\rho_1} + \frac{a_2}{\rho_2}\right)^2 + \frac{\beta_1}{\rho_1} + \frac{\beta_2}{\rho_2} = 0 \quad .$$

In exactly the same way, multiplying the first equation of (4.49) by ρ_1 and the second by ρ_2, we obtain, after adding:

(4.50a) $$(a_1 + a_2)^2 + \beta_1\rho_1 + \beta_2\rho_2 = 0 \quad .$$

Turning to relations (4.30) and (4.37), we see that the relations (4.50) and (4.50a) become meaningful if the expressions

$$r^2 = 1 + \frac{r\rho_1}{q}, \qquad -(\beta_1\rho_1 + \beta_2\rho_2) = D^2$$

are positive. From this we obtain the stability conditions

(4.51) $$a > qG^2 \quad ,$$ $$qG^2 < a < (q - p^2)G^2 + pE \quad .$$

On the other hand, simple addition of equations (4.49) yields still another inequality, the satisfaction of which also follows from the given modification of the theorem. Actually, since the expression

$$\frac{a_1^2}{\rho_1} + \frac{a_2^2}{\rho_2} + \frac{4a_1a_2}{\rho_1 + \rho_2} = 2F(a_1a_2)$$

represents a positive definite function of the variables a_1 and a_2, the expression

(4.52) $$-(\beta_1 + \beta_2)$$

should be positive. In this case it produces still one more inequality

(4.53) $$E > pG^2 \quad .$$

It is easy to show that inequality (4.53) is the direct consequence of inequalities (4.51).

Thus, in the second Bulgakov problem, the modified Lur'e theorem gives two inequalities (4.51) instead of one inequality (4.38), which we previously had.

Consequently, in this case, too, the solution obtained for the problem is less constructive than the solution (4.38).

It is easy to show that for this modification, inequalities (4.51) and (4.53) are of general character. Actually, term-by-term addition of (4.47) yields

$$\sum_{k=1}^{n+1} \beta_k + 2 \sum_{k=1}^{n+1} \sum_{i=1}^{n+1} \frac{a_k a_i}{\rho_k + \rho_i} = \sum_{k=1}^{n+1} \beta_k + 2F(a_1, \ldots, a_{n+1}) = 0 ,$$

where $F(a_1, \ldots, a_{n+1})$ is a positive definite function; consequently,

$$\sum_{k=1}^{n+1} \beta_k < 0 . \tag{4.54}$$

Next, multiplying each equation (4.47) by ρ_k and adding, we obtain

$$\sum_{k=1}^{n+1} \beta_k \rho_k + 2 \sum_{k=1}^{n+1} \sum_{i=1}^{n+1} \frac{\rho_k a_i a_k}{\rho_k + \rho_i} = 0 .$$

However, since

$$2 \sum_{k=1}^{n+1} \sum_{i=1}^{n+1} \frac{\rho_k a_k a_i}{\rho_k + \rho_i} = \sum_{k=1}^{n+1} \sum_{i=1}^{n+1} \frac{\rho_k a_k a_i}{\rho_k + \rho_i} + \sum_{k=1}^{n+1} \sum_{i=1}^{n+1} \frac{\rho_i a_k a_i}{\rho_k + \rho_i} = \sum_{k=1}^{n+1} \sum_{i=1}^{n+1} a_k a_i ,$$

the preceding relation can be written as

$$\sum_{k=1}^{n+1} \rho_k \beta_k + \left(\sum_{k=1}^{n+1} a_k \right)^2 = 0 .$$

From this we obtain

$$\sum_{k=1}^{n+1} \rho_k \beta_k < 0 . \tag{4.55}$$

Next, after dividing each (4.47) by $\rho_k \neq 0$ and adding terms, we get

$$\sum_{k=1}^{n+1} \frac{\beta_k}{\rho_k} + 2 \sum_{k=1}^{n+1} \frac{a_k}{\rho_k} \sum_{i=1}^{n+1} \frac{a_i}{\rho_k + \rho_i} = 0 .$$

Denoting

$$A_k = \frac{a_k}{\rho_k} \quad ,$$

we can write

$$\sum_{k=1}^{n+1} \frac{\beta_k}{\rho_k} + 2 \sum_{k=1}^{n+1} \sum_{i=1}^{n+1} \frac{A_k A_i \rho_i}{\rho_k + \rho_i} = 0 \quad ,$$

and thus, as before

$$2 \sum_{k=1}^{n+1} \sum_{i=1}^{n+1} \frac{A_k A_i \rho_i}{\rho_k + \rho_i} = \left(\sum A_k \right)^2 \quad ,$$

then

(4.56) $$\sum_{k=1}^{n+1} \frac{\beta_k}{\rho_k} < 0 \quad .$$

The advantage of (4.47) over (4.15) is their simplicity. But, apparently, they always yield less constructive solutions of the problem than do (4.15).

8. SECOND VARIANT OF THE LUR'E THEOREM

If we choose as the Lyapunov function not (4.17) but the function

$$V = F = \sum_{k=1}^{n+1} \sum_{i=1}^{n+1} \frac{a_k a_i}{\rho_k + \rho_i} x_k x_i$$

and calculate its derivative according to (4.11), we obtain

$$\frac{dV}{dt} = -\left(\sum_{k=1}^{n+1} a_k x_k \right)^2 + 2f(\sigma) \sum_{k=1}^{n+1} \sum_{i=1}^{n+1} \frac{a_k a_i x_k}{\rho_k + \rho_i} \quad .$$

This version can be readily obtained by elementary transformation of the right half of V. Adding to this part the term

$$- f(\sigma) \left[\sigma - \sum_{k=1}^{n+1} \gamma_k x_k \right] \equiv 0 \quad ,$$

we obtain

$$\frac{dV}{dt} = -\left(\sum_{k=1}^{n+1} a_k x_k\right)^2 - \sigma f(\sigma) + f(\sigma)\sum_{k=1}^{n+1}\left[\gamma_k + 2a_k \sum_{k=1}^{n+1} \frac{a_i}{\rho_k + \rho_i}\right] x_k \quad .$$

Requiring now that the relations

$$\gamma_k + 2a_k \sum_{i=1}^{n+1} \frac{a_i}{\rho_k + \rho_i} = 0 \qquad (k = 1, \ldots, n + 1) \quad , \tag{4.57}$$

be satisfied, we can guarantee the absolute stability of the control system.

The Lur'e theorem for this modification does not change, but instead of (4.15) it will now contain (4.57).

We can readily verify that, as in the preceding case, for these equations to be solvable it is essential* that inequalities (4.54), (4.55), and (4.56) be satisfied. The latter follows from the existing connection between the quantities γ_k and β_k, determined by formulas (2.17), and also from the structure of (4.57). In particular, these inequalities lead us to the already known solution in the case of the two Bulgakov problems.

9. STABILITY IN THE CASE OF A MULTIPLE ROOT

We shall consider here only that particular case in which (2.5) has only one multiple root ρ_1 of multiplicity ℓ. Here we shall employ the canonical equations (2.54).

As before, we are concerned with the stability of the obvious solution of (2.54). We can follow here two paths. First, we can use the procedure indicated in the beginning of this chapter in the proof of the Lur'e theorem. In this even we take, as before, the function (4.8) for the Lyapunov function and, calculating its total derivative, we obtain according to (2.54)**

---

* As follows from the considered examples, certain of the inequalities may be interrelated.

** In (2.54) we have $r = 1$.

$$\frac{dV}{dt} = -\sum_{k=1}^{s} \rho_k A_k x_k^2 - C_1(\rho_{s+1} + \rho_{s+2})x_{s+1}x_{s+2} - \cdots$$
$$\cdots - C_{n-s}(\rho_n + \rho_{n+1})x_n x_{n+1} -$$
$$-\left(\sum_{k=1}^{n+1} a_k x_k\right)^2 - (\sqrt{r}f(\sigma))^2 - 2\sqrt{r}f(\sigma)\sum_{k=1}^{n+1} a_k x_k +$$
$$+ f(\sigma)\sum_{k=1}^{\ell}\left[q_k A_k + \beta_k + 2\sqrt{r}a_k + 2a_k\sum_{i=1}^{n+1}\frac{a_i q_i}{\rho_k + \rho_i}\right]x_k +$$
$$+ f(\sigma)\sum_{k=\ell+1}^{s}\left[A_k + \beta_k + 2\sqrt{r}a_k + 2a_k\sum_{i=1}^{n+1}\frac{a_i}{\rho_k + \rho_i}\right]x_k +$$
$$+ f(\sigma)\sum_{\alpha=1}^{n-s+1}\left[C_\alpha + \beta_{s+\alpha} + 2\sqrt{r}a_{s+\alpha} + \right.$$
$$\left. + 2a_{s+\alpha}\sum_{i=1}^{n+1}\frac{a_i}{\rho_{s+\alpha} + \rho_i}\right]x_{s+\alpha} +$$
$$+ \sum_{k=1}^{\ell} A_k \varepsilon_{k-1} x_k x_{k-1} + \sum_{i=1}^{\ell}\sum_{k=1}^{\ell}\frac{a_k a_i \varepsilon_{i-1} x_k x_{i-1}}{\rho_k + \rho_i} .$$

In the calculation of this expression it was assumed that $\ell \leq s$ and $C_1 = C_2, \ldots, C_{n-s} = C_{n-s+1}$.

In order to obtain a sign-definite derivative, of sign opposite to that of V, it is enough to require that the relations

$$q_k A_k + \beta_k + 2\sqrt{r}a_k + 2a_k\sum_{i=1}^{n+1}\frac{q_i a_i}{\rho_k + \rho_i} = 0 \qquad (k = 1, \ldots, s).$$

$$\text{(4.58)} \qquad C_\alpha + \beta_{s+\alpha} + 2\sqrt{r}a_{s+\alpha} + 2a_{s+\alpha}\sum_{i=1}^{n+1}\frac{a_i}{\rho_k + \rho_i} = 0$$
$$(\alpha = 1, \ldots, n - s + 1) ,$$

be satisfied, in which we must assume that $q_i = 1$ when $i > \ell$.

Actually, in this case the sought expression for the function $\dot{V}$ will be

$$\frac{dV}{dt} = -\sum_{k=1}^{s} \rho_k A_k x_k^2 - C_1(\rho_{s+1} + \rho_{s+2})x_{s+1}x_{s+2} - \cdots$$

$$\cdots - C_{n-s}(\rho_n + \rho_{n+1})x_n x_{n+1} - \left[\sum_{k=1}^{n+1} a_k x_k + \sqrt{r}f(\sigma)\right]^2 +$$

$$+ \sum_{k=1}^{\ell} A_k \varepsilon_{k-1} x_k x_{k-1} + \sum_{i=1}^{\ell}\sum_{k=1}^{\ell} \frac{a_k a_i \varepsilon_{i-1} x_k x_{k-1}}{\rho_k + \rho_i} ,$$

and since the constant factors ε_{k-1} can always be chosen arbitrarily and as small as desired, the terms of the function dV/dt, containing these factors, cannot affect its sign at all; therefore, fulfillment of relations (4.58) is a sufficient condition for stability.

We have thus proved the following theorem: if the constants of the regulator are such that the system of quadratic equations (4.58), in which $\rho_1, \ldots, \rho_s, \beta_1, \ldots, \beta_s$, and $q_1, \ldots, q_s$ are real, while $\rho_{s+1}, \ldots, \rho_{n+1}$ and $\beta_{s+1}, \ldots, \beta_{n+1}$ are conjugate complex numbers, has a solution containing real roots $a_1, \ldots, a_s$ and conjugate complex roots $a_{s+1}, \ldots, a_{n+1}$, then the steady state of the system has asymptotic absolute stability.*

There is another possible way of solving this problem, through which the stability conditions can be obtained in a simpler form. For this purpose let us put

$$V = \frac{1}{2}\sum_{\alpha=1}^{\ell} B_\alpha x_\alpha^2 + \sum_{k=\ell+1}^{n+1}\sum_{i=\ell+1}^{n+1} \frac{a_k a_i x_k x_i}{\rho_k + \rho_i} + \Phi + \int_0^\sigma f(\sigma)d\sigma ,$$

where all the σ_α are positive numbers, and let us calculate dV/dt in accordance with (2.54). We have

---

* Let us note that in the case of a multiple root in the analysis of (4.58) we cannot put $A_1 = \cdots = A_s = C_1 = \cdots = C_{n-s} = 0$, as was done before, but we can assume them to be sufficiently small.

$$\frac{dV}{dt} = -\rho_1 \sum_{\alpha=1}^{\ell} B_\alpha x_\alpha^2 - \left(\sum_{k=\ell+1}^{n+1} a_k x_k \right)^2 - [\sqrt{r}f(\sigma)]^2 +$$

$$+ f(\sigma) \sum_{\alpha=1}^{\ell} (B_\alpha q_\alpha + \beta_\alpha) x_\alpha + \sum_{\alpha=1}^{\ell} \varepsilon_{\alpha-1} B_\alpha x_\alpha x_{\alpha-1} + \sum_{\alpha=1}^{\ell} \varepsilon_{\alpha-1} A_\alpha x_\alpha x_{\alpha-1} +$$

$$+ f(\sigma) \sum_{k=\ell+1}^{s} \left[\beta_k + 2a_k \sum_{i=\ell+1}^{n+1} \frac{a_i}{\rho_k + \rho_i} \right] x_k +$$

$$+ f(\sigma) \sum_{k=s+1}^{n+1} \left[\beta_k + 2a_k \sum_{i=\ell+1}^{n+1} \frac{a_i}{\rho_k + \rho_i} \right] x_k \quad .$$

If, as before, we add and subtract in turn to the right half of this equation the quantity

$$2\sqrt{r}f(\sigma) \sum_{k=\ell+1}^{n+1} a_k x_k \quad ,$$

and require the fulfillment of the following relations

$$B_\alpha q_\alpha + \beta_\alpha = 0 \qquad (\alpha = 1, \ldots, \ell) \quad ,$$

$$A_k + \beta_k + 2a_k\sqrt{r} + 2a_k \sum_{i=\ell+1}^{n+1} \frac{a_i}{\rho_k + \rho_i} = 0$$

(4.59) $$(k = \ell + 1, \ldots, s) \quad ,$$

$$C_k + \beta_k + 2a_k\sqrt{r} + 2a_k \sum_{i=\ell+1}^{n+1} \frac{a_i}{\rho_k + \rho_i} = 0$$

$$(k = s + 1, \ldots, n + 1) \quad ,$$

we then obtain a sign-definite derivative,

$$\frac{dV}{dt} = -\rho_1 \sum_{\alpha=1}^{\ell} B_\alpha x_\alpha^2 - \left[\sum_{k=\ell+1}^{n+1} a_k x_k + \sqrt{r}f(\sigma) \right]^2 +$$

$$+ \sum_{\alpha=1}^{\ell} \varepsilon_{\alpha-1} (A_\alpha + B_\alpha) x_\alpha x_{\alpha-1} - \sum_{k=1}^{s} \rho_k A_k x_k^2 -$$

$$- C_1(\rho_{s+1} + \rho_{s+2}) x_{s+1} x_{s+2} - \cdots - C_{n-s}(\rho_n + \rho_{n+1}) x_n x_{n+1} \quad .$$

which is negative everywhere. Consequently, conditions (4.59) are also sufficient for the absolute stability of the system in the case of the multiple root of (2.5); it differs from the conditions (4.58) of the Lur'e type only in the ℓ first relations. Relations (4.59) must be considered as equations, which must determine $n + 1$ constants: $B_1, \ldots, B_\ell, a_{\ell+1}, \ldots, a_n$.

Thus, we have proved the following theorem: if the constants of the regulator are such that the system of equations (4.59), in which $\rho_1, \ldots, \rho_s, \beta_1, \ldots, \beta_s$ and $q_1, \ldots, q_s$ are real and $\rho_{s+1}, \ldots, \rho_{n+1}$ and $\beta_{s+1}, \ldots, \beta_{n+1}$ are complex conjugate numbers, has a solution containing positive roots $B_1, \ldots, B_\ell$ real $a_{\ell+1}, \ldots, a_s$, and conjugate complex $a_{s+1}, \ldots, a_n$, then the steady state of the system has absolute asymptotic stability.

By way of an example of an application of these theorems, let us consider the second Bulgakov problem. Let us first consider the criterion (4.58). Here $n + 1 = \ell = 2$, and we have

$$A_1 + \beta_1 + 2\sqrt{r}a_1 + 2a_1 \left(\frac{q_1 a_1}{2\rho_1} + \frac{q_2 a_2}{2\rho_1}\right) = 0 ,$$

$$A_2 q_2 + \beta_2 + 2\sqrt{r}a_2 + 2a_2 \left(\frac{q_1 a_1}{2\rho_1} + \frac{q_2 a_2}{2\rho_1}\right) = 0 ,$$

where $q_1 = 1$ (2.98), and q_2 is an arbitrary number. If we employ this arbitrariness and put $q_2 = 1$ and $A_1 = A_2 = \varepsilon$, where ε is a sufficientl small number, then with an accuracy to within ε we arrive at the previous stability criterion, which was studied in Section 5 of this chapter. This conclusion holds also for the isodrome regulator, since according to (2.117) and (2.119) it is possible to put $q_1 = q_2 = 1$.

The criterion (4.59) yields the inequalities:

$$B_1 = -\beta_1 > 0 ,$$

$$B_2 = -\frac{\beta_2}{q_2} > 0 .$$

Taking (2.96) and (2.97) into account, we obtain

$$C_1(p_1 + b_{22}p_2 + \rho_1 p_2) > 0 ,$$

$$p_2 > C_1(p_1 + b_{22}p_2 + \rho_1 p_2) > 0 .$$

Obviously, the first inequality can always be satisfied by choosing the sign of C_1, and by virtue of the arbitrariness in the choice of $|C_1|$, the second inequality reduces to

$$E > pG^2 + \varepsilon^2 ,$$

where ε^2 is a sufficiently small positive number.

It is easy to verify that this inequality is bound to be satisfied if the stability of the system is assured by condition (4.51) when $\rho_1 \neq \rho_2$.

CHAPTER V: FORMULATION OF SIMPLIFIED STABILITY CRITERIA

1. FIRST CASE OF FORMULATION OF SIMPLIFIED STABILITY CRITERIA

In problems for which an analysis of (4.15) is difficult, it is advisable to try to find other, simpler stability criteria.

We shall indicate here one device* for finding such criteria, which can be applied to any system of (4.1), provided it is known beforehand that among the constants β_k there is at least one that is negative definite. Let, for example, such a constant be β_1; it corresponds to a real positive root ρ_1.

In this case the Lyapunov function should be taken in the form

$$V = \frac{1}{2} Kx_1^2 + F(a_2x_2, \ldots, a_{n+1}x_{n+1}) + \int_0^\sigma f(\sigma)d\sigma \tag{5.1}$$

where K is a positive constant; then the number of equations similar to (4.15) is reduced by one.

By way of an example, let us consider (2.133) for the control of a system subject to the action of constant disturbing forces.

It is clear from (2.132), that if there is no direct-connected load $(\rho_4 = 0)$, the constant β_4 must be negative definite. Actually

$$\beta_4 = \frac{p_1 b_{33}}{\rho_2 \rho_3} = - \frac{anr}{\sqrt{r}(pn+q)} \quad . \tag{5.2}$$

The Lyapunov function can be taken in this case in the form

$$V = \frac{1}{2} Kx_4^2 + \frac{a_2^2x_2^2}{2\rho_2} = \frac{a_3^2x_3^2}{2\rho_3} + \frac{2a_2a_3}{\rho_2 + \rho_3} x_2x_3 + \int_0^\sigma \Phi(\sigma)d\sigma \quad .$$

Its total derivative, calculated in accordance with (2.133) will be

$$\frac{dV}{dt} = (K + \beta_4)x_4\Phi(\sigma) - (a_2x_2 + a_3x_3)^2 - \Phi^2(\sigma) +$$
$$+ \left(\frac{a_2^2}{\rho_2} + 2a_2 + \frac{2a_2a_3}{\rho_2 + \rho_3} + \beta_2\right) x_2\Phi(\sigma) + \left(\frac{a_3^2}{\rho_3} + 2a_3 + \frac{2a_2a_3}{\rho_2 + \rho_3} + \beta_3\right) x_3\Phi(\sigma) \quad .$$

---

* M. Letov, Regulation of the Stationary State of a System Subjected to the Action of Constant Disturbing Forces, PMM, Volume XII, No. 2, 1948.

Obviously, by virtue of (5.2), the constant K can always be chosen from the condition $K = -\beta_4 > 0$, and consequently, the stability criterion reduces here to the fulfillment of the following two conditions:

$$\frac{a_2^2}{\rho_2} + 2a_2 + \frac{2a_2a_3}{\rho_2 + \rho_3} + \beta_2 = 0$$

$$\frac{a_3^2}{\rho_3} + 2a_3 + \frac{2a_2a_3}{\rho_2 + \rho_3} + \beta_3 = 0 .$$

In analogy with the second Bulgakov problem we arrive at two inequalities

$$(5.3) \qquad \begin{aligned} \Gamma^2 &= 1 - \frac{\beta_2}{\rho_2} - \frac{\beta_3}{\rho_3} > 0 , \\ D^2 &= \rho_2^2 + \rho_3^2 - \beta_2\rho_2 - \beta_3\rho_3 \pm 2\rho_2\rho_3\sqrt{\Gamma} > 0 . \end{aligned}$$

To calculate the left sides of these inequalities, it is necessary to employ formulas (2.132) and (2.124). We have

$$\frac{\beta_2}{\rho_2} + \frac{\beta_3}{\rho_3} = -\frac{1}{(\rho_2\rho_3)^2} [p_1\rho_2\rho_3 + p_1b_{33}(\rho_2 + \rho_3) - p_2b_{33}\rho_2\rho_3 + p_3\rho_2\rho_3] ,$$

$$(5.4) \qquad \beta_2\rho_2 + \beta_3\rho_3 = \frac{1}{\rho_3 - \rho_2} [p_1(\rho_3 - \rho_2) - p_2(\rho_3^2 - \rho_2^2) - p_2b_{33}(\rho_3 - \rho_2) + p_3(\rho_3 - \rho_2) ,$$

$$\rho_2^2 + \rho_3^2 = \left(\frac{p + n}{\sqrt{r}}\right)^2 - 2\,\frac{pn + q}{r} .$$

These relations permit calculation of the quantities Γ^2 and D^2. According to (5.3), we get

$$\Gamma^2 = \frac{r^2}{(pn+q)^2} \left[\left(\frac{pn + q}{r}\right)^2 + \frac{p_1}{r}(q - n^2) + \frac{n(pn+q)(E-pG^2)}{r} - \frac{qG^2(pn+q)}{r} \right] .$$

Introducing for brevity the dimensionless coefficients

$$(5.5)\qquad \alpha = \frac{k}{ai}\,,\quad \beta = \frac{U + i(E-pG^2)}{T\sqrt{ai}},\quad M = \frac{U}{T\sqrt{ai}}\,,\quad \chi = \frac{nT}{\sqrt{ai}}\,,\quad \theta = \frac{i}{\ell}$$

and subjecting the expression for r^2 to elementary transformation, we get

$$r^2 = \frac{p_1^2}{(\alpha+\chi M)^2}\left[(\alpha + \chi M)\left(\frac{i}{\ell}\alpha + \chi\beta\right) + \alpha - \chi^2\right] > 0 \quad .$$

Next, using formulas (5.3) and (5.4), we calculate D^2. We have

$$D^2 = \left(\frac{p + n}{\sqrt{r}}\right)^2 - 2\,\frac{pn + q}{r} - p_1 + \frac{p + n}{\sqrt{r}}\sqrt{r}(E - pG^2) -$$

$$- n(E - pG^2) + qG^2 \pm 2\,\frac{pn + q}{r}r > 0 \quad .$$

To interpret the resultant inequalities, we employ the symbols of (5.5) and write D^2 in the following form:

$$D^2 = p_1\left[(\chi + M)^2 - 2(\chi + M) - 1 + M\sqrt{\frac{r}{p_1}}(E - pG^2) + \right.$$

$$\left. + \frac{\alpha i G^2}{T^2} \pm 2\sqrt{(\alpha + \chi M)\ (\frac{i}{\ell}\alpha + \chi\beta)\ + \alpha - \chi^2}\,\right]$$

If we add the quantity M^2 to the right half of this equation and then subtract it, we can change this equation into

$$D^2 = p_1[\beta M - 1 + \chi^2 - (1 + \theta)\alpha \pm$$

$$\pm 2\sqrt{(\alpha + \chi M)(\theta\alpha+\chi\beta) + \alpha - \chi^2}\,] \quad .$$

Let us proceed to an examination of the inequalities $r^2 > 0$ and $D^2 > 0$. For this purpose, we denote

$$(5.6)\qquad \alpha = x,\quad \beta M = y$$

and examine the curve

$$(5.7)\qquad \theta x^2 + \frac{\chi}{M}xy + (1 + \chi M\theta)x + \chi^2 y - \chi^2 = 0 \quad ,$$

and also the curve

$$[y - 1 + \chi^2 - (1 + \theta)x]^2 = 4(\chi M + x)\quad \theta x + \frac{\chi}{M}y \quad + 4x - 4\chi^2 \quad .$$

The latter curve can be represented as

$$a_{11}x^2 + 2a_{12}xy + a_{22}y_2 + 2a_1x + 2a_2y + a_3 = 0 \quad , \tag{5.8}$$

where

$$a_{11} = (1 + \theta)^2 - 4\theta, \quad 2a_{12} = -2(1 + \theta) - 4\frac{\chi}{M}, \quad a_{22} = 1 \quad ,$$

$$2a_1 = -2(\chi^2 - 1)(1 + \theta) - 4\chi M\theta - 4 \quad ,$$

$$2a_2 = 2(\chi^2 - 1) - 4\chi^2, \quad a_3 = (1 + \chi^2)^2 \quad .$$

We shall first examine curve (5.8). It is easy to verify that it is a hyperbola. Let us consider that branch of the hyperbola located in the first quadrant. This hyperbola is tangent to the Oy axis at $y^* = 1 + \chi^2$ and intercepts the Ox axis at

$$x_1, x_2 = \frac{1}{(1-\theta)^2}\,[(1 + \theta)^2(\chi^2 - 1) + 2(1 + \chi M\theta) \pm$$

$$\pm \sqrt{[(1 + \theta)(\chi^2 - 1) + 2(1 + \chi M\theta)]^2 - (1 - \theta)^2(1 + \chi^2)^2}\;] \; .$$

For all regulators that contain no proportional feedback, (i.e., for which $\ell = \infty$ and $\theta = 0$), the intersection points x_1 and x_2 come together and become a point of tangency $x_1 = x_2 = 1 + \chi^2$; for all regulators with proportional feedback, $G^2 = 0$ and $\theta = 1$, we have $x_1 = \infty$, $x_2 = (1 + \chi^2)^2/r\chi(\chi + M)$; if $0 < \theta < 1$, we get $x_1 > x_2 > 0$. The region inside the hyperbola corresponds to the sought values of the regulator parameters, i.e., to the values for which $D^2 > 0$. When $\chi = 0$, the hyperbola (5.8) degenerates into the Bulgakov parabola.

Figure 7 gives the construction of the hyperbola for the case $\ell = \infty$.

It is easy to verify that when $\ell = \infty$, the curve (5.7) also is a hyperbola that intersects the coordinate axis at $x^* = \chi^2$, $y^* = 1$. The asymptotes of this curve are the straight lines

$$y = -\frac{M}{\chi}\,, \quad x = -\chi M \quad .$$

This hyperbola is tangent to the

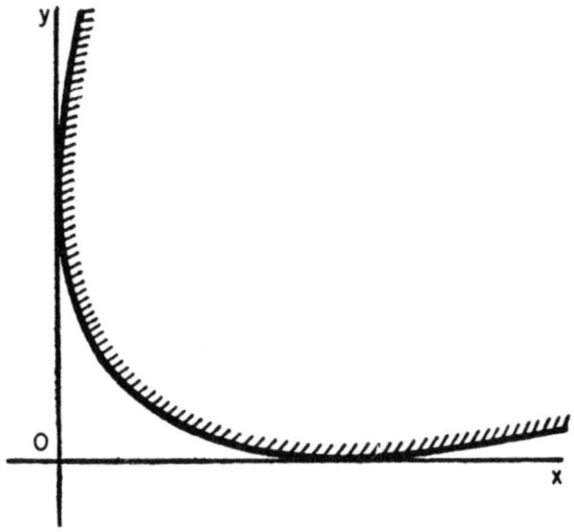

FIGURE 7

hyperbola (5.8) at a point having an abscissa

$$x = \frac{M}{2X}\left[\sqrt{\left[\frac{X}{M}(1 - X^2) + 1 + X^2\right]^2 + 4\frac{X^5}{M}} - \frac{X}{M}(1 - X^2) - 1 - X^2\right] .$$

The placement of the hyperbola (5.7) indicates that the boundary of the stability region, plotted with respect to the parameters x and y, is the hyperbola (5.8).

2. SECOND CASE OF FORMULATION OF SIMPLIFIED STABILITY CRITERIA

The Lur'e stability conditions considered above, like other conditions of the same type, reduce in the final analysis to the solvability of a finite number of quadratic equations with respect to certain unknown constants $a_1, \ldots, a_{n+1}$.

In analysis of control systems with many degrees of freedom, a study of these conditions can be made only by numerical analysis methods. In connection with this, it is preferable to formulate sufficient stability conditions that require no analysis of any equations whatever. We shall obtain conditions of this type in this section.

We shall make use of (4.1). Let us assume that all ρ_k are different real numbers. Let us consider the Lyapunov function

$$V = \frac{1}{2}\sum_{k=1}^{n+1} x_k^2 + R\int_0^\sigma f(\sigma)d\sigma \quad , \tag{5.9}$$

where R is a positive constant. Its total derivative, calculated in accordance with equations (4.1), is a quadratic form in the variables $x_1, \ldots, x_{n+1}$ and $f(\sigma)$:

$$\dot{V} = -\sum_{k=1}^{n+1} \rho_k x_k^2 - rRf^2(\sigma) + \sum_{k=1}^{n+1}(1 + R\beta_k)x_k f(\sigma) \quad . \tag{5.10}$$

To obtain an idea of the sign of this form, let us consider the determinant

(5.11)
$$\Delta = \begin{vmatrix} \rho_1 & 0 & \cdots & 0 & -\frac{1 + R\beta_1}{2} \\ 0 & \rho_2 & \cdots & 0 & -\frac{1 + R\beta_2}{2} \\ \cdots\cdots\cdots & & \cdots & 0 & \cdots\cdots \\ 0 & \cdots\cdots & \cdots & \rho_{n+1} & -\frac{1 + R\beta_{n+1}}{2} \\ -\frac{1 + R\beta_1}{2} & -\frac{1 + R\beta_2}{2} & \cdots & -\frac{1 + R\beta_{n+1}}{2} & rR \end{vmatrix},$$

and also a sequence of its diagonal minors

(5.12)
$$\Delta_1 = \rho_1, \quad \Delta_2 = \Delta_1\rho_2, \quad \ldots, \quad \Delta_{n+1} = \Delta_n\rho_{n+1} \quad .$$

From the theory of quadratic forms, it follows that a necessary and sufficient condition, for the form $-\dot{V}$ to be positive, is that the determinants (5.11) and (5.12) be positive. If the control system is inherently stable, i.e., $\rho_k > 0$ $(k = 1, \ldots, n + 1)$, then all the determinants in (5.12) are positive, and the only condition for the form $-\dot{V}$ to be positive is the inequality $\Delta > 0$. Expanding the determinant (5.11), we get

$$\Delta = rR \prod^{n+1} \rho_k - \frac{1}{4} \sum_{k=1}^{n+1} \rho_1, \ldots, \rho_{k-1}(1 + R\beta_k)^2\rho_{k+1}, \ldots, \rho_{n+1} > 0 \quad ,$$

and consequently the above condition can be written in the following final form

(5.13)
$$4rR > \sum_{k=1}^{n+1} \frac{(1 + R\beta_k)^2}{\rho_k} \quad .$$

If the roots ρ_k $(k = 1, \ldots, n + 1)$ are different, but one of these, say ρ_{n+1} vanishes, the Lyapunov function must be taken in the following form

(5.14)
$$V = \frac{1}{2} \sum_{k=1}^{n} x_k^2 + R \int_0^\sigma f(\sigma)d\sigma + \frac{1}{2} Kx_{n+1}^2 \qquad (K > 0)$$

and inequality (5.13) is retained, but the sign of the sum must be extended in it only over n terms.

An analogous comment must also be made concerning the two other cases, in which $\rho_{m+2} = \infty$ and $\rho_{m+1} = 0$. In both cases inequality (5.13) remains in force, but in the former the summation sign will extend over $m + 1$ components, and in the second case over m components.

Particular attention must be paid to that subcase of Case 3 in which $\rho_1 = 0$, and the remaining roots are different, real, and positive. Under these conditions, the Lyapunov function must be taken in the form

$$V = \frac{1}{2} Kx_1^2 + \sum_{k=2}^{m} x_k^2 + R \int_0^\sigma f(\sigma)d\sigma \quad , \tag{5.15}$$

where K is a positive constant. Its total derivative, calculated in accordance with (2.19a) is

$$\dot{V} = - \sum_{k=2}^{m} \rho_k x_k^2 - rRf^2(\sigma) + \sum_{k=2}^{m} (1 + R\beta_k)x_k f(\sigma) + \\ + (K + R\beta_1)x_1 f(\sigma) \quad .$$

Obviously, the conditions under which $\dot{V}$ will be negative definite assume in this case the following form

$$K = - R\beta_1 > 0 \quad , \tag{5.16}$$

$$4rR > \sum_{k=2}^{m} \frac{(1 + R\beta_k)^2}{\rho_k} \quad . \tag{5.17}$$

Finally, let us assume that among the roots of (2.5) there are real roots $\rho_1, \ldots, \rho_s$ and complex conjugate roots $\rho_{s+1}, \ldots, \rho_{n+1}$. Accordingly the quantities $\beta_1, \ldots, \beta_s$ and $x_1, \ldots, x_s$ will be real, and the quantities $\beta_{s+1}, \ldots, \beta_{n+1}$ and $x_{s+1}, \ldots, x_{n+1}$ will be complex conjugate pairs.

Let us put

$$\begin{aligned} x_{s+\alpha} &= u_{s+\alpha} + iv_{s+\alpha}, & x_{s+\alpha+1} &= u_{s+\alpha} - iv_{s+\alpha} \quad , \\ \beta_{s+\alpha} &= p_{s+\alpha} + iq_{s+\alpha}, & \beta_{s+\alpha+1} &= p_{s+\alpha} - iq_{s+\alpha} \quad , \\ \rho_{s+\alpha} &= a_{s+\alpha} + ib_{s+\alpha}, & \rho_{s+\alpha+1} &= a_{s+\alpha} - ib_{s+\alpha} \quad , \end{aligned} \tag{5.18}$$
$$(\alpha = 1, 3, \ldots, n + 1 - s)$$

and let us consider the function

$$V = \frac{1}{2}\sum_{k=1}^{s} x_k^2 + \sum_{\alpha=1,3,\ldots}^{n-s} x_{s+\alpha}x_{s+\alpha+1} + R\int_0^{\sigma} f(\sigma)d\sigma \quad . \tag{5.19}$$

According to (2.11), its total derivative is

$$\dot{V} = -\sum_{k=1}^{s} \rho_k x_k^2 - \sum_{\alpha=1,3,\ldots}^{n-s} (\rho_{s+\alpha} + \rho_{s+\alpha+1})x_{s+\alpha}x_{s+\alpha+1} + $$

$$+ \sum_{k=1}^{n+1} (1 + R\beta_k)x_k f(\sigma) - rRf^2(\sigma)$$

or, in accordance with (5.18),

$$\dot{V} = -\sum_{k=1}^{s} \rho_k x_k^2 - 2\sum_{\alpha=1,3}^{n-s} a_{s+\alpha}(u_{s+\alpha}^2 + v_{s+\alpha}^2) - rRf^2(\sigma) +$$

$$+ \sum_{k=1}^{s} (1 + R\beta_k)x_k f(\sigma) + 2\sum_{\alpha=1,3}^{n-s} [u_{s+\alpha} + R(p_{s+\alpha}u_{s+\alpha} - q_{s+\alpha}v_{s+\alpha})]f(\sigma) \quad .$$

As before, we obtain a quadratic form in the variables $x_1, \ldots, x_s$, u_{s+1} and $v_{s+1}, \ldots, f(\sigma)$. For this to be negative definite, it is necessary and sufficient that the determinant:

$$\begin{vmatrix}
\rho_1 & \cdots & 0 & 0 & \cdots & 0 & 0 & \cdots & 0 & \frac{1+R\beta_1}{2} \\
\vdots & & \vdots & \vdots & & \vdots & \vdots & \vdots & \vdots & \vdots \\
0 & \cdots & \rho_s & \cdots & \cdots & \cdots & \cdots & \cdots & \cdots & \frac{1+R\beta_s}{2} \\
0 & \cdots & \cdots & 2a_{s+1} & \cdots & \cdots & \cdots & \cdots & \cdots & 1 + Rp_{s+1} \\
\cdots & \cdots & \cdots & \cdots & \cdots & \cdots & \cdots & \cdots & \cdots & \cdots \\
0 & \cdots & \cdots & \cdots & \cdots & 2a_n & \cdots & \cdots & \cdots & 1 + Rp_n \\
0 & \cdots & \cdots & \cdots & \cdots & \cdots & 2a_{s+1} & \cdots & \cdots & - Rq_{s+1} \\
\cdots & \cdots & \cdots & \cdots & \cdots & \cdots & \cdots & \cdots & \cdots & \cdots \\
0 & \cdots & \cdots & \cdots & \cdots & \cdots & \cdots & \cdots & 2a_n & - Rq_n \\
\frac{1+R\beta_1}{2} & \cdots & \frac{1+R\beta_s}{2} & 1 + Rp_{s+1} & \cdots & 1 + Rp_n & - Rq_{s+1} & \cdots & - Rq_n & rR
\end{vmatrix} \tag{5.20}$$

and all its diagonal minors $\Delta_1, \ldots, \Delta_{n+1}$:

$$\Delta_1 = \rho_1, \ \ldots, \ \Delta_s = \Delta_{s-1}\rho, \ \ldots, \ \Delta_{n+1} = \Delta_n 2a_n \quad . \tag{5.21}$$

be positive. Obviously, by virtue of the condition $\operatorname{Re} \rho_k > 0$ $(k = 1, \ldots, n + 1)$, all the determinants (5.21) are positive. Consequently, the necessary and sufficient condition for the form $-\dot{V}$ to be positive definite is the inequality

$$\Delta > 0 \quad , \tag{5.22}$$

where Δ is the determinant (5.20).

Thus, it is possible to formulate the following theorem: In order to guarantee the absolute stability of the system of the form (4.1) when the roots ρ_k are simple and $\operatorname{Re} \rho_k > 0$ $(k = 1, \ldots, n + 1)$, it is sufficient to choose the regulator parameters such that the inequality (5.13) or (5.22) can be satisfied at $R > 0$. If, however, $\rho_1 = 0$, the suitable conditions are (5.16) and (5.17).

3. SPECIFIC PROBLEMS

As previously, the effectiveness of the given simplified stability criterion can be checked with specific examples.

Thus, in the case of the first Bulgakov problem it is necessary to employ inequalities (5.16) and (5.17). We have seen earlier than inequality (5.16) is satisfied in this case [see (4.23)]; inequality (5.17) assumes the following form

$$4R > \frac{(1+R\beta_2)^2}{\rho_2}$$

for the first problem, or, after elementary transformations

$$\dot{\beta}_2 < \frac{2\sqrt{R\rho_2} - 1}{R} \quad . \tag{5.23}$$

Let us now compare solution (4.23), obtained on the basis of the Lur'e theorem, with solution (5.23). There we have the inequality $\beta_2 < \rho_2$. Here, however, making use of the arbitrary choice of R, we attempt to select it in such a way that the upper limit of the values of β_2, determined by the right half of inequality (5.23), is greater, or at least equal to ρ_2. We have

$$\frac{2\sqrt{R\rho_2} - 1}{R} \geq \rho_2 \quad ,$$

or

$$(1 - \sqrt{R\rho_2})^2 \leq 0 \quad .$$

Obviously, it is necessary to take here the equal sign, and we find the most convenient value of R, which will be

$$R = \frac{1}{\rho_2} \quad .$$

But in this case the inequality (5.23) assumes the already known form (4.22), i.e.,

(5.24) $$\beta_2 < \rho_2 \quad .$$

Expanding inequality (5.24), we obtain the stability criterion (4.24), previously derived on the basis of the Lur'e theorem.

In the case of the second Bulgakov problem we have

(5.25) $$4R > \frac{(1+R\beta_1)^2}{\rho_1} + \frac{(1 + R\beta_2)^2}{\rho_2} \quad .$$

Denoting

$$\frac{1 + R\beta_1}{\rho_1} = u, \qquad \frac{1 + R\beta_2}{\rho_2} = v$$

and writing (5.25) in the new variables, we obtain a circle of radius $2\sqrt{R}$, which determines the stability region in the u, v plane.

4. GENERAL METHOD FOR FORMULATING SIMPLIFIED STABILITY CRITERIA

I. G. Malkin* proposed a general method for formulating simplified stability criteria, of which a particular case is the criterion of Section 2.

Let the ρ_k be real distinct roots, none of which vanishes.

Let us consider any real negative definite quadratic form W

* I. G. Malkin, Contribution to the Theory of Stability of Control Systems, PMM, Volume XV, No. 1, 1951.

(5.26)
$$W = -\frac{1}{2}\sum_{\alpha=1}^{n+1}\sum_{\beta=1}^{n+1} A_{\alpha\beta}x_\alpha x_\beta .$$

By Sylvester's rule, the coefficients $A_{\alpha\beta}$ of this form should satisfy the inequalities

$$\Delta_1 = A_{11} > 0, \quad \Delta_2 = \begin{vmatrix} A_{11} & A_{12} \\ A_{21} & A_{22} \end{vmatrix} > 0, \ \ldots$$

(5.27)
$$\ldots, \Delta_n = \begin{vmatrix} A_{11} & \cdots & A_{1,n+1} \\ \vdots & & \\ A_{n+1,1} & \cdots & A_{n+1,n+1} \end{vmatrix} > 0 .$$

On the other hand, let us consider the real quadratic form

(5.28)
$$F = \frac{1}{2}\sum_{\alpha=1}^{n+1}\sum_{\beta=1}^{n+1} B_{\alpha\beta}x_\alpha x_\beta ,$$

the coefficients of which will be determined by means of the inequalities

(5.29)
$$B_{\alpha\beta} = \frac{A_{\alpha\beta}}{\rho_\alpha + \rho_\beta} \qquad (\alpha, \beta = 1, \ldots, n+1) .$$

It is easy to show that the function F is positive definite everywhere.

This can be done by using a method similar to that employed to prove the theorem at the beginning of Section 2 of Chapter IV.

Turning now to (4.1), let us consider the positive definite function

$$V = F + \int_0^\sigma f(\sigma)d\sigma$$

and let us calculate its derivative with respect to time. On the basis of (4.1) we find

(5.30)
$$\frac{dV}{dt} = \frac{1}{2}\sum_{\alpha=1}^{n+1}\sum_{\beta=1}^{n+1}\Big[-(\rho_\alpha + \rho_\beta)x_\alpha x_\beta + f(\sigma)(x_\alpha + x_\beta)\Big]B_{\alpha\beta} + f(\sigma)\left[\sum_{\alpha=1}^{n+1}\beta_\alpha x_\alpha - f(\sigma)\right],$$

but according to (5.29) and (5.26) we have:

$$\frac{1}{2}\sum_{\alpha=1}^{n+1}\sum_{\beta=1}^{n+1} B_{\alpha\beta}(\rho_\alpha + \rho_\beta)x_\alpha x_\beta = -W ,$$

and consequently,

$$(5.31) \qquad \frac{dV}{dt} = W - f^2(\sigma) + f(\sigma)\sum_{\alpha=1}^{n+1}\left[\beta_\alpha + \frac{1}{2}\sum_{\beta=1}^{n+1}(B_{\alpha\beta} + B_{\beta\alpha})\right]x_\alpha .$$

Were all the coefficients in front of x_α equal to zero, the function dV/dt would be sign-definite and everywhere negative. Since this will not take place in the general case, to establish the sign of the quadratic form $-dV/dt$ in the variables $x_1, \ldots, x_{n+1}$ and $f(\sigma)$, let us write down its discriminant -- the determinant

$$(5.32) \qquad \Delta = \begin{vmatrix} A_{11} & \cdots & A_{1,n+1} & P_1 \\ A_{21} & \cdots & A_{2,n+1} & P_2 \\ \cdot & \cdot & \cdot & \cdot \\ A_{n+1,1} & \cdots & A_{n+1,n+1} & P_{n+1} \\ P_1 & \cdots & P_{n+1} & 1 \end{vmatrix} ,$$

where

$$(5.33) \qquad 2P_\alpha = \beta_\alpha + \frac{1}{2}\sum_{\beta=1}^{n+1}(B_{\alpha\beta} + B_{\beta\alpha}) \qquad (\alpha = 1, \ldots, n+1) .$$

The sequence of its diagonal minors is obviously here the sequence of the determinants (5.27). Since all these determinants are positive, the necessary and sufficient condition that the function (5.31) be negative definite everywhere is that the determinant (5.32) be positive. Consequently, the inequality

$$(5.34) \qquad \Delta > 0$$

is the only condition which assures absolute stability of the system. Several remarks must be made concerning this conclusion. First, let us note that all the specified particular canonical equations [Case 1, (2.12a); Case 2, (2.18); Case 3, (2.19a)] can be considered by the same method. Inequalities that would guarantee the stability of the system in each of these cases can be obtained from (5.32) and (5.33) by changing the number of columns and rows of the determinant (5.32) and the summation indices. Thus, in Case 1 we find

$$\Delta = \begin{vmatrix} A_{11} & \cdots & A_{1n} & P_1 \\ \vdots & & \vdots & \vdots \\ A_{n1} & \cdots & A_{nn} & P_n \\ P_1 & \cdots & P_n & 1 \end{vmatrix} > 0 \quad , \tag{5.35}$$

where

$$2\,P_\alpha = \beta_\alpha + \frac{1}{2}\sum_{\beta=1}^{n}(B_{\alpha\beta} + B_{\beta\alpha}) \qquad (\alpha = 1, \ldots, n) \quad .$$

Analogously, in Case 2

$$\Delta = \begin{vmatrix} A_{11} & \cdots & A_{1,m+1} & P_1 \\ \vdots & & \vdots & \vdots \\ A_{m+1} & \cdots & A_{m+1,m+1} & P_{m+1} \\ P_1 & \cdots & P_{m+1} & 1 \end{vmatrix} > 0 \quad , \tag{5.36}$$

where

$$2\,P_\alpha = \beta_\alpha + \frac{1}{2}\sum_{\beta=1}^{m+1}(B_{\alpha\beta} + B_{\beta\alpha}) \qquad (\alpha = 1, \ldots, m+1) \quad ,$$

and in Case 3

$$\Delta = \begin{vmatrix} A_{11} & \cdots & A_{1m} & P_1 \\ \vdots & & \vdots & \vdots \\ A_{m1} & \cdots & A_{mm} & P_m \\ P_1 & \cdots & P_m & 1 \end{vmatrix} > 0 \quad , \tag{5.37}$$

where

$$2\,P_\alpha = \beta_\alpha + \frac{1}{2}\sum_{\beta=1}^{m}(B_{\alpha\beta} + B_{\beta\alpha}) \qquad (\alpha = 1, \ldots, m) \quad .$$

Finally, if Case 3 includes a subcase characterized by the vanishing of the root ρ_1, then the Lyapunov function must be taken in the form

$$V = \frac{1}{2} Sx_1^2 + \frac{1}{2} \sum_{\alpha=2}^{m} \sum_{\beta=2}^{m} B_{\alpha\beta} x_\alpha x_\beta + \int_0^\sigma f(\sigma) d\sigma ,$$

where S is a positive number, and the $B_{\alpha\beta}$ are determined by the formulas (5.29), with α and β varying from 2 to m. Then the stability conditions assume the following form

$$(5.38) \qquad \Delta = \begin{vmatrix} A_{22} & \cdots & A_{2m} & P_2 \\ \vdots & & \vdots & \vdots \\ A_{m2} & \cdots & A_{mm} & P_m \\ P_2 & \cdots & P_m & 1 \end{vmatrix} > 0 ,$$

where

$$2 P_\alpha = \beta_\alpha + \frac{1}{2} \sum_{\beta=2}^{m} (B_{\alpha\beta} + B_{\beta\alpha}) \qquad (\alpha = 2, \ldots, m) ,$$

and

$$S = - \beta_1 > 0 .$$

It remains for us to consider only the case of complex ρ_k. Omitting derivations, we employ here the construction of the function suggested by I. G. Malkin. Thus, let there be simple complex roots $\rho_1 = a + ib$ and $\rho_2 = a - ib$. Introducing the variables

$$x_1 = u + iv, \quad x_2 = u - iv$$

we can always reduce (4.1) to a form containing only positive quantities. We can convert to the old variables in accordance with the formulas

$$u = \frac{x_1 + x_2}{2}, \quad v = \frac{x_1 - x_2}{2i} .$$

Let us now assume that we have chosen some form $W(u, v, x_3, \ldots, x_{n+1})$. Changing to the variables u and v with these formulas, we obtain a form in the variables $x_1, \ldots, x_{n+1}$. Then, using roots (5.28) and (5.29), it is possible to make up a form F which will depend on $x_1, \ldots, x_{n+1}$ and which will contain complex coefficients. Finally, using the direct-transformation formulas to change to real variables we obtain a function $F(u, v, x_3, \ldots, x_{n+1})$ that is suitable for our problem.

Thus, we have proved the following Malkin theorem. In order to guarantee the absolute stability of a system of the type (4.1), in which the roots ρ_k are simple and $\operatorname{Re} \rho_k > 0$ $(k = 1, \ldots, n + 1)$, it is sufficient that the parameters of the regulator be such that it is possible to choose the numbers $A_{\alpha\beta}$ to satisfy conditions (5.27) and (5.34) or conditions (5.35), (5.36), (5.37), and (5.38).

This theorem remains in force also when two of the roots ρ_k are equal and non-zero.

5. CERTAIN MODIFICATIONS OF THE MALKIN METHOD

Turning to the formula (5.31), we see that by choosing the coefficients in such a way as to satisfy the equations

$$\beta_\alpha + \frac{1}{2} \sum_{\beta=1}^{n+1} (B_{\alpha\beta} + B_{\beta\alpha}) = 0 \qquad (\alpha = 1, \ldots, n + 1) , \tag{5.39}$$

we can make $\dot{V}$ a negative definite function. From (5.29) we have

$$\beta_\alpha + \frac{1}{2} \sum_{\beta=1}^{n+1} \frac{A_{\alpha\beta} + A_{\beta\alpha}}{\rho_\alpha + \rho_\beta} = 0 \qquad (\alpha = 1, \ldots, n + 1) , \tag{5.40}$$

or

$$\beta_\alpha + \sum_{\beta=1}^{n+1} \frac{A_{\alpha\beta}}{\rho_\alpha + \rho_\beta} = 0 \qquad (\alpha = 1, \ldots, n + 1) . \tag{5.41}$$

If the choice of the quantities $A_{\alpha\beta}$ is based only on (5.27), we can obtain a more general stability criterion. Actually, let us assume that we have solved equations (5.41) with respect to the unknowns $A_{\alpha\beta}$ and that we have found

$$A_{\alpha\beta} = A^*_{\alpha\beta} (\beta_1, \ldots, \beta_{n+1}) \qquad (\alpha, \beta = 1, \ldots, n + 1) . \tag{5.42}$$

This is always possible if the ρ_α are different numbers, although such a solution will not always be single-valued. Then, the stability criterion becomes

$$A^*_{11} > 0, \quad \begin{vmatrix} A^*_{11} & A^*_{12} \\ A^*_{21} & A^*_{22} \end{vmatrix} > 0, \ \ldots$$
$$\begin{vmatrix} A^*_{11} & \cdots & A^*_{1,n+1} \\ \cdot & \cdot\ \cdot\ \cdot & \cdot \\ A^*_{n+1,1} & \cdots & A^*_{n+1,n+1} \end{vmatrix} > 0 . \tag{5.43}$$

Inequalities (5.43) will contain the parameters of the regulator and certain of the numbers A^*_{ki}.

Thus, we have proved the following theorem: In order to guarantee absolute stability of a system of the type (4.1), for which the roots of (1.2) are simple and $\operatorname{Re} \rho_k > 0$ $(k = 1, \ldots, n+1)$, it is sufficient to choose the parameters of the regulator in accordance with inequalities (5.43) in which the numbers $A^*_{\alpha\beta}$ serve as solutions of the system of (5.41).

By way of an example, let us consider the second Bulgakov problem. Thus, let ρ_1 and ρ_2 be real roots of (2.87), and let β_α be determined by the formulas (2.92). We have

$$\frac{A_{11}}{2\rho_1} + \frac{A_{12}}{\rho_1 + \rho_2} = -\beta_1 ,$$

$$\frac{A_{12}}{\rho_1 + \rho_2} + \frac{A_{22}}{2\rho_2} = -\beta_2 .$$

From this we find

$$A^*_{11} = -2\rho_1\beta_1 - \frac{2\rho_1}{\rho_1 + \rho_2} A_{12} ,$$

$$A^*_{22} = -2\rho_2\beta_2 - \frac{2\rho_2}{\rho_1 + \rho_2} A_{12} ,$$

$$A^*_{12} = A_{12} .$$

In these equalities, the number A_{12} remains arbitrary and, obviously, can always be so chosen as to satisfy the conditions $A^*_{11} > 0$ and $A^*_{22} > 0$ for all regulator constants. For this it is necessary to put

$$\text{(5.44)} \qquad \begin{aligned} A_{12} &< -(\rho_1 + \rho_2)\beta_1 , \\ A_{12} &< -(\rho_1 + \rho_2)\beta_2 . \end{aligned}$$

On the other hand, we have a discriminant of the form

$$A^*_{11}A^*_{22} - A^{*2}_{12} = 4\rho_1\rho_2\beta_1\beta_2 + \frac{4\rho_1\rho_2(\beta_1+\beta_2)}{\rho_1 + \rho_2} A_{12} + \left[\frac{4\rho_1\rho_2}{(\rho_1+\rho_2)} - 1 \right] A^2_{12} > 0 .$$

An elementary transformation changes this inequality to

$$4\rho_1\rho_2\beta_1\beta_2 + \frac{4\rho_1\rho_2(\beta_1+\beta_2)}{\rho_1 + \rho_2} A_{12} - \left(\frac{\rho_1 - \rho_2}{\rho_1 + \rho_2}\right)^2 A_{12}^2 > 0 \quad .$$

To analyze this inequality, let us consider the equation

$$\left(\frac{\rho_1 - \rho_2}{\rho_1 + \rho_2}\right)^2 x^2 - \frac{4\rho_1\rho_2(\beta_1+\beta_2)}{\rho_1 + \rho_2} x - 4\rho_1\rho_2\beta_1\beta_2 = 0 \quad , \tag{5.45}$$

whose roots are

$$x_1, x_2 = \left[\frac{2\rho_1\rho_2(\beta_1+\beta_2)}{\rho_1 + \rho_2} \pm \right.$$
$$\left. \pm \sqrt{\frac{4\rho_1^2\rho_2^2(\beta_1+\beta_2)^2}{(\rho_1+\rho_2)^2} + 4\rho_1\rho_2\beta_1\beta_2 \left(\frac{\rho_1 - \rho_2}{\rho_1 + \rho_2}\right)^2} \,\right] \left(\frac{\rho_1 + \rho_2}{\rho_1 - \rho_2}\right)^2 \quad .$$

In order for a real A_{12} to exist such as to satisfy this inequality, it is necessary and sufficient that

$$\rho_1\rho_2(\beta_1 + \beta_2)^2 + \beta_1\beta_2(\rho_1 - \rho_2)^2 > 0 \quad . \tag{5.46}$$

But, according to (2.92), we find

$$\beta_1\beta_2 = - \frac{1}{(\rho_2-\rho_1)^2} [p_1^2 - (\rho_1 + \rho_2)p_1p_2 + \rho_1\rho_2p_2^2]$$

and

$$\beta_1 + \beta_2 = \frac{1}{\rho_2 - \rho_1} [- p_1 + \rho_1p_2 + p_1 - \rho_2p_2] = - p_2 \quad .$$

Consequently, we have

$$- p_1^2 + (\rho_1 + \rho_2)p_1p_2 > 0 \quad .$$

Using (2.82) and (2.87), it is possible to express the left half of this inequality in terms of the initial parameters. Performing this operation, we have

$$p_1[p(E - pG^2) - p_1] > 0 \quad ,$$

where $p_1 = a - qG^2$. Consequently, we should have*

$$a > qG^2 \quad , \tag{5.47}$$
$$pE + qG^2 > a + p^2G^2 \quad .$$

Inequalities (5.47) are the same as (4.51), obtained with the aid of the modified Lur'e theorem. If these are satisfied, the roots of (5.45) are sure to be real, and consequently, this guarantees the existence of a real form W, negative everywhere, for the given problem. Thus, we find

$$A_{11}^*A_{22}^* - A_{12}^2 = -\left(\frac{\rho_1 - \rho_2}{\rho_1 + \rho_2}\right)^2 (A_{12} - x_1)(A_{12} - x_2) > 0 \quad ,$$

and therefore it is possible to choose any value of A_{12} lying in the interval

$$x_2 < A_{12} < x_1 \quad . \tag{5.48}$$

If inequalities (5.44) and (5.48) are compatible, the stability is guaranteed by the satisfaction of conditions (5.47).

Let us now consider the problem of the isodrome regulator. From (2.113) we obtain

$$\beta_1\beta_2 = \frac{1}{(\rho_2-\rho_1)^2}\left[-p_1^2 + (\rho_1 + \rho_2)p_1p_2 - \rho_1\rho_2p_2^2 + \right.$$
$$\left. + \frac{b_{31}p_1p_3(\rho_1+\rho_2)}{\rho_1\rho_2} - \frac{b_{31}p_2p_3(\rho_1^2+\rho_2^2)}{\rho_1\rho_2} + \frac{b_{31}^2p_3^2}{\rho_1\rho_2}\right] ,$$

$$\beta_1 + \beta_2 = \frac{b_{31}p_3}{\rho_1\rho_2} - p_2 \quad .$$

Substituting these expressions in (5.46) and transforming the latter, we get

$$[(\rho_1 + \rho_2)p_2]\left[p_1 - \frac{b_{31}(\rho_1+\rho_2)}{\rho_1\rho_2}p_2\right] > 0 \quad .$$

This inequality breaks up into two subsidiary inequalities. According to (2.104) they can be written

---

* It is impossible to assume the inverse inequalities, for they contradict the physical meaning of the problem and the conditions previously obtained.

$$a > qG^2 + \frac{p\sqrt{r}}{qN} \cdot \frac{E - pG^2}{a - qG^2} ,$$

$$pE + qG^2 > a + p^2G^2 .$$

Comparison of these inequalities with (5.47) shows that the introduction of the isodrome coefficient N has changed only the first inequality of (5.47) by a certain term that depends on N.

6. REALIZABILITY OF SOLUTIONS OBTAINED BY THE LYAPUNOV METHOD

It was noted above that the Lyapunov direct method gives, generally speaking, only the sufficient conditions for stability. Consequently, it may happen that the sufficient conditions obtained in this manner are quite different from the necessary conditions, and their practical realization will involve greater design difficulties.

Thus, the problem arises of the realizability of the solutions of problems in the theory of automatic control, obtained by the Lyapunov direct method.

The principal importance of this problem is emphasized by the fact that upon superficial examination the appearance of these difficulties may stem from the application of the method, and not from the physical nature of the problem being solved.

At the present time, it is not possible to investigate this problem in general form; it was considered partially by Lur'e, who, in particular, has shown that one of the inequalities of the type $r^2 > 0$ is the Hurwitz inequality for a suitably linearized control system.

In this section we consider the realization of the solutions obtained, using specific examples. The solution method we choose in these examples is that of direct comparison of the sufficient conditions with the necessary conditions obtained by the Lyapunov method (which is based on the study of the first-approximation equations).

Let us denote

$$\left[\frac{df^*(\sigma)}{d\sigma}\right]_{\sigma=0} = h \qquad (0 < h < \infty)$$

and let us consider the Bulgakov second problem.

In correspondence with the first-approximation method, we must consider the characteristic equation of the system. Turning to (2.81) we find

$$T^2\lambda^3 + \left[U + h\,\frac{T^2 + \ell G^2}{\ell}\right]\lambda^2 + \left[k + \frac{U + \ell E}{\ell}\,h\right]\lambda + h\,\frac{k + a\ell}{\ell} = 0 \quad .$$

Among all the Hurwitz conditions, the only one of consequence is

$$\left[U + h\,\frac{T^2 + \ell G^2}{\ell}\right]\left[k + \frac{U + \ell E}{\ell}\,h\right] - T^2\,\frac{k + a\ell}{\ell}\,h > 0 \quad .$$

In the space of the regulator parameters, the equation

$$(5.49) \qquad h^2\,\frac{U + \ell E}{\ell}\cdot\frac{T^2 + \ell G^2}{\ell} + \left[k\,\frac{T^2 + \ell G^2}{\ell} + \frac{U(U+\ell E)}{\ell} - T^2\,\frac{k + a\ell}{\ell}\right]h + kU = 0$$

defines a family of curves that depends on the parameter $h (0 < h < \infty)$; each curve of this family bounds a corresponding stability region. We find the envelope of this family, according to the following general rule. If a family of curves that depend on a single parameter h is specified by means of an equation $F(x, y, h) = 0$, then to obtain the equation of the envelope it is necessary to eliminate the parameter h from the equations

$$F(x, y, h) = 0 \quad \text{and} \quad \frac{\partial F(x,y,h)}{\partial h} = 0 \qquad (0 < h < \infty) \ .$$

In our case, this operation leads to the equation

$$\left[k\,\frac{T^2 + \ell G^2}{\ell} + \frac{U(U+\ell E)}{\ell} - \frac{T^2(k+a\ell)}{\ell}\right]^2 = 4KU\,\frac{T^2 + \ell G^2}{\ell}\,\frac{U + \ell E}{\ell} \quad .$$

If we use the symbols of (4.39) and (4.40), this equation can be rewritten as

$$(5.50) \qquad (\xi + \eta - 1)^2 = 4\xi\eta \quad .$$

It can be seen that curve (5.50) is the same as curve (4.41), which bounds the region of absolute stability, obtained by the Lyapunov direct method.

By way of another example, let us consider the problem of the isodrome regulator.*

---

* A. M. Letov. Contribution to the Theory of the Isodrome Regulator, PMM, Volume XII, No. 4, 1948.

In accordance with (2.99), we find the characteristic equation

$$T^2\lambda^4 + \left[U + \frac{h}{r}\right]\lambda^3 + \left[k + \frac{U + \ell E}{\ell} h\right]\lambda^2 + \frac{k + a\ell}{\ell} h\lambda + \frac{h}{N} = 0 \quad .$$

If the only Hurwitz inequality that is of importance here is written in terms of h, we obtain:

$$\left[\frac{U + \ell E}{r\ell} \cdot \frac{k + a\ell}{\ell} - \frac{h}{r^2 N}\right] h^2 + \left[\left\{\frac{U(U+\ell E)}{\ell} + \frac{k}{r} - \frac{T^2(k+a\ell)}{\ell}\right\} \frac{k + a\ell}{\ell} - \frac{2U}{rN}\right] h + \frac{kU(k+a\ell)}{\ell} - \frac{U^2}{N} > 0 \quad .$$

Using the symbols of (4.39) and (4.43), the family of curves that outline the stability regions at various values of $h > 0$ will be represented by the equation

$$(5.51) \qquad (\beta - \rho)\frac{h^2}{(rsT)^2} + (\beta M + \alpha - 1 - 2\rho M)\frac{h}{sTr} + M(\alpha - \rho M) = 0.$$

The envelope of this family is the curve

$$(5.52) \qquad (x + y - 1)^2 = 4xy \quad ,$$

where x and y are determined from (4.44). It is easy to see that curve (5.52) coincides with curve (4.46), which bounds the region of absolute stability of the system.

Finally, let us consider the problem of the control of the steady state of a system, subject to the action of constant disturbing forces.* We shall confine ourselves to the case when the constant χ of the actuator is zero. Turning to (2.120) we find

$$T^2\lambda^4 + \left[U + nT^2 + \frac{h}{r}\right]\lambda^3 + \left[k + nU + \left(\frac{U + \ell E}{\ell} + \frac{n}{r}\right)h\right]\lambda^2 + \left[\frac{n(U+\ell E)}{\ell} + \frac{k + a\ell}{\ell}\right]h\lambda + anh = 0 \quad .$$

For the sake of brevity, we assume $\ell = \infty$ in addition (absence of a proportional feedback in the regulator). Then, denoting

---

* A. M. Letov, Regulation of the Stationary State of a System Subjected to the Action of Constant Disturbing Forces, PMM, Volume XII, No. 2, 1948.

$$h \frac{G^3}{T^2 \sqrt{a}} = H, \qquad \frac{4\lambda}{\sqrt{a}} = \mu \tag{5.53}$$

and using (5.5), we obtain the following characteristic equation for the roots of the first approximation

$$\begin{aligned} \mu^4 + (M + \chi + H)\mu^3 + [\alpha + \chi M + (\beta + \chi)H]\mu^2 + \\ + (1 + \chi\beta)H\mu + \chi H = 0 \quad . \end{aligned} \tag{5.54}$$

Among all the Hurwitz inequalities, the only one of importance here is the next to the last, which is rewritten

$$\begin{aligned} [(A + BH)(C + H) - \chi H(B + D)]\chi H(B + D) - \\ - \chi H(C + H)^2 > 0 \quad , \end{aligned} \tag{5.55}$$

where

$$A = \alpha + \chi M, \quad B = \chi + \beta, \quad C = \chi + M, \quad D = \frac{1}{\chi} - \chi$$

The stability regions obtained for various values of H and determined from the first-approximation equations, are bounded by curves in the family

$$a_0 H^2 + a_1 H + a_2 = 0 \quad , \tag{5.57}$$

where

$$\begin{aligned} a_0 &= B(B + D) - 1 \quad , \\ a_1 &= (B + D)(A + BC) - \chi(B + D)^2 - 2C \quad , \\ a_2 &= AC(B + D) - C^2 \quad . \end{aligned} \tag{5.58}$$

The equation of the envelope of the family (5.57) is

$$a_1^2 = 4a_0 a_2 \quad ,$$

or, in greater detail,

$$\begin{aligned} [(B + D)(A + BC) - \chi(B + D)^2 - 2C]^2 = \\ = 4[B(B + D) - 1][AC(B + D) - C^2] \quad . \end{aligned}$$

It is now necessary to carry out elementary transformation of the last relation. We find:

$$\begin{aligned} (B + D)^2(A + BC)^2 + \chi^2(B + D)^4 - 2\chi(B + D)^3(A + BC) \\ + 4\chi C(B + D)^2 = 4ABC(B + D)^2 \quad . \end{aligned}$$

Finally, after cancelling $(B + D)^2$ we obtain the two equations

(5.59) $$(B + D)^2 = 0$$

and

(5.6) $$(A + BC)^2 + x^2(B + D)^2 - 2x(B + D)(A + BC) + 4xC = 4ABC \quad .$$

Let us consider the first equation. According to the notation of (5.56) and (5.6), it yields

(5.61) $$y = -\frac{M}{x} \quad .$$

This equation is represented by a straight line in the xy plane and plays no role whatever in the determination of the stability-region boundaries.

Let us now consider (5.60). It can be simplified to

$$[\alpha - \beta(x + M) - x^2]^2 + x\left(\beta + \frac{1}{x}\right)[1 + x\beta - 2\alpha - 4xM -$$
$$- 2\beta(x + M) - 2x^2] + 4x(x + M) = 0 \quad .$$

If we use the symbols of (5.6), we find

(5.62) $$a_{11}x^2 + 2a_{12}xy + a_{22}y^2 + 2a_1x + 2a_2y + a_3 = 0 \quad ,$$

where the coefficients a are determined by the following formulas

(5.63) $$a_{11} = 1, \quad 2a_{12} = -\frac{2(x+M)}{M} - 2\frac{x}{M}, \quad a_{22} = 1 \quad ,$$
$$2a_1 = -2 - 2x^2, \quad 2a_2 = -2(1 + x^2) \quad ,$$
$$a_3 = (1 = x^2)^2 \quad .$$

Comparison of (5.63) with (5.8) for $\theta = 0$ shows that the envelope of the family (5.57) is the same as the hyperbola (5.8), the outline of the absolute-stability region.

Thus, for each of the three problems, the following conclusion is correct: the curve that outlines the absolute-stability region, plotted in accordance with the Lyapunov theorem, is the envelope of the family of the curves of the single parameter $0 < h < \infty$, which bound the stability region that is deduced from the first-approximation equations. There are many more such examples.*

---

* P. V. Bromberg, Ustoichivost' i avtokolebaniya impul'snykh sistem regulirovaniya (Stability and Self-Oscillations of Pulsed Control Systems), Oborongiz, 1953.

Two important consequences result from this conclusion.

First, if we solve the problem of stability by using the first-approximation equations and specify that the system be stable for all $h > 0$ (for any servo motor), we obtain the same stability region as the absolute-stability region obtained from the Lur'e theorem.

Second, the contour L of the absolute-stability region, obtained by the first Lyapunov method according to the Lur'e theorem, approaches in the limit the contour L' of the stability region that can be obtained from the first-approximation equations and gives us the necessary and sufficient conditions for the stability. The meaning of "approaching in the limit" is that any other contour L'', plotted in accordance with the Lyapunov direct method, will be separated from the contour L' by the contour L, plotted by the Lur'e theorem.

The realizability of other solutions obtained by the Lyapunov direct method in accordance with the simplified stability criteria, can be investigated by comparing these solutions with the solution afforded by the Lur'e theorem.

The above considerations lead to one mathematical problem. We have seen that if the conditions of the Lur'e theorem are satisfied, the control system is stable for all functions $f(\sigma)$ of class (A), and, in particular, for all linear functions $f(\sigma) = h\sigma$.

The question then arises as to whether the inverse is also correct. That is to say, if the system is stable for all functions $f(\sigma) = h\sigma$ $(0 < h < \infty)$, is it correct to state that the Lur'e stability conditions are satisfied and the system is stable for any function $f(\sigma)$ of class (A)?

The above examples allow us to expect that this question can be answered in the affirmative. We shall show that the inverse of the Lur'e theorem holds for a third-order system.

Assume that we have the canonical equations

$$x_1 = -\rho_1 x_1 + f(\sigma)$$
$$x_2 = -\rho_2 x_2 + f(\sigma)$$
$$\dot{\sigma} = \beta_1 x_1 + \beta_2 x_2 - f(\sigma) \quad .$$

We know that for any $f(\sigma)$ of class (A) system stability is ensured by fulfillment of the inequalities $\Gamma^2 > 0$, $D^2 > 0$, (4.31), and (4.36). We now put $f(\sigma) = h\sigma$, $0 < h < \infty$. The characteristic equation of the thus linearized system is

$$\lambda^3 + a_1\lambda^2 + a_2\lambda + a_3 = 0$$

where

$$a_1 = \rho_1 + \rho_2 + h$$

$$a_2 = \rho_1\rho_2 + h(\rho_1 + \rho_2 - \beta_1 - \beta_2)$$

$$a_3 = h\rho_1\rho_2 r^2 \ .$$

This leads to the Hurwitz inequalities

$$r^2 > 0$$

$$(\rho_1 + \rho_2 = h)\ [\rho_1 + \rho_2 + h(\rho_1 + \rho_2 - \beta_1 - \beta_2)] > h\rho_1\rho_2 r^2 \ .$$

The first inequality gives half the solution to the problem. The second can be expressed as a polynomial in h

$$b_o h^2 + b_1 h + b_2 = 0$$

where

$$b_o = \rho_1 + \rho_2 - \beta_1 - \beta_2$$

$$b_1 = \rho_1\rho_2 + (\rho_1 + \rho_2)(\rho_1 + \rho_2 - \beta_1 - \beta_2) - \rho_1\rho_2 r^2$$

$$b2 = \rho_1\rho_2(\rho_1 + \rho_2) \ .$$

For any fixed h, the last equation yields the boundary of the stability region in the particular coordinates chosen. When $0 < h < \infty$ this equation represents a single-parameter family of curves the envelope of which is $b_1^2 = 4b_o b_2$, or

$$[\rho_1\rho_2 + (\rho_1 + \rho_2)(\rho_1 + \rho_2 - \beta_1 - \beta_2) - \rho_1\rho_2 r^2]^2 =$$

$$= 4\rho_1\rho_2(\rho_1 + \rho_2)(\rho_1 + \rho_2 - \beta_1 - \beta_2) \ .$$

An elmentary transformation of the quantity in the square brackets yields

$$[\rho_1 + \rho_2)^2 - \beta_1\rho_1 - \beta_2\rho_2]^2 =$$

$$= 4\rho_1\rho_2(\rho_1 + \rho_2)(\rho_1 + \rho_2 - \beta_1 - \beta_2) \ .$$

Finally, separating the sum of the squares from $(\rho_1 + \rho_2)^2$ and using the terms containing $2\rho_1\rho_2$ to simplify the right half of this relation, we obtain as the end result

$$[\rho_1^2 + \rho_2^2 - \beta_1\rho_1 - \beta_2\rho_2]^2 = 4\rho_1^2\rho_2^2 r^2$$

which indeed yields the second half of the solution.

7. ANOTHER METHOD OF FORMULATING SIMPLIFIED STABILITY CRITERIA

We have seen above, using several examples, that the stability criteria afforded by the Lur'e theorem are sufficiently structurally realizable. However, their derivation involves a complicated analysis of systems of algebraic quadratic equations, which in general cannot be written out in full if n is greater than 3.

This difficulty arises frequently also in the methods described above for the formulation of the simplified criteria.

We describe below a method for formulating the simplified criteria, which will always permit their analysis when written out in full.

For this purpose, let us turn to (4.1) and retain all the assumptions concerning the character of the quantities ρ_k $(k = 1, \ldots, n)$ that enter into these equations.

Let us assume that $2q = n + 1$ is even, and consider the function

$$V = \sum_{i=1}^{n} F_i(a_i x_i,\ a_{i+1}x_{i+1}) + \int_0^{\sigma} f(\sigma)d\sigma \quad , \tag{5.64}$$

where

$$F_i(a_i x_i,\ a_{i+1}x_{i+1}) = \frac{a_i^2 x_i^2}{2\rho_i} + \frac{a_{i+1}^2 x_{i+1}^2}{2\rho_{i+1}} + \frac{2a_i a_{i+1} x_i x_{i+1}}{\rho_i + \rho_{i+1}} \quad , \tag{5.65}$$

and the number i assumes only odd values, i.e., $1, 3, \ldots, n$.

For any $i = 1, 3, \ldots, n$ the function F_i becomes sign-definite, and for all non-zero values x_i, x_{i+1} it assumes only positive values. Consequently, the function (5.64) is positive definite in the variables $x_1, \ldots, x_{n+1}$ and σ everywhere except at the origin, where it vanishes.

Let us calculate its total derivative. We have

$$\frac{dV}{dt} = - \sum_{i=1}^{n} (a_i x_i + a_{i+1} x_{i+1})^2 - rf^2(\sigma) +$$

$$(5.66) \qquad + f(\sigma) \sum_{i=1}^{n} \left[\beta_i + \frac{a_i^2}{\rho_i} + \frac{2a_i a_{i+1}}{\rho_i + \rho_{i+1}} \right] x_i +$$

$$+ f(\sigma) \sum_{i=1}^{n} \left[\beta_{i+1} + \frac{a_{i+1}^2}{\rho_{i+1}} + \frac{2a_i a_{i+1}}{\rho_i + \rho_{i+1}} \right] x_{i+1} \quad ,$$

where $i = 1, 3, 5, \ldots, n$. The expression obtained can be modified by adding to its right side a term

$$2 \sum_{i=1}^{n} (a_i x_i + a_{i+1} x_{i+1}) \left(\sqrt{\frac{r}{f}}\, f(\sigma) \right) \quad ,$$

and then subtracting this term. Then, considering that $q = (n + 1)/2$ and making elementary transformations, we have

$$\frac{dV}{dt} = - \sum_{i=1}^{n} \left[a_i x_i + a_{i+1} x_{i+1} + \sqrt{\frac{r}{q}}\, f(\sigma) \right]^2 +$$

$$(5.67) \qquad + f(\sigma) \sum_{i=1}^{n} \left[\beta_i + \frac{a_i^2}{\rho_i} + \frac{2a_i a_{i+1}}{\rho_i + \rho_{i+1}} \; 2 \sqrt{\frac{r}{q}}\, a_i \right] x_i +$$

$$+ f(\sigma) \sum_{i=1}^{n} \left[\beta_{i+1} + \frac{a_{i+1}^2}{\rho_{i+1}} + \frac{2a_i a_{i+1}}{\rho_i + \rho_{i+1}} + 2 \sqrt{\frac{r}{q}}\, a_{i+1} \right] x_{i+1} .$$

From (5.66) and (5.67) it is seen that for $\dot{V}$ to be sign-definite it is sufficient to satisfy the relations

$$\beta_i + \frac{a_i^2}{\rho_i} + \frac{2a_i a_{i+1}}{\rho_i + \rho_{i+1}} = 0 \quad , \qquad (i = 1, 3, \ldots, n)$$

(5.68)

$$\beta_{i+1} + \frac{a_{i+1}^2}{\rho_{i+1}} + \frac{2a_i a_{i+1}}{\rho_i + \rho_{i+1}} = 0$$

or

$$\beta_i + 2\sqrt{\frac{r}{q}}\, a_i + \frac{a_i^2}{\rho_i} + \frac{2a_i a_{i+1}}{\rho_i + \rho_{i+1}} = 0 \ ,$$

(5.69) $\qquad (i = 1, 3, \ldots, n)$

$$\beta_{i+1} + 2\sqrt{\frac{r}{q}}\, a_{i+1} + \frac{a_{i+1}^2}{\rho_{i+1}} + \frac{2a_i a_{i+1}}{\rho_i + \rho_{i+1}} = 0 \ .$$

In the former case we have:

(5.70)
$$\frac{dV}{dt} = - \sum_{i=1}^{n} (a_i x_i + a_{i+1} x_{i+1})^2 - rf^2(\sigma) \ ,$$

and in the latter case:

(5.70a)
$$\frac{dV}{dt} = - \sum_{i=1}^{n} [a_i x_i + a_{i+1} x_{i+1} + \sqrt{\frac{r}{f}}\, f(\sigma)]^2 \ .$$

According to (5.70) and (5.70a), the function $\dot{V}$ can be only negative or zero. Consequently, fulfillment of (5.68) or (5.69) guarantees the absolute stability of the system. Naturally, were we to evaluate here the function Φ as in the proof of the Lur'e theorem, we would recognize that we deal with asymptotic stability.

The quadratic systems (5.68) and (5.69) differ from those used in connection with the Lur'e theorem in that they form q systems of quadratic equations that are not interrelated by the unknown a_i. As a consequence, such systems can be analyzed although written out in full.

Analogous systems of equations could be obtained by putting instead of (5.64)

(5.71)
$$V = \sum_{i=1}^{n} F_i(a_i x_i, a_{i+1} x_{i+1}) \qquad (i = 1, 3, \ldots, n) \ .$$

As in the case of (5.64), V is here a positive definite function in the variables $x_1, \ldots, x_{n+1}$. Its total derivative, calculated according to (4.1) is

$$\frac{dV}{dt} = -\sum_{i=1}^{n}(a_i x_i + a_{i+1} x_{i+1})^2 + f(\sigma)\sum_{i=1}^{n}\left[\frac{a_i^2}{\rho_i} + \frac{2a_i a_{i+1}}{\rho_i + \rho_{i+1}}\right] x_i +$$

$$+ f(\sigma)\sum_{i=1}^{n}\left[\frac{a_{i+1}^2}{\rho_{i+1}} + \frac{2a_i a_{i+1}}{\rho_i + \rho_{i+1}}\right] x_{i+1} \ .$$

Adding here the expression

$$f(\sigma)\sum_{k=1}^{n+1} \gamma_k x_k - \sigma f(\sigma) = f(\sigma)\left(\sum_{k=1}^{n+1} \gamma_k x_k - \sigma\right) = 0$$

and requiring that the following relations be satisfied

$$\gamma_i + \frac{a_i^2}{\rho_i} + \frac{2a_i a_{i+1}}{\rho_i + \rho_{i+1}} = 0 \ ,$$

(5.72) $$(i = 1, 3, \ldots, n) \ ,$$

$$\gamma_{i+1} + \frac{a_{i+1}^2}{\rho_{i+1}} + \frac{2a_i a_{i+1}}{\rho_i + \rho_{i+1}} = 0 \ ,$$

we obtain a sign-definite derivative

$$\frac{dV}{dt} = -\sum_{i=1}^{n}(a_i x_i + a_{i+1} x_{i+1})^2 - \sigma f(\sigma) \ ,$$

which assumes negative or zero values. Consequently, relations (5.72) are also stability conditions.

Thus, we have proved the following theorem: A system has absolute asymptotic steady state stability if the constants of the system are such that $q = (n + 1/2$ pairs of independent systems of equations (5.68), (5.69), and (5.72), in which $\rho_1, \ldots, \rho_s, \beta_1, \ldots, \beta_s$, and $\gamma_1, \ldots, \gamma_s$ are real, and $\rho_{s+1}, \ldots, \rho_{n+1}, \beta_{s+1}, \ldots, \beta_{n+1}$, and $\gamma_{s+1}, \ldots, \gamma_{n+1}$ are complex conjugate pairs, admit of at least one solution that contains real roots $a_1, \ldots, a_s$ and complex conjugate pairs of roots $a_{s+1}, \ldots, a_{n+1}$.

This theorem has been proved only for the case of even $n + 1$; if $n + 1$ is odd, it is possible to proceed in the following manner.

It is first necessary to subject to a thorough analysis all the quantities β_k and to find among them those that assume only negative values. Let us assume that such a value is β_{n+1}. Then, the V function can be chosen to be

$$V = \frac{1}{2} Kx_{n+1}^2 + \sum_{i=1}^{n-1} F_i(a_i x_i, a_{i+1} x_{i+1}) + \int_0^\sigma f(\sigma) d\sigma$$

or

$$V = \frac{1}{2} Kx_{n+1}^2 + \sum_{i=1}^{n-1} F_i(a_i x_i, a_{i+1} x_{i+1}) \quad ,$$

where K is any positive number, and $i = 1, 3, \ldots, n - 1$.

If, however, none of the numbers β_k are known to be negative, the V function is chosen to be

$$V = F(a_1 x_1, a_2 x_2, a_3 x_3) + \sum_{i=4}^{n} F_i(a_i x_i, a_{i+1} x_{i+1}) + \int_0^\sigma f(\sigma) d\sigma \quad ,$$

or

$$V = F(a_1 x_1, a_2 x_2, a_3 x_3) + \sum_{i=4}^{n} F_i(a_i x_i, a_{i+1} x_{i+1}) \quad ,$$

where $i = 4, 6, \ldots, n$. In this case one set of three equations with three unknowns a_1, a_2, and a_3 can be found among the systems of equations of type (5.68), (5.69), and (5.72), each containing two unknowns a_i, a_{i+1}.

Thus, for example, an investigation of a fifth-order control system reduces to the analysis of one inequality and of two individual systems, each of which consists of two quadratic equations (or to an analysis of one system of three* and one system of two quadratic equations).

It is obvious that in the case of a control system of second or third order, we obtain, according to this theorem, the same result given by the Lur'e theorem.

Let us proceed to a consideration of the applications. We consider the second Bulgakov problem in the case of non-ideal sensing elements.

---

* An analysis of a system of three equations was made by Lur'e in his book Nekotorye nelineǐnye zadachi teorii avtomaticheskovo regulirovaniya (Certain Nonlinear Problems in the Theory of Automatic Control), Gostekhizdat, 1951.

Here $n + 1 = 4$ and $q = 2$, and consequently we have two systems of quadratic equations. Let us first consider the system of type (5.68). We have

$$(5.73)\qquad \begin{aligned} \beta_1 + \frac{a_1^2}{\rho_1} + \frac{2a_1a_2}{\rho_1 + \rho_2} &= 0 \quad , \\ \beta_2 + \frac{a_2^2}{\rho_2} + \frac{2a_1a_2}{\rho_1 + \rho_2} &= 0 \quad . \end{aligned}$$

We have already seen that for (5.73) to be solvable it is necessary and sufficient to satisfy the two inequalities

$$(5.74)\qquad \frac{\beta_1}{\rho_1} + \frac{\beta_2}{\rho_2} < 0$$

and

$$(5.75)\qquad \beta_1\rho_1 + \beta_2\rho_2 < 0 \quad .$$

At the same time, for the system (5.73), it is necessary to satisfy the inequality

$$(5.76)\qquad \beta_1 + \beta_2 < 0 \quad ,$$

which is readily seen to follow from (5.74) and (5.75). Actually, let us multiply (5.74) by $\rho_1\rho_2 > 0$ and then add it to inequality (5.75). Then taking the common factor outside the parenthesis, we have

$$(\rho_1 + \rho_2)(\beta_1 + \beta_2) < 0 \quad ,$$

from which the above statement follows.

An analogous pair of inequalities is also obtained for the quantities β_3, ρ_3 and β_4, ρ_4:

$$(5.77)\qquad \frac{\beta_3}{\rho_3} + \frac{\beta_4}{\rho_4} < 0 \quad ,$$

$$(5.78)\qquad \beta_3\rho_2 + \beta_4\rho_4 < 0 \quad .$$

Had we taken (5.69) as the initial equations, then, repeating the arguments already known from Section 5 of Chapter IV, and taking into account that in our problem $q = 2$, we would have obtained the necessary and sufficient conditions for the solvability of (5.69) in the form of two systems of inequalities:

$$\Gamma_1^2 = \frac{1}{2} - \frac{\beta_1}{\rho_1} - \frac{\beta_2}{\rho_2} > 0 \ ,$$

(5.79)

$$D_1^2 = \frac{1}{2}(\rho_1 + \rho_2)^2 - \rho_1\rho_2 - \beta_1\rho_1 - \beta_2\rho_2 \pm \sqrt{2}\,\rho_1\rho_2\Gamma_1 > 0 \ ,$$

$$\Gamma_2^2 = \frac{1}{2} - \frac{\beta_3}{\rho_3} - \frac{\beta_4}{\rho_4} > 0 \ ,$$

(5.80)

$$D_2^2 = \frac{1}{2}(\rho_3 + \rho_4)^2 - \rho_3\rho_4 - \beta_3\rho_3 - \beta_4\rho_4 \pm \sqrt{2}\,\rho_3\rho_4\Gamma_2 > 0 \ .$$

Comparing inequalities (5.79) and (5.80) with inequalities (5.74), (5.75), (5.77), and (5.78), we see that (5.79) and (5.80) are less stringent and therefore give a more acceptable stability criterion than the other inequalities. Let us analyze these inequalities, and for this purpose express the right sides in terms of the initial data. We then obtain for inequalities (5.79)

$$\frac{\beta_1}{\rho_1} + \frac{\beta_2}{\rho_2} = \left[-\frac{p_1}{\rho_1\rho_2} + \frac{b_{42}p_3(b_{44}-b_{22})}{\Delta_3(\rho_1)\Delta_3(\rho_2)} \right](-n_2) + \frac{(-n_2)b_{41}p_3}{\rho_1\rho_2} \cdot \frac{(\rho_1+\rho_2)(\rho_3+\rho_4) - \rho_3\rho_4 - \rho_1^2 - \rho_2^2 - \rho_1\rho_2}{\Delta_3(\rho_1)\Delta_3(\rho_2)} \ ,$$

$$\beta_1\rho_1 + \beta_2\rho_2 = \left[p_1 + \frac{b_{42}p_3(b_{44}b_{21}-b_{43}b_{22})}{\Delta_3(\rho_1)\Delta_3(\rho_2)} \right](-n_2) + \frac{(-n_2)b_{41}p_3(\rho_3\rho_4-\rho_1\rho_2)}{\Delta_3(\rho_1)\Delta_3(\rho_2)} \ .$$

In analogous manner we obtain for the first two inequalities

$$\frac{\beta_3}{\rho_3} + \frac{\beta_4}{\rho_4} = p_3\left[\frac{n_4}{\rho_3\rho_4} + \frac{b_{42}(b_{22}-b_{44})}{\Delta_1(\rho_3)\Delta_1(\rho_4)} \right] ,$$

$$\beta_3\rho_3 + \beta_4\rho_4 = p_3\left[-n_4 + \frac{b_{42}(\rho_1\rho_2b_{44}-\rho_3\rho_4b_{22})}{\Delta_1(\rho_3)\Delta_1(\rho_4)} \right] .$$

Noting that

$$r^{*2}T_s^4\Delta_3(\rho_1)\Delta_3(\rho_2) = (1 - qT_s^2)^2 - (p - qH)(H - pT_s^2) ,$$

$$p_3b_{42}(b_{22} - b_{44}) = \frac{ET^2 - UG^2}{T^2T_s^2\sqrt{r^*}} \cdot \frac{H - pT_s^2}{T_s^2\sqrt{r^*}} ,$$

$$p_3b_{42}[\rho_3\rho_4b_{22} - \rho_1\rho_2b_{44}] = \frac{ET^2 - UG^2}{T^2T_s^2\sqrt{r^*}} \cdot \frac{qH - p}{T_s^2r^*\sqrt{r^*}} ,$$

$$(5.81) \qquad p_3b_{41}(\rho_3\rho_4 - \rho_1\rho_2) = -\frac{qG^2}{r^*T_s^2} \cdot \frac{1 - qT_s^2}{r^*T_s^2} ,$$

$$p_3b_{41}[b_{22}b_{44} - \rho_3\rho_4 - \rho_1^2 - \rho_2^2 - \rho_1\rho_2] =$$

$$= -\frac{qG^2}{r^*T_s^2} \cdot \frac{qT_s^2 + pH - 1 - p^2T_s^2}{r^*T_s^2} ,$$

$$p_3b_{41}[\rho_1\rho_2 + \rho_3\rho_4 + \rho_3^2 + \rho_4^2 - b_{22}b_{44}] =$$

$$= -\frac{qG^2}{r^*T_s^2} \cdot \frac{H^2 - T_s^2 - pHT_s^2 + qT_s^4}{r^*T_s^4} ,$$

we write the equations in the following form

$$\frac{\beta_1}{\rho_1} + \frac{\beta_2}{\rho_2} = (-n_2)\left[-\frac{p_1}{\rho_1\rho_2} + \frac{ET^2 - UG^2}{T^2T_s^2\sqrt{r^*}} \cdot \frac{pT_s^2 - H}{T_s^2\sqrt{r^*}\Delta_3(\rho_1)\Delta_3(\rho_3)} - \right.$$
$$\left. - \frac{qG^2(qT_s^2 - pH - 1 - p^2T_s^2)}{\rho_1\rho_2r^{*2}T_s^4\Delta_3(\rho_1)\Delta_3(\rho_2)}\right] ,$$

$$\beta_1\rho_1 + \beta_2\rho_2 = (-n_2)\left[p_1 - \frac{ET^2 - UG^2}{T^2T_s^2\sqrt{r^*}} \cdot \frac{p - qH}{T_{sr^*}^2\sqrt{r^*}}\,\frac{1}{\Delta_3(\rho_1)\Delta_3(\rho_2)} - \right.$$
$$\left. - \frac{qG^2(1 - qT_s^2)}{r^{*2}T_s^4\Delta_3(\rho_1)\Delta_3(\rho_2)}\right] ,$$

$$\frac{\beta_3}{\rho_3} + \frac{\beta_4}{\rho_4} = \left[-\frac{\ell G^2}{T^2} + (-n_2)\frac{ET^2 - UG^2}{T^2T_s^2\sqrt{r^*}} \cdot \frac{H - pT_s^2}{T_s^2\sqrt{r^*}\Delta_3(\rho_1)\Delta_3(\rho_2)} \right] +$$

$$+ (-n_2)\frac{qG^2(H^2-T_s^2-pHT_s^2+qT_s^4)}{r^*T_s^4\Delta_3(\rho_1)\Delta_3(\rho_2)} ,$$

$$\beta_3\rho_3 + \beta_4\rho_4 = \left[-\frac{\ell G^2}{r^*T_s^2T^2} + \frac{ET^2 - UG^2}{T^2T_s^2\sqrt{r^*}} \cdot \frac{p - qH}{T_{sr*}^2\sqrt{r^*}\Delta_3(\rho_1)\Delta_3(\rho_2)} \right] .$$

The computations yield the stability criterion for the control system in an explicit form that contains only the initial parameters:

$$r_1^2 = \frac{1}{2} + (-n_2)r^* \left[\frac{p_1}{q} - \frac{(ET^2-UG^2)(pT_s^2-H)}{T^2T_{sr*}^4{}^2\Delta_3(\rho_1)\Delta_3(\rho_2)} + \frac{qG^2(qT_s^2+pH-p^2T_s^2-1)}{qT_{sr*}^4{}^2\Delta_3(\rho_1)\Delta_3(\rho_2)} \right] > 0 ,$$

(5.82)

$$D_1^2 = \frac{p^2 - 2q}{2r^*} - (-n_2)\left[p_1 - \frac{(ET^2-UG^2)(p-qH)}{T^2T_{sr*}^4{}^2\Delta_3(\rho_1)\Delta_3(\rho_2)} - \frac{qG^2(1-qT_s^2)}{T_{sr*}^4{}^2\Delta_3(\rho_1)\Delta_3(\rho_2)} \right] \pm$$

$$\pm \sqrt{2}\,\frac{q}{r^*}\,r_1 > 0 ,$$

$$r_2^2 = \frac{1}{2} + \frac{\ell G^2}{T^2} - \frac{(ET^2-UG^2)(H-pT_s^2)}{T^2T_{sr*}^4\Delta_3(\rho_1)\Delta_3(\rho_2)} > 0 ,$$

(5.83)

$$D_2^2 = \frac{H^2 - 2T_s^2}{T^2T_{sr*}^2} + \frac{\ell G^2}{T^2T_{sr*}^2} - \frac{(ET^2-UG^2)(p-qH)}{T^2T_{sr*}^4{}^2\Delta_3(\rho_1)\Delta_3(\rho_2)} +$$

$$\pm \frac{\sqrt{2}}{r^*T_s^2}\,r_2 > 0 .$$

If the parameters of the regulated object are specified and if the parameters H and T_s^2 of the sensing device are specified, inequalities (5.82) and (5.83) limit the choice of regulator parameters E, G^2, a, and ℓ.

Let us consider the particular case of $H = T_s = 0$. Taking into account inequalities (5.81) we find

$$r_1^2 = \frac{1}{2} + \frac{r^*(-n_2)}{q} (p_1 - qG^2) > 0 \quad ,$$

$$D_1^2 = \frac{p^2 - 2q}{2r^*} - (-n_2)\left[p_1 - \frac{ET^2 - UG^2}{T^2} p - qG^2 \right] \pm \sqrt{2}\, \frac{q}{r^*} r_1 > 0 \quad .$$

Noting that $r^* (-n_2) = r$ and using (2.136) and (2.144), we reduce these inequalities to

$$\frac{k + a\ell}{k} \frac{T^2}{T^2 + \ell G^2} > \frac{1}{2} \quad ,$$

(5.84)
$$\left[\frac{U}{T^2} \frac{U + \ell E}{T^2 + \ell G^2} - \frac{k + a\ell}{T^2 + \ell G^2} - \frac{u^2}{2T^4} \pm \right.$$

$$\left. \pm\, 2\sqrt{\frac{k(k+2a\ell) - \frac{k\ell G^2}{T^2}}{2T^2(T^2+\ell G^2)}} \right] > 0 \quad .$$

Subject to the condition assumed, $(k > 0)$, both inequalities are satisfie for sufficiently large $a\ell$ and $E\ell$, and for sufficiently small G^2. For the two other inequalities, (5.83), we find

(5.85)
$$\lim_{H, T_s \to 0} r_2^2 = \frac{1}{2} + \frac{\ell G^2}{T^2} > 0, \qquad \lim_{H, T_s \to 0} T_s^4 D_2^2 = 0 \quad ,$$

and consequently these inequalities no longer contribute to the formulation of the stability criteria for this limiting case.

Returning to inequalities (5.82), we can now state that fulfillment of conditions (5.84) for the stability of the system, applicable to ideal sensing devices, guarantees the stability of this system, even in the case of non-ideal sensing devices, provided the constants H and T_s are sufficiently small. The same argument holds for the first inequality of (5.38). As to the second inequality of (5.83), it cannot, by virtue of the last equation of (5.85) play any substantial role in the formulation of the stability criterion.

Let us study these inequalities in greater detail. To simplify the calculations, we restrict ourselves to sufficiently small values of H and T_s. Subject to this assumption, we can put

$$r^{*2}T_s^4\Delta_3(\rho_1)\Delta_3(\rho_2) \approx 1, \quad qT_s^2 + pH - p^2T_s^2 - 1 \approx -1, \quad 1 - qT_s^2 \approx 1 \quad ,$$

so that these inequalities in a simpler approximate form become

$$\frac{k + a\ell}{k} \frac{T^2}{T^2 + \ell G^2} - \left(E - \frac{UG^2}{T^2} \right) \frac{(pT_s^2 - H)\ell}{T^2 + \ell G^2} > \frac{1}{2} ,$$

(5.86)

$$\left[\frac{U}{T^2} \cdot \frac{U + \ell E}{T^2 + \ell G^2} - \frac{2k + a\ell + \frac{k\ell G^2}{T^2}}{T^2 + \ell G^2} \pm \right.$$

$$\left. \pm 2 \sqrt{\frac{k(k+a\ell)}{T^2(T^2+\ell G^2)} - \left(E - \frac{UG^2}{T^2} \right) \frac{(pT_s^2-H)k^2\ell}{T^4(T^2+\ell G^2)}} \right] -$$

$$- (- n_2)qH \quad E - \frac{UG^2}{T^2} \quad > 0 ,$$

(5.87)

$$r_2^2 = \frac{1}{2} + \frac{\ell G^2}{T^2} - r^*(H - pT_s^2)(E - pG^2) > 0 ,$$

$$D_2^2 = \frac{1}{2r^* T_s^4} \left[H^2 - 2T_s^2 + \frac{2\ell G^2 T_s^2}{T^2} \pm 2 \sqrt{2} T_s^2 r_2 \right] > 0 .$$

We have already discussed the inequalities of (5.86). Let us turn now to inequalities (5.87). If the constants H and T_s^2 are sufficiently small, we can assume that the value of r_2^2 remains greater than $\sqrt{2}/2$. In this case, obviously, we have

$$D_2^2 = \frac{1}{2r^* T_s^2} \left[H^2 + 2T_s^2(\sqrt{2} r_2 - 1) + \frac{2\ell G^2 T_s^2}{T^2} \right] > 0 ,$$

and the inequality is indeed satisfied. We can therefore conclude that the solution (5.84) of the second Bulgakov problem for the case of ideal sensing devices, remains valid also when the sensing devices are non-ideal, provided, however, their parameters H and T_s remain sufficiently small.

The above arguments show that a stability criterion based on the theorem can be analyzed when written out in full for a control system with any degrees of freedom.

8. STABILITY IN INDIRECT CONTROL

Let us attempt to solve the problem of the stability of an indirectly controlled machine, making use of the canonical equations (2.180) and (2.187).

Since we wish to obtain a solution in explicit form, we shall employ the simplified stability criterion detailed in the preceding section of this chapter.

Let us first consider the equation (2.180).* In this case $q = 2$ and $r = \beta$. The stability criterion will be formulated on the basis of (5.69). Here we can choose arbitrary combinations of the quantities $\beta_1, \ldots, \beta_4$. Accordingly, we formulate a stability criterion which contains the pair β_1, β_2, and the pair β_3, β_4. Then the stability criterion of the system (2.180) can be written in the form of two pairs of inequalities

$$\Gamma_1^2 = \frac{r}{2} - \frac{\beta_1}{\rho_1} - \frac{\beta_2}{\rho_2} > 0 \quad ,$$

(5.88)
$$D_1^2 = \frac{r}{2}\,[(\rho_1 + \rho_2)^2 - 2\rho_1\rho_2] - \beta_1\rho_1 - \beta_2\rho_2 \pm \sqrt{2}\,\sqrt{r\rho_1\rho_2}\,\Gamma_1 > 0 \quad .$$

$$\Gamma_2^2 = \frac{r}{2} - \frac{\beta_3}{\rho_3} - \frac{\beta_4}{\rho_4} > 0 \quad ,$$

(5.89)
$$D_2^2 = \frac{r}{2}\,[(\rho_3 + \rho_4)^2 - 2\rho_3\rho_4] - \beta_3\rho_3 - \beta_4\rho_4 \pm \sqrt{2}\,\sqrt{r\rho_3\rho_4}\,\Gamma_2 > 0 \quad .$$

For the first inequalities we have, in accordance with (2.179), the following relations

$$\frac{\beta_1}{\rho_1} + \frac{\beta_2}{\rho_2} = -\frac{b_{23}b_{35}[\rho_1+\rho_2-\rho_2]}{\rho_1\rho_2(\rho_3-\rho_1)(\rho_3-\rho_2)} \,, \quad \beta_1\rho_1 + \beta_2\rho_2 = -\frac{b_{23}b_{35}\rho_3}{(\rho_3-\rho_1)(\rho_3-\rho_2)} \,,$$

$$\frac{\beta_3}{\rho_3} + \frac{\beta_4}{\rho_4} = \frac{b_{23}b_{35}}{\rho_3(\rho_3-\rho_1)(\rho_3-\rho_2)} + \frac{h_4\rho_4 - h_5b_{45}}{\rho_4} \,,$$

$$\beta_3\rho_3 + \beta_4\rho_4 = \frac{b_{23}b_{35}\rho_3}{(\rho_3-\rho_1)(\rho_3-\rho_2)} + \rho_4(h_4\rho_4 - h_5b_{45}) \quad .$$

---

* By virtue of $\rho_5 = \beta_5 = 0$, the fifth equation of (2.180) can be omitted.

The relations obtained can be expressed in terms of the initial parameters with the aid of (2.175) and (2.155). We have

$$(\rho_3 - \rho_1)(\rho_3 - \rho_2) = n^2 - \frac{T_\alpha T_k}{T_r^2}\, n + \frac{\delta T_\alpha^2}{T_r^2} \quad ,$$

$$\rho_1 + \rho_2 - \rho_3 = \frac{T_\alpha T_k}{T_r^2} - n \quad ,$$

$$h_4\rho_4 - b_{45} = - \frac{(\gamma - \beta) T_\alpha}{T_i} \quad ,$$

thanks to which we obtain

$$\frac{\beta_1}{\rho_1} + \frac{\beta_2}{\rho_2} = - \frac{nT_r^2 - T_\alpha T_k}{\delta(n^2T_r^2 - nT_\alpha T_k + \delta T_\alpha^2)} \, , \quad \beta_1\rho_1 + \beta_2\rho_2 = \frac{nT_\alpha^2}{n^2T_r^2 - nT_\alpha T_k + \delta T_\alpha^2} \quad ,$$

$$\frac{\beta_3}{\rho_3} + \frac{\beta_4}{\rho_4} = - \frac{T_\alpha^2}{n(n^2T_r^2 - nT_\alpha T_k + \delta T_\alpha^2)} + \beta - \gamma \quad ,$$

$$\beta_3\rho_3 + \beta_4\rho_4 = - \frac{nT_\alpha^2}{n^2T_r^2 - nT_\alpha T_k + \delta T_\alpha^2} + \frac{(\beta - \gamma)T_\alpha^2}{T_r^2} \quad .$$

Now inequalities (5.88) and (5.89) can be rewritten

$$\Gamma_1^2 = \frac{\beta}{2} + \frac{nT_r^2 - T_\alpha T_k}{\delta(n^2T_r^2 - nT_\alpha T_k + \delta T_\alpha^2)} > 0 \quad ,$$

$$D_1^2 = \frac{\beta}{2}\left[\frac{T_\alpha^2 T_k^2}{T_r^4} - \frac{2\delta T_\alpha^2}{T_r^2}\right] - \frac{nT_\alpha^2}{n^2T_r^2 - nT_\alpha T_k + \delta T_\alpha^2} \pm \sqrt{2}\,\sqrt{\beta}\,\frac{\delta T_\alpha^2}{T_r^2}\,\Gamma_1 > 0 \quad ,$$

(5.90) $$\Gamma_2^2 = \frac{\beta}{2} + \frac{T_\alpha^2}{n(n^2T_r^2 - nT_\alpha T_k + \delta T_\alpha^2)} - \beta + \gamma > 0 \quad ,$$

$$D_2^2 = \frac{\beta}{2}\left[\left(n + \frac{T_\alpha}{T_i}\right)^2 - 2n\,\frac{T_\alpha}{T_i}\right] +$$

$$+ \frac{nT_\alpha^2}{n^2T_r^2 - nT_\alpha T_k + \delta T_\alpha^2} - \frac{(\beta - \gamma)T_\alpha^2}{T_i^2} \pm \sqrt{2}\,\sqrt{\beta}\,\frac{nT_\alpha}{T_i}\,\Gamma_2 > 0 \quad .$$

Thus, if the parameters of the control system (2.180) satisfy inequalities (5.90), absolute stability of the system is guaranteed.

Let us consider a particular case, in which we assume that the machine has no self-equalization (i.e., $n = 0$). We find

$$(5.91)\quad \begin{aligned} \Gamma_1^2 &= \frac{\beta}{2} - \frac{T_k}{\delta^2 T_\alpha} > 0 \ , \\ D_1^2 &= \left[\frac{\beta}{2}\left(\frac{T_k^2}{T_r^2} - 2\delta\right) \pm \sqrt{2\delta}\,\sqrt{\beta\Gamma_1} \right] \frac{T_\alpha^2}{T_r^2} > 0 \ , \\ \Gamma_2^2 &= +\infty < 0 \ , \\ D_2^2 &= \left(\gamma - \frac{\beta}{2}\right)\left(\frac{T_\alpha}{T_1}\right)^2 > 0. \end{aligned}$$

The first and fourth inequalities impose certain limitations on the choice of the isodrome coefficient β, if γ and $T_k/\delta^2 T_a$ are specified. The second inequality imposes limitations on the parameters of the sensing element, T_k and T_r, but is sure to be satisfied if the sensing element has viscous friction and a sufficiently small mass. Obviously, this inequality cannot be satisfied in the absence of a dashpot and if $T_r \neq 0$.

Let us return now to (2.187).

We first call attention to the fact that as $\alpha \longrightarrow 0$, (2.158) changes continuously into (2.156). On the other hand, (2.182) makes it possible to write

$$(5.92)\quad \begin{aligned} \rho_5 - \rho_4 &= \frac{1}{\alpha}\sqrt{\left(\frac{T_a}{T_1}\right)^2 (1 - 4\alpha)} \ , \\ \lim_{\alpha\to 0} \rho_4 &= \frac{T_a}{T_1}, \qquad \lim_{\alpha\to 0} \rho_5 = +\infty \ . \end{aligned}$$

Consequently,

$$(5.93)\quad \begin{aligned} \lim_{\alpha\to 0} \beta_4 &= \frac{T_a}{T_1}(\beta - \gamma) \ , \\ \lim_{\alpha\to 0} \beta_5 &= -\infty \ . \end{aligned}$$

On the other hand, (2.187) with $s = 5$ yields

$$\lim_{\alpha \to 0} x_5 = - \lim_{\alpha \to 0} \frac{\dot{x}_5 - f(\sigma)}{\rho_5} = 0 .$$

From this we obtain

(5.94) $$\lim_{\alpha \to 0} [- \rho_5 x_5 + f(\sigma)] = 0 ,$$

thanks to which

(5.94a) $$\lim_{\alpha \to 0} \beta_5 x_5 = \lim_{\alpha \to 0} \frac{\frac{T_a}{T_i}\left(\beta - \frac{\gamma}{\rho_5}\right)(- \rho_5 x_5)}{\frac{T_a}{T_i}\sqrt{1 - \alpha}} = - \beta f(\sigma) .$$

Consequently, as $\alpha \longrightarrow 0$, (2.187) go continuously into (2.180).

Let us write the stability criterion of the system, making use at the same time of the limiting relation (5.94a).

We choose as the V function for this problem the positive definite function

$$V = \frac{a_1^2 x_1^2}{2\rho_1} + \frac{a_2^2 x_2^2}{2\rho_2} + \frac{2a_1 a_2 x_1 x_2}{\rho_1 + \rho_2} + \frac{a_3^2 x_3^2}{2\rho_3} + \frac{a_4^2 x_4^2}{2\rho_4} + \frac{2a_3 a_4 x_3 x_4}{\rho_3 + \rho_4} +$$

$$+ \frac{1}{2} A_5 x_5^2 + \frac{1}{2} \sum_{k=1}^{4} A_k x_k^2 + \int_0^\sigma f(\sigma) d\sigma .$$

It is assumed that the ρ_k are real numbers and that all A_k are positive. Let us assume that α is sufficiently small, and write the last equation of (2.187) in the form

$$\dot{\sigma} = \sum_{k=1}^{4} \beta_k x_k - \beta f(\sigma) + [\beta_5 x_5 + \beta f(\sigma)] .$$

Calculating the total derivative of the function V and requiring that the following relations be satisfied

$$A_1 + \beta_1 + 2\sqrt{\frac{\beta}{2}}\, a1 + \frac{a_1^2}{\rho_1} + \frac{2a_1a_2}{\rho_1 + \rho_2} = 0 \ ,$$

(5.95)

$$A_2 + \beta_2 + 2\sqrt{\frac{\beta}{2}}\, a_2 + \frac{a_2^2}{\rho_2} + \frac{2a_1a_2}{\rho_1 + \rho_2} = 0 \ ,$$

$$A_3 + \beta_3 + 2\sqrt{\frac{\beta}{2}}\, a_3 + \frac{a_3^2}{\rho_3} + \frac{2a_3a_4}{\rho_3 + \rho_4} = 0 \ ,$$

(5.96)

$$A_4 + \beta_4 + 2\sqrt{\frac{\beta}{2}}\, a_4 + \frac{a_4^2}{\rho_4} + \frac{2a_3a_4}{\rho_3 + \rho_4} = 0 \ ,$$

(5.97) $$A_5 + \beta_5 = 0 \ ,$$

we can write

$$\dot{V} = - [a_1x_1 + a_2x_2 + \sqrt{\frac{\beta}{2}}\, f(\sigma)]^2 -$$

$$- [a_3x_3 + a_4x_4 + \sqrt{\frac{\beta}{2}}\, f(\sigma)]^2 - \varepsilon f^2(\sigma) = - W - \varepsilon f^2(\sigma) \ .$$

Here W is a function in the variables $x_1, \ldots, x_4, f(\sigma)$, which is everywhere positive definite. If α is sufficiently small ε is also a sufficiently small number that vanishes simultaneously with α. In this case relations (5.95), (5.96), and (5.97) are the sufficient conditions for the absolute asymptotic stability of the system. Let us consider these conditions.

Obviously, by virtue of (5.93), relation (5.97) is always satisfied. In the analysis of (5.95) and (5.96) the comments made with respect to similar equations in the proof of the Lur'e theorem, also apply to the quantities $A_1, \ldots, A_4$. Then, for (5.95) and (5.96) to be solvable it is enough to require that inequalities (5.88) and (5.89) be satisfied, with $\beta_1, \ldots, \beta_4$ being determined by (2.186). Noting that β_1 and β_2, obtained from (2.186), are the same as β_1 and β_2 obtained from (2.179), we conclude that inequalities (5.88) remain valid for this problem. We obtain for the inequalities (5.89)

$$\frac{\beta_3}{\rho_3} + \frac{\beta_4}{\rho_4} = - \frac{T_a^2}{n(n^2T_r^2 - nT_aT_k + \delta T_a^2)} - \frac{\beta - \gamma\left(1 + \frac{\varepsilon T_a}{T_1} + \ldots\right)}{\sqrt{1 - 4\alpha}} \ ,$$

$$\beta_3\rho_3 + \beta_4\rho_4 = -\frac{nT_a^2}{n^2T_r^2 - nT_aT_k + \delta T_a^2} -$$

$$-\frac{T_a^2}{T_i^2}\frac{\beta - \gamma}{\sqrt{1-4\alpha}} - \frac{\left[\beta\left(\varepsilon - \frac{T_a}{T_i}\right) + \frac{\gamma T_a}{T_i} - \frac{\beta T_a}{T_i}\right]}{\sqrt{1-4\alpha}}\varepsilon ,$$

where ε is a sufficiently small number that vanishes together with α.

Thus, the two last inequalities in the stability criterion assume the form

$$\Gamma_2^2 = \frac{\beta}{2} + \frac{T_a^2}{n(n^2T_r^2 - nT_aT_k + \delta T_a^2)} -$$

$$-\frac{\beta - \gamma\left(1 + \frac{\varepsilon T_i}{T_a} + \dots\right)}{\sqrt{1-4\alpha}} > 0 ,$$

(5.98)

$$D_2^2 = \frac{\beta}{2}\left[n^2 + \left(\frac{T_a}{T_i}\right)^2\right] + \frac{nT_a^2}{n^2T_r^2 - nT_aT_k + \delta T_a^2} -$$

$$-\frac{T_a^2}{T^2}(\beta - \gamma) \pm 2\sqrt{\frac{\beta}{2}}\frac{nT_a}{T_i}\Gamma_2 + v(\varepsilon) > 0 .$$

It is seen from inequalities (5.98) that if $\varepsilon(\alpha)$ is sufficiently small, a machine stable when the mass of the isodrome is zero remains stable when this mass is sufficiently small but not zero.

An analogous argument can be made also with respect to the Bulgakov problem, if the relay time constant T_i is sufficiently small. For this purpose, it is necessary to choose s proportional to T_i. Actually, turning to formulas (2.158a) and (2.192), we have

$$\lim_{T_i \to 0} \beta_1 = -\frac{p_1 - \rho_1 p_2}{\rho_2 - \rho_1}, \qquad \lim_{T_i \to 0} \beta_2 = \frac{p_1 - \rho_2 p_2}{\rho_2 - \rho_1} .$$

Since when $T_i \longrightarrow 0$

$$\lim \rho_3 = \lim \frac{1}{T_i\sqrt{r}} = +\infty ,$$

$$\lim_{T_i \to 0} \beta_3 = \lim_{T_i \to 0}\left[-\rho_3 - n_2p_2 + \frac{p_3h_4b_{24}(b_{31} - \rho_3 b_{32})}{\rho_3^2 + b_{22}\rho_3 - b_{21}}\right] = -\infty ,$$

then

$$\lim_{T_1 \to 0} x_3 = - \lim_{T_1 \to 0} \frac{\dot{x}_3 - f(\sigma)}{\rho_3} = 0 \quad .$$

From this we obtain

$$\lim_{T_1 \to 0} [- \rho_3 x_3 + f(\sigma)] = 0 \quad .$$

The limiting equation we need,

$$\lim_{T_1 \to 0} \beta_3 x_3 = - \lim_{T_1 \to 0} \left[1 + \frac{n_2 p_2}{\rho_3} - \frac{p_3 h_4 b_{24}(b_{31} - \rho_3 b_{32})}{\rho_3(\rho_3^2 + b_{22}\rho_3 - b_{21})} \right] \rho_3 x_3 = - f(\sigma)$$

permits, as in the preceding example, formulation of the Lyapunov function for such a problem, making use of the equation for the variable σ, written in the form

$$\dot{\sigma} = \beta_1 x_1 + \beta_2 x_2 - f(\sigma) + [\beta_3 x_3 + f(\sigma)] \quad .$$

Let us assume that ρ_1 and ρ_2 are real. Let

$$V = \frac{a_1^2 x_1^2}{2\rho_1} + \frac{a_2^2 x_2^2}{2\rho_2} + \frac{2a_1 a_2 x_1 x_2}{\rho_1 + \rho_2} + \frac{1}{2} A_3 x_3^2 + \int_0^\sigma f(\sigma) d\sigma + \frac{1}{2} A_1 x_1^2 + \frac{1}{2} A_2 x_2^2$$

represent a positive definite function (A_1, A_2, and A_3 are positive numbers). If the constants β_1, β_2, and β_3 satisfy the relations

$$A_1 + \frac{a_1^2}{\rho_1} + 2a_1 + \frac{2a_1 a_2}{\rho_1 + \rho_2} = 0 \quad ,$$

$$A_2 + \frac{a_2^2}{\rho_2} + 2a_2 + \frac{2a_1 a_2}{\rho_1 + \rho_2} = 0 \quad ,$$

$$A_3 + \beta_3 = 0 \quad ,$$

then the total derivative dV/dt will have the form

$$\frac{dV}{dt} = - [a_1 x_1 + a_2 x_2 + f(\sigma)]^2 + A_1 \rho_1 x_1^2 - A_2 \rho_2 x_2^2 - \varepsilon f^2(\sigma) \quad .$$

An examination of these equations leads us to a stability criterion that reduces to the satisfaction of three inequalities. One of these is $\beta_3 < 0$. As follows from the preceding, it is always satisfied. The two other inequalities are

$$\Gamma^2 > 0, \quad D^2 > 0 \quad .$$

It is easy to verify that at sufficiently small T_1 the limiting relations written above are sure to be satisfied, provided inequalities (4.33) and (4.36) are satisfied, and the control system is stable when the relay has no time delay.

In conclusion, let us note another important premise of importance to the theory of automatic control.

Quite frequently, when some problem is posed, it becomes necessary to idealize the system and to neglect the so-called "small" or "parasitic" parameters.

It has been shown here by several examples that the method proposed for solving the fundamental problem of the theory of automatic control makes it possible to clarify the cases in which such parameters can actually be neglected.

Naturally, the stability criterion for these systems could have been derived also for the case in which the parasitic parameters are no longer small.

CHAPTER VI: INHERENTLY UNSTABLE CONTROL SYSTEMS

1. GENERALIZATION OF THE LUR'E THEOREM

As was already mentioned, in those cases when the regulated object is unstable with the regulator disconnected, the roots ρ_k include at least one root for which $\text{Re}\ \rho_k < 0$. This circumstance makes it quite difficult to formulate the Lyapunov function for such a problem. In all such cases, and also in those cases when more than one ρ_k vanishes, it is recommended that forms (3.1) and (3.6) of the canonical transformation of the initial equations be used; this form has already been considered. Thus, we have:

$$(6.1)\qquad \begin{aligned} \dot{x}_s &= -r_s x_s + \sigma \qquad (s = 1, \ldots, m)\ , \\ \dot{\sigma} &= \sum_{k=1}^{m} \bar{\beta}_k x_k - \bar{\rho}\sigma - f(\sigma)\ , \end{aligned}$$

where r_s are the roots of the equation

$$(6.2)\qquad \Delta(r) = |\bar{b}_{k\alpha} + r\delta_{k\alpha}| = 0\ .$$

We assume that $\text{Re}\ r_k \geq 0$ $(k = 1, \ldots, m)$. As was noted earlier, this requirement is satisfied for systems that are stable in the presence of an ideal regulator. Conditions of such stability reduce to the fulfillment of the Hurwitz inequality

$$(6.3)\qquad \Delta_1 \geq 0, \ldots, \Delta_n \geq 0\ ,$$

written for the equation that results from (6.2) by replacing r with $-r$.

Let us consider the positive definite function

$$(6.4)\qquad V = \sum_{k=1}^{m} \sum_{i=1}^{m} \frac{a_k a_i x_k x_i}{r_k + r_i} + \frac{x^2}{2}\sigma^2 + \Phi(x_1, \ldots, x_m)\ ,$$

where $\chi^2 > 0$, $a_1, \ldots, a_s$ are real, and $a_{s+1}, \ldots, a_m$ are complex conjugate pairs. Its total derivative is calculated in accordance with (6.1) and equals

$$\dot{V} = -\sum_{k=1}^{m}\sum_{i=1}^{m} a_k a_i x_k x_i + 2\sigma \sum_{k=1}^{m}\sum_{i=1}^{m} \frac{a_k a_i x_k}{r_k + r_i} - \sum_{k=1}^{s} r_k A_k x_k^2 +$$

(6.5)

$$+ \chi^2\sigma \sum_{k=1}^{m} \bar{\beta}_k x_k - \bar{\rho}\chi^2\sigma^2 - \chi^2\sigma f(\sigma) - C_1(r_{s+1} + r_{s+2})x_{s+1}x_{s+2} - \cdots .$$

But since

$$\sum_{k=1}^{m}\sum_{i=1}^{m} a_k a_i x_k x_i = \left(\sum_{k=1}^{m} a_k x_k\right)^2 ,$$

then, adding to (6.5) the expression

$$\sigma^2 + 2\sigma \sum_{k=1}^{m} a_k x_k$$

and then subtracting it, we can write

$$\dot{V} = -\left[\sum_{k=1}^{m} a_k x_k + \sigma\right]^2 - (\chi^2\rho - 1)\sigma^2 - \chi^2\sigma f(\sigma) -$$

$$- \sum_{k=1}^{s} r_k A_k x_k^2 - C_1(r_{s+1} + r_{s+2})x_{s+1}x_{s+2} - \cdots$$

$$\cdots + \sigma \sum_{k=1}^{s}\left[2\sum_{i=1}^{m} \frac{a_k a_i}{r_k + r_i} + \chi^2\bar{\beta}_k + 2a_k + A_k\right] x_k +$$

$$+ \sigma \sum_{\alpha=1}^{m-s}\left[2\sum_{i=1}^{m} \frac{a_{s+\alpha} a_i}{r_{s+\alpha} + r_i} + \chi^2\bar{\beta}_{s+\alpha} + 2a_{s+\alpha} + C_\alpha\right] x_{s+\alpha} .$$

Using expressions (6.5) and (6.6), we can obtain various forms of the stability conditions.

Thus, if we require the fulfillment of the relation

(6.7)
$$A_k + \chi^2\bar{\beta}_k + 2a_k \sum_{i=1}^{m} \frac{a_i}{r_k + r_i} = 0 \qquad (k = 1, \ldots, s) ,$$

$$C_\alpha + \chi^2\bar{\beta}_{s+\alpha} + 2a_{s+\alpha} \sum_{i=1}^{m} \frac{a_i}{r_{s+\alpha} + r_i} = 0 \qquad (\alpha = 1, \ldots, n - s),$$

(6.8)
$$\bar{\rho} > 0 ,$$

then (6.5) can be written as

$$\dot{V} = -\left(\sum_{k=1}^{m} a_k x_k \right)^2 - \chi^2[\sigma f(\sigma) + \bar{\rho}\sigma^2] -$$

$$- \sum_{k=1}^{s} r_k A_k x_k^2 - C_1(r_{s+1} + r_{s+2})x_{s+1}x_{s+2} - \cdots .$$

On the other hand, if we require the following relations to be satisfied

(6.10)
$$A_k + \chi^2\bar{\beta}_k + 2a_k + 2a_k \sum_{i=1}^{m} \frac{a_i}{r_k + r_i} , \qquad (k = 1, \ldots, s) ,$$

$$C_\alpha + \chi^2\bar{\beta}_{s+\alpha} + 2a_{s+\alpha} + 2a_{s+\alpha} \sum_{i=1}^{m} \frac{a_i}{r_{s+\alpha} + r_i} = 0 \qquad (\alpha = 1, \ldots, m - s) ,$$

(6.11)
$$\chi^2\rho - 1 > 0 ,$$

we may rewrite (6.6) as

(6.12)
$$\dot{V} = -\left[\sum_{k=1}^{m} a_k x_k + \sigma \right]^2 - [\chi^2\bar{\rho} - 1]\sigma^2 - \chi^2\sigma f(\sigma) -$$

$$- \sum_{k=1}^{s} r_k A_k x_k^2 - C_1(r_{s+1} + r_{s+2})x_{s+1}x_{s+2} -$$

$$- C_3(r_{s+3} + r_{s+4})x_{s+3}x_{s+4} - \cdots .$$

Relations (6.3), (6.7), and (6.8) or else relations (6.3), (6.10), and (6.11) can be taken to be the sufficient conditions for the absolute stability of the system. However, the examples show that these conditions are frequently incompatible, owing to the impossibility of satisfying conditions (6.8), or, respectively, (6.11).* In such cases, the matter can be rectified by imposing addition requirements on the function $f(\sigma)$.

Let us assume that

$$h \leq \left[\frac{df}{d\sigma} \right]_{\sigma=0}$$

is a fixed number, and

$$\varphi(\sigma) = f(\sigma) - h\sigma \tag{6.13}$$

satisfies the conditions mentioned in Section 2 of Chapter I [i.e., the function $f(\sigma)$ belongs to subclass (A')]. Then, writing (6.9) in the form**

$$\dot{V} = -\left(\sum_{k=1}^{m} a_k x_k \right)^2 - \chi^2(\rho + h)\sigma^2 - \chi^2\sigma\varphi(\sigma) - \ldots ,$$

and (6.12) in the form

$$\dot{V} = -\left[\sum_{k=1}^{m} a_k x_k + \sigma \right]^2 - [\chi^2(\bar{\rho} + h) - 1]\sigma^2 - \chi^2\sigma\varphi(\sigma) - \ldots$$

and replacing conditions (6.8) and (6.11) by

$$\bar{\rho} + h > 0 \tag{6.14}$$

and respectively

---

* I. G. Malkin noted that this incompatibility always occurs in the case of an inherently unstable control system. Actually, in this case it is impossible to attain stability with any arbitrary servo motor, i.e., with any $f(\sigma)$ of class (A), for the first-approximation Lyapunov theorems become subject to instability at

$$\left[\frac{df}{d\sigma} \right]_{\sigma=0} = \varepsilon ,$$

where ε is a positive number, as small as desired.

** Terms of the type $C_1(\rho_{s+1} + \rho_{s+2})x_{s+1}x_{s+2}$ are omitted.

$$x^2(\bar{\rho} + h) - 1 > 0 \quad , \tag{6.15}$$

we see that in each of these cases $\dot{V}$ can assume only negative or zero values. Thus, the fulfillment of conditions (6.3), (6.7), and (6.14) or else of (6.3), (6.10), and (6.15) gives sufficient ground for concluding that the system is stable.

For specified parameters of the regulated object and for specified regulator parameters, specified in accordance with (6.3) and (6.7) [or else in accordance with (6.3) and (6.10)], relation (6.14) [or (6.15)] serves as a lower limit on the average slope of the characteristic of the actuator of the regulator. When the velocity $f(\sigma)$ of the actuator and the intersection of the curve $f(\sigma)$ with the straight line $h\sigma$ passes through $\sigma = \bar{\sigma}$ (Figure 2), then, if the conditions (6.3), (6.7), and (6.14) [or else (6.3), (6.10), and (6.15)] are satisfied, the estimate of the stability of the equilibrium position can be extended to include only those initial disturbances for which σ does not exceed $\bar{\sigma}$. It can be said that in this case we have a unique, conditional stability in the large.

Thus we have proven the following theorem:* If the constants of the regulator are chosen to satisfy inequalities (6.3) and (6.14) [or (6.3) and (6.15)] and are such that the system of equations (6.7) [or (6.11)] (in which $r_1, \ldots, r_s$ and $\bar{\beta}_1, \ldots, \bar{\beta}_s$ are real and $r_{s+1}, \ldots, r_m$ and $\bar{\beta}_{s+1}, \ldots, \bar{\beta}_m$ are complex conjugate pairs), has at least one solution containing real roots $a_1, \ldots, a_s$ and complex conjugate pairs $a_{s+1}, \ldots, a_n$, then the steady state of the system has absolute asymptotic stability.

2. SOLUTION OF THE SECOND BULGAKOV PROBLEM FOR $k < 0$

The simplest example, illustrating the above, is the second Bulgakov problem in the case when $k < 0$. In this case it is impossible to construct a Lyapunov function by the Lur'e method detailed above.

Turning to the canonical equations (3.27), we will have, in accordance with the theorem proved, the following sufficient conditions for the stability ($x^2 = 1$):

$$\begin{aligned} &\bar{\beta}_1 + 2a_1 + \frac{a_1^2}{} + \frac{2a_1a_2}{r_1 + r_2} = 0 \quad , \\ &\bar{\beta}_2 + 2a_2 + \frac{a_2^2}{} + \frac{2a_1a_2}{r_1 + r_2} = 0 \quad , \end{aligned} \tag{6.16}$$

* A. M. Letov. Inherently unstable control systems, PMM, Vol. XIV, No. 2, 1950.

$$\bar{\rho} + h - 1 > 0 \quad . \tag{6.17}$$

(6.16) have the same form as (4.28), pertaining to the same problem for $k > 0$. The difference between the two lies in the difference between the numbers r_1, r_2, $\bar{\beta}_1$ and $\bar{\beta}_2$, and the numbers ρ_1, ρ_2, β_1, and β_2. As before, the conditions for the solvability of the above equations will be

$$r^2 = 1 - \frac{\bar{\beta}_1}{r_1} - \frac{\bar{\beta}_2}{r_2} > 0 \quad , \tag{6.18}$$

$$D^2 = r_1^2 + r_2^2 - \bar{\beta}_1 r_1 - \bar{\beta}_2 r_2 \pm 2 r_1 r_2 \Gamma > 0 \quad , \tag{6.19}$$

and in the condition (6.19) it is enough to take any one sign ahead of Γ. Henceforth we shall choose the positive sign. First let us consider the case when there is no proportional feedback (i.e., when $\ell = \infty$). According to (3.21) and (3.26), we find

$$r_1 + r_2 = \frac{E}{G}, \qquad r_1 r_2 = a \quad ,$$

$$\frac{\bar{\beta}_1}{r_1} + \frac{\bar{\beta}_2}{r_2} = \frac{\bar{p}_1}{r_1 r_2}, \qquad \bar{\beta}_1 r_1 + \bar{\beta}_2 r_2 = -\bar{p}_1 + \frac{E}{G}\bar{p}_2 \quad .$$

But formulas (3.20) and (2.82) yield for $\ell = \infty$

$$\bar{p}_1 = -\frac{a(E-pG^2)}{G}, \qquad \bar{p}_2 = a - qG^2 - \frac{E(E-pG^2)}{G^2} \quad ,$$

thanks to which, conditions (6.18) and (6.19) become

$$r^2 = 1 + \frac{E - pG^2}{G} > 0 \quad , \tag{6.20}$$

$$\begin{aligned} D^2 = \left(\frac{E}{G}\right)^3 + \left(\frac{E}{G}\right)^2 + (ap + qE)G \, + \\ + 2a\left(\Gamma - 1 - \frac{E}{G}\right) - pG\left(\frac{E}{G}\right)^2 > 0 \quad . \end{aligned} \tag{6.21}$$

Let us show that for all $p > 0$ and for all q it is possible to find an infinite set of values of the parameters E, G, and a, comprising region B of the absolute stability of the system (3.19).

For this purpose, putting $E/G = x$ and $G = y$ and relaxing inequality (6.21) by omitting the positive term $2a\Gamma$, we plot in the xy plane the following curves

$$y = \frac{1}{p} x + \frac{1}{p} , \tag{6.22}$$

$$F(xy) = qxy^2 + p(a - x^2)y + x^3 + x^2 - 2a(x + 1) = 0 . \tag{6.23}$$

The first of these is a straight line with a slope $k = 1/p$; the second is a third degree curve.

From (6.23) we get

$$y_1, y_2 = \frac{p(x^2-a) \pm \Delta(x)}{2qx} , \tag{6.24}$$

where

$$[\Delta(x)]^2 = p^2(x^2 - a)^2 - 4qx(x + 1)(x^2 - 2a) . \tag{6.25}$$

By varying x from $-\infty$ to $+\infty$ we can obtain from (6.24) all the points of the curve (6.23). By the usual methods of differential geometry, it is easy to establish: (1) that the curve (6.23) has no singular points, and (2) that the curve (6.23) has three asymptotes, determined by the equations

$$x = 0 ,$$

$$y = \frac{p - \sqrt{p^2 - 4q}}{2q} x + \frac{1}{\sqrt{p^2 - 4q}} ,$$

$$y = \frac{p + \sqrt{p^2 - 4q}}{2q} x - \frac{1}{\sqrt{p^2 - 4q}} .$$

Next, investigating the signs of the polynomial (6.25), it is easy to establish that its roots x_1, x_2, x_3 and x_4 lie, respectively, in the intervals $(-\sqrt{2a}, -\sqrt{a})$, $(-\sqrt{a}, -1)$, $(0, \sqrt{a})$, $(\sqrt{a}, \sqrt{2a})$ (it is assumed here that $a > 1$). This makes it possible to conclude that the curve (6.23) has vertical tangents, whose equations are $x = x_1$, $x = x_2$, $x = x_3$, and $x = x_4$. In the intervals (x_1, x_2), (x_3, x_4) the curve has not a single point. Finally, the direct investigation of expression (6.24) makes it possible to verify that the curve passes through the following points:

(1) $y_1 = 0, \quad y_2 = -\frac{p}{q}\sqrt{\frac{a}{2}} > 0 \quad \text{if} \quad x = -\sqrt{2a}$,

(2) $y_1 = 0, \quad y_2 = \frac{p}{q}(a - 1) < 0 \quad \text{if} \quad x = -1$,

(3) $y_1 = \frac{2}{p}, \quad y_2 = \pm\infty \quad \text{if} \quad x = 0$,

(4) $y_1 = 0, \quad y_2 = \frac{p}{q}\sqrt{\frac{a}{2}} < 0 \quad \text{if} \quad x = +\sqrt{2a}$.

The above is enough for plotting the curve $F(x, y) = 0$ (Figure 8).

Conditions (6.20) and (6.21) are satisfied in those points of the plane xy towards which the cross hatched portion of the curve (6.22) is facing.

Let us now return to condition (6.17). In the absence of the quantity h from the last part of this inequality, it contradicts conditions (6.20).

Assuming $h \neq 0$, we plot the straight line

(6.26) $$y = \frac{1}{p}x + \frac{1 - h}{p} .$$

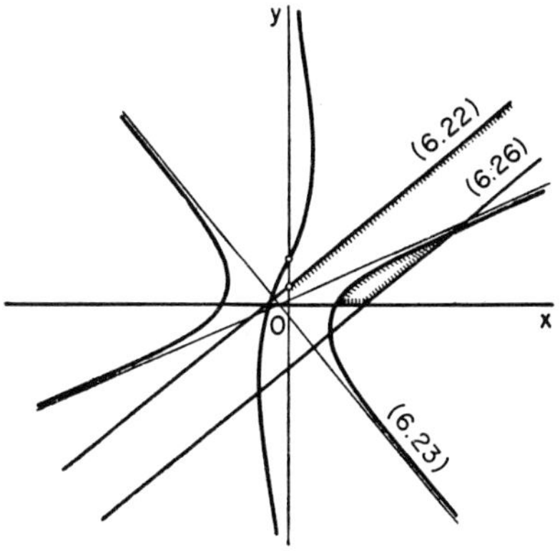

FIGURE 8

The lines (6.22) and (6.23) are parallel; when $h = 0$ they coincide. In order to obtain a region of absolute stability for the system (3.19) on the xy plane, it is sufficient to choose so large a value of h as to make the line (6.26) intersect the curve (6.23). The area obtained in this manner is indicated by cross hatching (Figure 8).

It is clear from this analysis that the limiting value $h = h^*$, for which the line (6.26) is tangent to the curve (6.23), gives the value of the minimum slope of the actuator characteristic. Therefore, when $h < h^*$, the absolute stability of the system cannot be guaranteed.

Analogous constructions can be made also for the case of a regulator containing proportional feedback $(G = 0)$. From (3.21) and (3.26) we obtain

$$r_1 + r_2 = \frac{U + \ell E}{T\sqrt{\ell}}, \quad r_1 r_2 = \frac{k + a\ell}{\ell}, \quad \frac{\bar{\beta}_1}{r_1} + \frac{\bar{\beta}_2}{r_2} = -\frac{E\sqrt{\ell}}{T} ,$$

$$\bar{\beta}_1 r_1 + \bar{\beta}_2 r_2 = \frac{k + a\ell}{T\sqrt{\ell}} E + \frac{U + \ell E}{T\sqrt{\ell}} \left[a - \frac{(U+\ell E)E}{T^2} \right] ,$$

$$r = \frac{\ell}{T^2}, \quad \bar{\rho} = -\frac{E\sqrt{\ell}}{T} .$$

Obviously, the condition

$$\Gamma^2 = 1 + \frac{E\sqrt{\ell}}{T} > 0 \tag{6.27}$$

is always satisfied. Next, condition (6.17) can be written in the form

$$h > 1 + \frac{E\sqrt{\ell}}{T} . \tag{6.27a}$$

As in the preceding case, the number h is here bounded from below by the positive definite number Γ^2.

Finally, the last condition of (6.19) can be readily reduced to the form

$$\left(\frac{U + \ell E}{T\sqrt{\ell}} \right)^2 \Gamma^2 - \frac{a(U+\ell E)}{T\sqrt{\ell}} - \frac{k + a\ell}{\ell} (1 \pm \Gamma)^2 > 0 . \tag{6.28}$$

The last half of inequality (6.28) can always be made positive by choosing a sufficiently large value of the parameter of the regulator E, since the left of (6.28) is a polynomial in E with a positive coefficient for the highest-order term. The analysis performed is based on an examination of the stability conditions, expressed in the form (6.3), (6.10), and (6.15). It is equally possible to consider conditions taken in the form (6.3), (6.7), and (6.14). Condition (6.3) is already known; for conditions (6.7) and (6.14) we have

$$\bar{\beta}_1 + \frac{a_1^2}{r_1} + \frac{2a_1 a_2}{r_1 + r_2} = 0, \quad \bar{\beta}_2 + \frac{a_2^2}{r_2} + \frac{2a_1 a_2}{r_1 + r_2} = 0 , \tag{6.29}$$

$$\bar{\rho} + h > 0 . \tag{6.30}$$

The conditions for the solvability of (6.29) are as follows:*

---

* See Section 8 of Chapter IV.

$$\frac{\bar\beta_1}{r_1} + \frac{\bar\beta_2}{r_2} < 0, \qquad \bar\beta_1 r_1 + \bar\beta_2 r_2 < 0 \ . \tag{6.31}$$

Let us consider inequalities (6.31). According to the derivations detailed in this section, we obtain in the absence of proportional feedback

$$\frac{E - pG^2}{G} > 0 \ , \tag{6.32}$$

$$\left(\frac{E}{G}\right)^2 \frac{E - pG^2}{G} - \frac{E}{G}(a - qG^2) - \frac{a(E-pG^2)}{G} > 0 \ . \tag{6.33}$$

Let us compare inequality (6.32) with (6.20). It is clear that if (6.32) is satisfied, then (6.20) is satisfied to an even greater extent; consequently, (6.32) is the more stringent inequality.

Let us compare inequality (6.32) with (6.20). It is clear that if (6.32) is satisfied, then (6.20) is satisfied to an even greater extent; consequently, (6.32) is the more stringent inequality.

By comparison of inequalities (6.33) and (6.21) we see that when $\Gamma - 1 \geq 0$, the satisfaction of (6.33) involves the satisfaction of (6.12). If, however, $\Gamma - 1 < 0$, then, the satisfaction of (6.21) when $(E/G)^2 + 2a(\Gamma - 1) < 0$ guarantees the satisfaction of (6.33). Inequality (6.30) assumes the form

$$h - \frac{E - pG^2}{G} > 0 \tag{6.34}$$

and its satisfaction guarantees the satisfaction of inequality (6.17). In the case of a regulator containing proportional feedback, inequalities (6.31) are written as follows

$$\frac{E\sqrt{\ell}}{T} > 0 \ , \tag{6.35}$$

$$\frac{(U+\ell E)^2 E}{T^3\sqrt{\ell}} - \frac{a(U+\ell E)}{T\sqrt{\ell}} - \frac{k + a\ell}{T\sqrt{\ell}} E > 0 \ , \tag{6.36}$$

and inequality (6.15) is the same as (6.27a). Inequality (6.35), like inequality (6.27), is obvious. Inequality (6.36), by virtue of the fact that $\Gamma - 1 > 0$, is more stringent than inequality (6.28). The above comparison of the two stability criteria does not disclose, for this example, any substantial difference between them.

Condition (6.30) can be written as

$$h - \sqrt{r}(E - pG^2) > 0 \quad . \tag{6.37}$$

In the case of the absence of proportional feedback

$$h > \frac{E}{G} + pG \quad .$$

In the presence of a proportional feedback

$$h > \frac{\sqrt{\ell}}{T} E \quad .$$

It would be of interest to compare the above solution with that given by the Lyapunov first-approximation stability theorems. For this purpose, putting $f(\sigma) = h\sigma$ in (3.27), let us set up the characteristic equation

$$\begin{aligned} &\lambda^3 + (r_1 + r_2 + h - p_2)\lambda^2 + [r_1 r_2 + (r_1 + r_2)(h - p_2) - \beta_1 - \beta_2]\lambda + \\ &+ r_1 r_2 (h - p_2) - \beta_1 r_2 - \beta_2 r_1 = 0 \quad . \end{aligned} \tag{6.38}$$

We consider only the case of the absence of proportional feedback (i.e., when $\ell = \infty$). In this case, (6.38) assumes the following form

$$\lambda^3 + (h + py)\lambda^2 + (xh - qy^2)\lambda + ah = 0 \quad ,$$

where x and y assume their previous values. Now let x and y be fixed coordinates of any point in the plane, and let h_1 and h_2 be the roots of the equation

$$xh^2 + (qy^2 + pxy - a)h + pqy^3 = 0 \quad .$$

The stability conditions of the linearized system are

$$\begin{aligned} &(h + py)(hx + qy^2) - ah = x(h - h_1)(h - h_2) > 0 \quad , \\ &hx + qy^2 > 0 \quad . \end{aligned} \tag{6.39}$$

Considering only the region of positive values of x and y, we have $h_1 > 0$ and $h_2 < 0$.

Let us show that if the foregoing stability conditions are satisfied, then conditions (6.39) are all the more satisfied. For this purpose we insert into inequalities (6.39) the lower value of h, which is determined by the condition (6.26), i.e.,

$$h^* = 1 + x - py \; .$$

When $h > h^*$, inequalities (6.39) are assuredly satisfied if they are satisfied when $h = h^*$. Making the above substitution in (6.39), and changing from inequalities to equalities, we obtain the two limiting curves

$$f(x, y) = (1 + x)x - pxy + qy^2 = 0$$

and

$$\Phi(x, y) = F(x, y) + f(x, y) + ax + a = 0 \quad ,$$

where $F(x, y)$ is determined by equality (6.23). The first of these curves is a hyperbola passing through the origin; its asymptotes are also the asymptotes of the curve (6.23). Obviously, the region Q, where $f(x, y) > 0$, contains that region which we plotted in Figure 8. Inside this region, the second inequality $\Phi(x, y) > 0$ is everywhere satisfied.

3. OTHER FORMS OF STABILITY CRITERIA

The above solution was based on the assumption that the inherently-unstable system becomes stable if an ideal regulator $(h = \infty)$ is introduced into the system.

In an attempt to expand the region of stability, we sought a solution $h_{min} = h^*$ that would guarantee system stability for all $h > h^*$ in the case of a non-ideal regulator.

In this section, we illustrate another approach to the solution of the problem. We assume that the linearized system (with a linearization coefficient h) is stable. Subject to this condition, we seek stability criteria for the initial control system. The statement and solution of this problem are due to Lur'e.*

Thus, assume a system of equations for disturbed motion in the form (1.33). Taking into account everything said above, we rewrite these equations

$$\dot{\eta}_k = \sum_{\alpha=1}^{m} bk_\alpha \eta_\alpha + n_k \eta_{m+1} \qquad (k = 1, \ldots, m) \quad ,$$

(6.40)
$$\dot{\eta}_{m+1} = - \rho_{m+1} \eta_{m+1} + h \left[\sum_{\alpha=1}^{m} p_\alpha \eta_\alpha - r\eta_{m+1} \right] + \varphi(\sigma) \quad ,$$

* A. I. Lur'e, Nekotorye nelineǐnye zadachi teorii avtomaticheskovo regulirovaniya (Certain Nonlinear Problems in the Theory of Automatic Control), Gostekhizdat, 1951.

where

(6.41)
$$\eta_{m+1} = \xi ,$$
$$\sigma = \sum_{\alpha=1}^{m} p_\alpha \eta_\alpha - r\eta_{m+1} .$$

We introduce $m + 1$ new independent variables, determined by the equations

(6.42)
$$x_s = \sum_{\alpha=1}^{m+1} C_\alpha^{(s)} \eta_\alpha \qquad (s = 1, \ldots, m + 1) ,$$

where $C_\alpha^{(s)}$ are constants, so far undetermined. Differentiating (6.42) and then eliminating the variables $\dot\eta_k$ in accordance with (6.40), we obtain

$$\dot x_s = \sum_{\alpha=1}^{m} C_\alpha^{(s)} \left[\sum_{\beta=1}^{m} b_{\alpha\beta}\eta_\beta + \eta_\alpha \eta_{m+1} \right] +$$
$$+ C_{m+1}^{(s)} \left[- \rho_{m+1}\eta_{m+1} + h \left(\sum_{\alpha=1}^{m} p_\alpha \eta_\alpha - r\eta_{m+1} \right) + \varphi(\sigma) \right] .$$

To transform the equations in the new variables into a canonical form, it is necessary to subject the transformation coefficients to the relations

(6.43)
$$\mu C_\beta^{(s)} = \sum_{\alpha=1}^{m} C_\alpha^{(s)} b_{\alpha\beta} + hp_\beta C_{m+1}^{(s)} \qquad (\beta = 1, \ldots, m) ,$$
$$\mu C_{m+1}^{(s)} = \sum_{\alpha=1}^{m} C_\alpha^{(s)} n_\alpha - (hr + \rho_{m+1}) C_{m+1}^{(s)} ,$$

(6.44)
$$C_{m+1}^{(s)} = 1 \qquad (s = 1, \ldots, m + 1) .$$

If such a transformation really exists, then the quantity σ can be expressed as a linear function of the new variables, i.e., it will have the form

(6.45)
$$\sigma = \sum_{k=1}^{m} \gamma_k x_k + \gamma_{m+1} x_{m+1} .$$

Here γ_k are certain numbers, for which special formulas will be given later.

Let us first consider (6.43). We set up the determinant

$$D(\mu) = \begin{vmatrix} b_{11} - \mu & b_{21} & \cdots & b_{m1} & hp_1 \\ b_{12} & b_{22} - \mu & \cdots & b_{m2} & hp_2 \\ \cdots & \cdots & \cdots & \cdots & \cdots \\ b_{1m} & b_{2m} & \cdots & b_{mm} - \mu & hp_m \\ n_1 & n_2 & \cdots & n_m & -(hr + \rho_{m+1} + \mu) \end{vmatrix} \tag{6.46}$$

and let $\mu_1, \ldots, \mu_{m+1}$ be the roots of the equation

$$D(\mu) = 0 \ . \tag{6.47}$$

Unlike the roots ρ_k of (2.5), the roots μ_k are influenced by the parameters of the regulator, owing to the separation of the linear part $h\sigma$ in the function $f(\sigma)$.

It is natural to require that all roots μ_k of (6.47) satisfy the condition

$$\operatorname{Re} \mu_k < 0 \qquad (k = 1, \ldots, m + 1) \ . \tag{6.48}$$

The assumption (6.48) is connected with the statement of this problem, according to which the stability of the given system is assured when $f(\sigma) = h\sigma$. The condition for stability is the satisfaction of the Hurwitz inequalities

$$\Delta_1 > 0, \ \ldots, \ \Delta_{m+1} > 0 \tag{6.49}$$

for (6.47).

Let B_1 be the stability region, plotted in the space of the parameters of the regulator, in accordance with inequalities (6.49). It is natural to assume that, since we consider inherently unstable systems, a certain lower limit is imposed on the choice of the positive number h in the entire region B_1.

Thus, the canonical form of the sought equations is

$$\dot{x}_s = \mu_s x_s + \varphi(\sigma) \qquad (s = 1, \ldots, m + 1) \ ,$$

$$\sigma = \sum_{s=1}^{m+1} \gamma_s x_s \ . \tag{6.50}$$

Later on we shall need an expression for $\dot{\sigma}$. From (6.50) we find

$$\dot{\sigma} = \sum_{k=1}^{m+1} \gamma_k[\mu_k x_k + \varphi(\sigma)] \quad .$$

If we denote

$$\beta_k = \mu_k\gamma_k, \quad r^* = -\sum_{k=1}^{m+1} \gamma_k \qquad (k = 1, \ldots, m + 1) , \tag{6.51}$$

we have

$$\dot{\sigma} = \sum_{k=1}^{m+1} \beta_k x_k - r^*\varphi(\sigma) \quad . \tag{6.52}$$

Thus, we can use either (6.50) or

$$\begin{aligned} \dot{x}_s &= \mu_s x_s + \varphi(\sigma) \qquad (s = 1, \ldots, m + 1) , \\ \dot{\sigma} &= \sum_{k=1}^{m+1} \beta_k x_k - r^*\varphi(\sigma) \quad . \end{aligned} \tag{6.53}$$

Here, as in the preceding cases, it is possible to obtain general formulas for the coefficients of the direct and inverse transformations. Let, as before, $D_{ik}(\mu_s)$ be the minors of the elements in the i'th row and k'th column of the determinant (6.46). Considering the system of linear homogeneous equations (6.43), we can write its solution in the form

$$C_k^{(s)} = A_i^{(s)} D_{ik}(\mu_s) \qquad (k, s = 1, \ldots, m + 1) , \tag{6.54}$$

where $A_i^{(s)}$ is proportionality coefficient, which can be determined from (6.44). Taking this equation into account, we obtain

$$A_i^{(s)} = \frac{1}{D^i{}_{,m+1}(\mu_s)} \cdot \qquad (s = 1, \ldots, m + 1) . \tag{6.55}$$

hence

$$C_k^{(s)} = \frac{D_{ik}(\mu_s)}{D_{im+1}(\mu_s)} \qquad (k, s = 1, \ldots, m) . \tag{6.56}$$

Formulas (6.56) yield the direct-transformation coefficients. To obtain the inverse transformation, we must solve the equations

$$\sum_{k=1}^{m+1} D_{1k}(\mu_s)\eta_k = \frac{x_s}{A_1^{(s)}} \qquad (s = 1, \ldots, m+1) \tag{6.57}$$

with respect to the unknowns η_k. This can be done by repeating all the steps listed in the solution of (2.3). As a result we obtain

$$\eta_k = -\sum_{j=1}^{m+1} \frac{D_{k,m+1}(\mu_j)x_j}{D'(\mu_j)} \qquad (k = 1, \ldots, m+1), \tag{6.58}$$

where $D'(\mu_j)$ is the derivative of the determinant (6.46), calculated for $\mu = \mu_j$. In the case of simple μ_j $(j = 1, \ldots, m+1)$, all the $D'(\mu_j)$ are different from zero, and the foregoing inverse transformation becomes feasible.

We can now obtain formulas for the quantities γ_k, β_k, and r^*. According to (6.41) we find

$$\sigma = \sum_{j=1}^{m+1} \frac{rD_{m+1,m+1}(\mu_j) - \sum_{k=1}^{m} p_k D_{k,m+1}(\mu_j)}{D'(\mu_j)} - x_j . \tag{6.59}$$

Comparison of (6.50) and (6.59) gives the first group of formulas

$$\gamma_j = \frac{rD_{m+1,m+1}(\mu_j) - \sum_{k=1}^{m} p_k D_{k,m+1}(\mu_j)}{D'(\mu_j)}, \tag{6.60}$$

and also the second group

$$(6.61) \quad \begin{aligned} \beta_j &= \frac{rD_{m+1,m+1}(\mu_j) - \sum_{k=1}^{m} p_k D_{km+1}(\mu_j)}{D'(\mu_j)} \mu_j , \\ r^* &= \sum_{j=1}^{m+1} \frac{rD_{m+1,m+1}(\mu_j) - \sum_{k=1}^{m} p_k D_{km+1}(\mu_j)}{D'(\mu_j)} . \end{aligned}$$

Formulas (6.60) and (6.61) express all the constants of the canonical transformation in terms of the initial constants of the system. Starting with (6.50) and (6.53), we can propose several solutions for the stability problem.

Let us first consider the Lur'e solution. Let $\mu_1, \ldots, \mu_s$ be the real roots of (6.47), and $\mu_{s+1}, \ldots, \mu_{m+1}$ be the complex conjugate pairs of roots. By way of the Lyapunov function, we assume the sign-definite function

$$V = F = -\sum_{k=1}^{m+1}\sum_{i=1}^{m+1} \frac{a_k a_i x_k x_i}{\mu_k + \mu_i}, \tag{6.62}$$

where $a_1, \ldots, a_s$ are real, and $a_{s+1}, \ldots, a_{m+1}$ are complex conjugate pairs. By virtue of condition (6.48), this function is everywhere positive. Its total derivative, calculated according to (6.50), is

$$\frac{dV}{dt} = -\sum_{k=1}^{m+1}\sum_{i=1}^{m+1} a_k a_i x_k x_i - 2\varphi(\sigma)\sum_{k=1}^{m+1}\sum_{i=1}^{m+1} \frac{a_k a_i x_k}{\mu_k + \mu_i}. \tag{6.63}$$

Let us add to the right half of this expression the quantity

$$\varphi(\sigma)\sum_{k=1}^{m+1} \gamma_k x_k - \sigma\varphi(\sigma) = 0$$

and choose the coefficients a_k in accordance with the conditions

$$\gamma_k - 2\sum_{i=1}^{m+1} \frac{a_k a_i}{\mu_k + \mu_i} = 0 \qquad (k = 1, \ldots, m + 1). \tag{6.64}$$

Then the derivative dV/dt becomes

$$\frac{dV}{dt} = -\left(\sum_{k=1}^{m+1} a_k x_k\right)^2 - \sigma\varphi(\sigma)$$

and will be negative definite no matter what the function $f(\sigma)$ of subclass (A') might be. Relations (6.64) were derived by Lur'e. They must be considered as $m + 1$ equations in the unknowns a_k.

Let B_2 be the region in the space of the regulator parameters in which there is at least one solution of (6.64), having real roots $a_1, \ldots, a_s$ and complex conjugate roots $a_{s+1}, \ldots, a_n$. We denote by B that portion of the region B_2, which belongs simultaneously to region B_1. In particular, it may turn out that $B = B_2 \in B_1$. Then, for any point in the region B, the stability of the control system is determined by the Lyapunov function (6.62).

Thus, it is possible to formulate the following result: If in region B_1 of the stability of a linear system [i.e., in the region (6.49)] there is a subregion $B \in B_1$, in every point of which the system (6.64) has at least one solution containing real roots $a_1, \ldots, a_s$ and complex conjugate pairs $a_{s+1}, \ldots, a_{n+1}$, then the control system (6.40) is absolutely stable.

The absolute-stability conditions can be obtained also in a different form. For this purpose, instead of the Lyapunov function (6.62), one must assume

$$V = F + \int_0^\sigma \varphi(\sigma)d\sigma \quad . \tag{6.65}$$

The total derivative of V, calculated in accordance with (6.53), is

$$\frac{dV}{dt} = -\left(\sum_{k=1}^{m+1} a_k x_k\right)^2 - r^*\varphi^2(\sigma) + \varphi(\sigma)\sum_{k=1}^{m+1}\left[\beta_k - 2\sum_{i=1}^{m+1}\frac{a_k a_i}{\mu_k + \mu_i}\right]x_k \quad . \tag{6.66}$$

As before, adding to the right half of (6.66) the expression

$$2\varphi(\sigma)\sum_{k=1}^{m+1} a_k x_k \sqrt{r^*}$$

and then subtracting it, we can write

$$\frac{dV}{dt} = -\left[\sum_{k=1}^{m+1} a_k x_k + \sqrt{r^*}\,\varphi(\sigma)\right]^2 + \varphi(\sigma)\sum_{k=1}^{m+1}\left[\beta_k + 2a_k\sqrt{r^*} - 2a_k\sum_{i=1}^{m+1}\frac{a_i}{\mu_k + \mu_i}\right]x_k \quad . \tag{6.67}$$

(Naturally, it must be assumed that $r^* > 0$.) For such a choice of the function V, the equations that determine the choice of the constants a_k are written

$$\beta_k - 2a_k \sum_{i=1}^{m+1} \frac{a_i}{\mu_k + \mu_i} = 0 \qquad (k = 1, \ldots, m+1) \tag{6.68}$$

[on the basis of (6.66)], or else in the form

$$\beta_k + 2\sqrt{r^*}\, a_k - 2a_k \sum_{i=1}^{m+1} \frac{a_i}{\mu_k + \mu_i} = 0 \qquad (k = 1, \ldots, m+1) \tag{6.69}$$

[on the basis of (6.67)]. Thus, we formulate the following result: If in the stability region B_1 of a linear system [i.e., in the region (6.49)] there is a subregion $B \in B_1$, for each point of which the system (6.68) or (6.69) has at least one solution, containing real roots $a_1, \ldots, a_s$ and complex conjugate root pairs $a_{s+1}, \ldots, a_{m+1}$, and $r^* > 0$, then the system (6.40) is absolutely stable.

Let us see what solution is obtained for the second Bulgakov problem with this method.

We start with (2.83). Here $n_1 = 0$, $n_2 = -1$, and $r = 1$. Let us set up the determinant (6.46)

$$D(\mu) = \begin{vmatrix} -\mu & b_{21} & hp_1 \\ 1 & b_{22} - \mu & hp_2 \\ 0 & -1 & -(h+\mu) \end{vmatrix} . \tag{6.70}$$

From this we find the characteristic equation of the linearized problem

$$\mu^3 + (h - b_{22})\mu^2 + (hp_2 - b_{21} - hb_{22})\mu + h(p_1 - b_{21}) = 0 . \tag{6.71}$$

Condition (6.48) is satisfied if the system parameters obey the inequalities

$$a - qG^2 + \frac{g}{r} > 0 ,$$

$$\left(h + \frac{p}{\sqrt{r}}\right)\left[h\sqrt{r}(E - pG^2) + \frac{ph}{\sqrt{r}} + \frac{g}{r}\right] > h\left[a - qG^2 + \frac{g}{r}\right] . \tag{6.72}$$

Let us consider these inequalities. According to (2.82) we obtain

$$a - qG^2 + \frac{q}{r} = a + \frac{qT^2}{\ell} ,$$

and the first inequality of (6.72) becomes

(6.73) $$\frac{k + a\ell}{\ell} > 0 .$$

It is easy to verify that it coincides with inequality (3.22) (since $\ell > 0$ always). On the other hand

$$hE\sqrt{r} - hpG^2\sqrt{r} + \frac{hp}{\sqrt{r}} + \frac{q}{r} = hE\sqrt{r} + \frac{q}{r} + \frac{hpT^2}{\sqrt{\ell(T^2+LG^2)}} ,$$

and the second inequality (6.72) becomes

(6.74) $$\left(h + \frac{p}{\sqrt{r}}\right)\left[hE\sqrt{r} + \frac{q}{r} + \frac{hpT^2}{\sqrt{\ell(T^2+\ell G^2)}}\right] > \frac{h(k+a\ell)}{\ell} .$$

Obviously, (6.74) is satisfied when $k < 0$ provided, in turn, the following inequality is satisfied

(6.75) $$hE\sqrt{r} + \frac{q}{r} + \frac{hpT^2}{\sqrt{\ell(T^2+\ell G^2)}} > 0 .$$

The first, necessary conditions for stability are thus (6.73), (6.74), and (6.75).

To formulate the remaining conditions, we must carry out the series of calculations discussed previously in general form. Let us put $i = 1$. The minors of interest to us in the determinant (6.70) are

$$D_{13} = -1, \quad D_{12} = h + \mu, \quad D_{11} = hp_2 - (h + \mu)(b_{22} - \mu) ,$$

$$D_{33} = \mu^2 - b_{22}\mu - b_{21}, \quad D_{23} = -\mu .$$

Next, differentiating the left half of (6.71), we obtain

$$D'(\mu) = 3\mu^2 + 2(h - b_{22})\mu + hp_2 - b_{21} - hb_{22} .$$

Expressing now the coefficients of (6.71) in terms of its root, we obtain the derivative of determinant (6.70):

$$D'(\mu_1) = (\mu_2 - \mu_1)(\mu_1 - \mu_3) ,$$

$$D'(\mu_2) = (\mu_3 - \mu_2)(\mu_2 - \mu_1) ,$$

$$D'(\mu_3) = (\mu_1 - \mu_3)(\mu_3 - \mu_2) .$$

It is now possible to calculate the quantities γ_k and β_k. For this purpose, we use formulas (6.60) and (6.61). We have the first group of quantities

$$\gamma_1 = \frac{D_{33}(\mu_1) - p_1 D_{13} - p_2 D_{23}}{D'(\mu_1)} = \frac{D_{33}(\mu_1) + p_1 + \mu_1 p_2}{D'(\mu_1)} ,$$

$$\gamma_2 = \frac{D_{33}(\mu_2) - p_1 D_{13} - p_2 D_{23}}{D'(\mu_2)} = \frac{D_{33}(\mu_2) + p_1 + \mu_2 p_2}{D'(\mu_2)} ,$$

$$\gamma_3 = \frac{D_{33}(\mu_3) - p_1 D_{13} - p_2 D_{23}}{D'(\mu_3)} = \frac{D_{33}(\mu_3) + p_1 + \mu_3 p_2}{D'(\mu_3)} .$$

Using (6.51), we obtain the second group of quantities

(6.76)
$$\beta_1 = \frac{D_{33}(\mu_1) + p_1 + \mu_1 p_2}{D'(\mu_1)}\mu_1, \qquad \beta_2 = \frac{D_{33}(\mu_2) + p_1 + \mu_2 p_2}{D'(\mu_2)}\mu_2 ,$$
$$\beta_3 = \frac{D_{33}(\mu_3) + p_1 + \mu_3 p_2}{D'(\mu_3)}\mu_3, \qquad r^* = -(\gamma_1 + \gamma_2 + \gamma_3) .$$

We write the conditions for absolute stability, using the stability criteri[on] in the form (6.64) and in the form (6.68). For this purpose, we consider the equations

(6.77)
$$\gamma_1 - \frac{a_1^2}{\mu_1} - \frac{2a_1 a_2}{\mu_1 + \mu_2} - \frac{2a_1 a_3}{\mu_1 + \mu_3} = 0 ,$$
$$\gamma_2 - \frac{a_2^2}{\mu_2} - \frac{2a_1 a_2}{\mu_1 + \mu_2} - \frac{2a_2 a_3}{\mu_2 + \mu_3} = 0 ,$$
$$\gamma_3 - \frac{a_3^2}{\mu_3} - \frac{2a_1 a_3}{\mu_1 + \mu_3} - \frac{2a_2 a_3}{\mu_2 + \mu_3} = 0 .$$

(6.77) differ from (4.47) only in that the former contains the terms γ_k and $-\mu_k$ while the latter contains the terms β_k and ρ_k. Therefore, denoting

$$\Gamma^2 = \frac{\gamma_1}{\mu_1} + \frac{\gamma_2}{\mu_2} + \frac{\gamma_3}{\mu_3} > 0 \quad ,$$

(6.78)

$$D^2 = \gamma_1\mu_1 + \gamma_2\mu_2 + \gamma_3\mu_3 > 0 \quad ,$$

we have, according to (4.55) and (4.56)

$$a_1 + a_2 + a_3 = \pm D \quad ,$$

$$\frac{a_1}{\mu_1} + \frac{a_2}{\mu_2} + \frac{a_3}{\mu_3} = \pm \Gamma \quad .$$

From this we obtain

$$a_1 = \frac{\mu_1}{\mu_1 - \mu_2}\left[\pm D \pm \mu_2\Gamma - \frac{\mu_3 - \mu_2}{\mu_3} a_3\right] ,$$

(6.79)

$$a_2 = -\frac{\mu_2}{\mu_1 - \mu_2}\left[\pm D \pm \mu_1\Gamma + \frac{\mu_1 - \mu_3}{\mu_3} a_3\right] .$$

Inserting the values of (6.79) into the last equation of (6.77), and making elementary simplifications, we obtain an equation for the constants a_3

$$(\mu_1 - \mu_3)(\mu_2 - \mu_3)a_3^2 + 2\mu_3\ (\pm \mu_3 D \mp \mu_1\mu_2\Gamma)a_3 -$$

(6.80)

$$- \gamma_3(\mu_1 + \mu_3)(\mu_2 + \mu_3) = 0 \quad .$$

No matter what the roots of (6.71), (6.80) will always have real coefficients. If μ_1, μ_2, and μ_3 are real numbers, this statement is quite obvious. If, however, μ_1 and μ_2 are complex conjugate numbers, then the products

$$(\mu_1 - \mu_3)(\mu_2 - \mu_3) \quad \text{and} \quad (\mu_1 + \mu_3)(\mu_2 + \mu_3)$$

are also real and, furthermore, they are positive. Consequently, a_3 will be a real number if $\gamma_3 > 0$. In the opposite case it is necessary to demand that the discriminant of (6.80) be positive.

Thus, the stability criterion for the system deduced from (6.64) is of the form

$$(6.81) \qquad \Gamma^2 > 0, \quad D^2 > 0, \quad \gamma_3 > 0$$

or

$$(6.82) \qquad \begin{gathered} \Gamma^2 > 0, \quad D^2 > 0 \ , \\ \mu_3^2[\pm D\mu_3 \mp \mu_1\mu_2\Gamma]^2 + \gamma_3(\mu_1^2 - \mu_3^2)(\mu_2^2 - \mu_3^2) > 0 \ . \end{gathered}$$

In both cases (6.79) determine real or complex conjugate quantities a_1 and a_2. Naturally, these stability conditions must not contradict conditions (6.73), (6.74), and (6.75).

Our analysis could be repeated, with equal success, on the basis of (6.68). However, if we bear in mind the already noted analogy between (6.64) and (6.68), and denote

$$(6.83) \qquad \begin{gathered} \bar{\Gamma}^2 = \frac{\beta_1}{\mu_1} + \frac{\beta_2}{\mu_2} + \frac{\beta_3}{\mu_3} > 0 \ , \\ \bar{D}^2 = \beta_1\mu_1 + \beta_2\mu_2 + \beta_3\mu_3 > 0 \ , \end{gathered}$$

then the stability conditions can be presented in the form of the inequaliti

$$(6.84) \qquad \bar{\Gamma}_2 > 0, \quad \bar{D}^2 > 0, \quad \beta_3 > 0$$

or the inequalities

$$(6.85) \qquad \begin{gathered} \bar{\Gamma}^2 > 0, \quad \bar{D}^2 > 0 \ , \\ \mu_3^2[\pm \bar{D}\mu_3 \mp \mu_1\mu_2\bar{\Gamma}]^2 + \beta_3(\mu_1^2 - \mu_3^2)(\mu_2^2 - \mu_3^2) > 0 \ . \end{gathered}$$

We note that to investigate inequalities (6.81) and (6.82), or inequalities (6.82) and (6.84) in general form involves considerable computational difficulties. Even more substantial difficulties arise in the derivation of a stability criterion on the basis of (6.69).

The above analysis shows that the solution of this problem, obtained in Section 2 of this chapter, involves less computational difficulty than the solution obtained here.

4. FORMULATION OF SIMPLIFIED STABILITY CRITERIA

We can obtain various simplified stability criteria for inherentl unstable systems. Thus, for example, if it is known with certainty that th quantity $\bar{\beta}_1$ is negative, it is possible to employ the device described in Section 1 of Chapter V. In this case, instead of the V-function (6.4), it is necessary to take

$$V = \frac{1}{2} Kx_1^2 + \sum_{k=2}^{n} \sum_{i=2}^{n} \frac{a_k a_i x_k x_i}{r_k + r_i} + \frac{x^2}{2} \sigma^2 ,$$

where $k > 0$, thanks to which the number of quadratic equations (6.7) [(6.10)] diminishes by one, and the missing equation is replaced by inequality $\bar{\beta}_1 < 0$. This simplification can be attained for any other value of $\bar{\beta}_k$ known with certainty to be negative.

Let us assume now that r_k are real numbers. The function

$$V = \frac{1}{2} \sum_{k=1}^{m} x_k^2 + \frac{1}{2} \sigma^2$$

is positive definite everywhere, and its derivative, calculated according to (6.1) is

$$\dot{V} = - \sum_{k=1}^{m} r_k x_k^2 - (\bar{\rho} + h)\sigma^2 - \sigma\varphi(\sigma) + \sum_{k=1}^{m} (1 + \beta_k) x_k \sigma .$$

Let us consider the discriminant of the form $-[\dot{V} + \sigma\varphi(\sigma)]$, i.e., the determinant

$$\Delta = \begin{vmatrix} r_1 & 0 & \cdots & 0 & -\frac{1}{2}(1 + \beta_1) \\ 0 & r_2 & \cdots & 0 & -\frac{1}{2}(1 + \beta_2) \\ \vdots & \vdots & & \vdots & \vdots \\ 0 & \cdot & \cdots & r_m & -\frac{1}{2}(1 + \beta_m) \\ -\frac{1}{2}(1 + \beta_1) & -\frac{1}{2}(1 + \beta_2) & \cdots & -\frac{1}{2}(1 + \beta_m) & (\bar{\rho} + h) \end{vmatrix}$$

and all its diagonal minors

$$\Delta_1 = r_1, \quad \Delta_2 = \Delta_1 r_2, \quad \ldots, \quad \Delta_m = \Delta_{m-1} r_m .$$

But by virtue of (6.3) all r_k are positive numbers. Consequently, a sufficient condition for the function $\dot{V}$ to be sign-definite is the fulfillment of merely one inequality

$$\Delta = \Delta_m \left[\bar{\rho} + h - \frac{1}{4} \sum_{k=1}^{m} \frac{(1+\beta_k)^2}{r_k} \right] > 0 \quad . \tag{6.87}$$

Consequently, the sufficient conditions for stability reduce to the fulfillment of inequalities (6.3) and (6.87).

If the various roots r_k of (6.2) include one vanishing root or two complex conjugate roots, the simplified stability criterion can be obtained by repeating, word for word, the corresponding analysis given in Section 2 of Chapter V. It is possible, in equal measure, to apply to the systems under investigation here the method for obtaining the simplified criteria, proposed by I. G. Malkin. Let us assume that all r_k are real numbers. Let us consider the sign-definite form, which is everywhere negative,

$$W = -\frac{1}{2} \sum_{\alpha=1}^{m} \sum_{\beta=1}^{m} A_{\alpha\beta} x_\alpha x_\beta \quad . \tag{6.88}$$

The coefficients $A_{\alpha\beta}$ of the form satisfy the Sylvester inequalities

$$\Delta = A_{11} > 0, \quad \Delta_2 = \begin{vmatrix} A_{11} & A_{12} \\ A_{21} & A_{22} \end{vmatrix} > 0, \ \ldots, \ \Delta_m = \begin{vmatrix} A_{11} & \cdots & A_{1m} \\ \cdot & \cdots & \cdot \\ A_{m1} & \cdots & A_{mm} \end{vmatrix} > 0. \tag{6.89}$$

Next, as in Section 4 of Chapter V, it is possible to show that the form

$$F = \frac{1}{2} \sum_{\alpha=1}^{m} \sum_{\beta=1}^{m} B_{\alpha\beta} x_\alpha x_\beta \quad , \tag{6.90}$$

whose coefficients are

$$B_{\alpha\beta} = \frac{A_{\alpha\beta}}{r_\alpha + r_\beta} \qquad (\alpha, \beta = 1, \ldots, m) \quad , \tag{6.91}$$

is positive definite everywhere. Let us now take by way of the V-function the following:

$$V = F + \frac{1}{2} \sigma^2 \ . \tag{6.92}$$

Its derivative, calculated from (6.1), is

(6.93)
$$\dot{V} = W - (\bar{\rho} + h)\sigma^2 + \sigma \sum_{\alpha=1}^{m} \left[\beta_\alpha + \frac{1}{2} \sum_{\beta=1}^{m} (B_{\alpha\beta} + B_{\beta\alpha}) \right] x_\alpha - \sigma\varphi(\sigma) \quad .$$

The discriminat of the form $-[\dot{V} + \sigma\varphi(\sigma)]$ is

(6.94)
$$\Delta = \begin{vmatrix} A_{11} & \cdots & A_{1m} & \bar{P}_1 \\ \cdot & \cdot & \cdot & \cdot \\ A_{m1} & \cdots & A_{mm} & \bar{P}_m \\ \bar{P}_1 & \cdots & \bar{P}_m & (\bar{\rho} + h) \end{vmatrix} ,$$

where

(6.95)
$$2\bar{P}_\alpha = \beta_\alpha + \frac{1}{2} \sum_{\beta=1}^{m} (B_{\alpha\beta} + B_{\beta\alpha}) \qquad (\alpha = 1, \ldots, m) .$$

From this it is seen that by virtue of (6.89), the stability criterion of this system reduces merely to one inequality, $\Delta > 0$.

Naturally, all the above can be repeated in suitable form also as regards complex roots r_k.

Starting with (6.93), it is possible to obtain still another stability criterion in the form of m equations in the coefficients $A_{\alpha\beta}$ and one equality in h:

(6.96)
$$\beta_\alpha + \sum_{\beta=1}^{m} \frac{A_{\alpha\beta}}{r_\alpha + r_\beta} = 0 \qquad (\alpha = 1, \ldots, m) ,$$

(6.97)
$$\bar{\rho} + h > 0 \quad .$$

The system will be stable when a solution of (6.96), satisfying inequalities (6.89), exists. Applying this criterion to the solution of the second Bulgakov problem for $k < 0$, we obtain one inequality, resulting from (6.96) and (6.89):

(6.98)
$$r_1 r_2 (\beta_1 + \beta_2)^2 + \beta_1\beta_2 (r_1 - r_2)^2 > 0 \quad .$$

This is analogous to inequality (5.46). Let us make use of (3.20), (3.21), and (3.26) and represent inequality (6.98) in the form

$$r_1 r_2 \bar{p}_2^2 + (\bar{p}_1 - r_1\bar{p}_2)(-\bar{p}_1 + r_2\bar{p}_2) > 0$$

or, after transformation

$$- \bar{p}_1^2 + (r_1 + r_2)\bar{p}_1\bar{p}_2 > 0 \quad .$$

The latter inequality breaks up into two. The first is the simplest

$$(6.99) \qquad \sqrt{r}\,(E - pG^2)\left[a - qG^2 + \frac{q}{r}\right] = \frac{\sqrt{r}}{\ell}\,(E - pG^2)(k + a\ell) > 0 \quad ,$$

and the second, on the basis of the formulas given above, reduces to

$$(6.100) \qquad \frac{U + \ell E}{\sqrt{\ell(T^2+\ell G^2)}}\left[2 - qG^2 - \frac{(U+\ell E)(E-pG^2)}{T^2 + \ell G^2}\right] - \frac{(k+a\ell)(E-pG^2)}{\sqrt{\ell(T^2+\ell G^2)}} > 0 \ .$$

In addition to these inequalities, we must also bear in mind inequality (6.97), which in this case is the same as (6.30). In the case of a regulator not containing proportional feedback all three inequalities assume the following simple form:

$$(6.101) \qquad E > pG^2, \ ap - qE > \frac{E^2}{G^4}\,(E - pG^2), \ h > \frac{E - pG^2}{G} \quad .$$

5. STABILITY INVESTIGATION BASED ON EQUATIONS (3.57) AND (3.60)

A solution of the first fundamental problem in control theory can be obtained from (3.57) and (3.60), no matter what the initial control system may be. It must be remembered that this solution, which will now be given here, is best used only in two cases:

(1) When the control system is inherently unstable.

(2) When, along with a solution of the stability problem, we are interested also in the problem of transient response.

The use of this method in the second case will be discussed later.

The canonical form of (3.57) and (3.60) has a particularly clear geometrical interpretation. As already noted, the new variables ζ_k and ζ are by definition the direction cosines of the radius vector R of the representative point M. The time variation of these variables does not influence the stability of the obvious solution of the system.

Consequently, the first fundamental problem in the control theory is solved uniquely by investigating only the first equation of the system (3.57) and (3.60).

Since the function $f(R\zeta)$ belongs to class (A) or to subclass (A'), to answer the question of interest to us it is enough to consider only the abbreviated equation

(6.102) $$\dot{R} = -WR ,$$

where W is meant to be the function (3.55) or the function (3.62). The stability of this control system corresponds to any positive definite function W. Let us introduce for brevity the symbols

$$B_{sr} = -\frac{\sqrt{a_{ss}a_{rr}}}{\Delta_m^2}\sum_{\alpha,\beta=1}^{m}\left[\sum_{k=1}^{m}\bar{a}_{k\beta}\bar{b}_{k\alpha}\right]\Delta_{\alpha s}\Delta_{\beta r}$$

(6.103) $$(s, r = 1, \ldots, m) ,$$

$$Q_s = -\frac{\sqrt{a_{ss}}}{2S\Delta_n}\left[S^2\sum_{\alpha=1}^{m}\bar{p}_\alpha\Delta_{\alpha s} + \sum_{\alpha,\beta=1}^{m}a_{\alpha\beta}\bar{n}_\alpha\Delta_{\beta s}\right]$$

[coefficients of the form W (3.55)] and

$$B_{rs} = \frac{\sqrt{a_{rr}a_{ss}}}{\Delta^2}\sum_{\alpha=1}^{m}\sum_{\beta=1}^{m}\frac{r_\alpha + r_\beta}{2}a_{\alpha\beta}\Delta_{r\beta}\Delta_{s\alpha} ,$$

(6.104) $$Q_k = -\frac{\sqrt{a_{kk}}}{2\Delta\sqrt{S}}\sum_{\alpha=1}^{m}\left[S\beta_\alpha + \sum_{\beta=1}^{m}a_{\alpha\beta}\right]\Delta_{k\alpha}$$

[coefficients of the form W (3.62)]. We purposely use here identical symbols for the coefficients of these forms, so as to obtain a single formulation for the solution of the problem, regardless of the form of its initial equations.

Let us write the Sylvester inequalities for the function W. We have

$$\begin{vmatrix} B_{11} & \cdots & B_{1k} \\ B_{21} & \cdots & B_{2k} \\ \cdots & \cdots & \cdots \\ B_{k1} & \cdots & B_{kk} \end{vmatrix} > 0 \qquad (k = 1, \ldots, m) ,$$

(6.105) $$\begin{vmatrix} B_{11} & \cdots & B_{1m} & Q_1 \\ B_{21} & \cdots & B_{2m} & Q_2 \\ \cdots & \cdots & \cdots & \cdots \\ B_{m1} & \cdots & B_{mm} & Q_m \\ Q_1 & \cdots & Q_m & \bar{\rho} + h \end{vmatrix} > 0 .$$

Thus, we have proved the following theorem: If the parameters of the regulator are chosen in accordance with inequalities (6.105), then the control system is absolutely stable, no matter what the real, positive-definite form of F^2 in (3.49) may be.

It must be borne in mind that if the symbols (6.104) are used in inequalities (6.105), it is necessary to consider only the case of real r_α. If, however, among $r_1, \ldots, r_m$ there are also complex conjugate numbers, analysis can be repeated, using the Hermitian metric of the unitary variable space, defining the square of the length of the vector of the representative point M to the origin. However, the example given below will show how to avoid these arguments and to reduce the problem to the case under consideration.

Let us return now to the second Bulgakov problem, defined by (3.27).

We shall use the form (3.61) and the symbols (6.104). Let us assume that r_1 and r_2 are real numbers. For the system (3.27), r_1, r_2, β_1, and β_2 are determined by the following equations

$$r_1 + r_2 = \frac{U + \ell E}{\sqrt{\ell(T^2 + \ell G^2)}}, \quad r_1 r_2 = \frac{k + a\ell}{\ell},$$

$$\beta_1 = \frac{1}{r_2 - r_1}(\bar{p}_1 - r_1\bar{p}_2), \quad \beta_2 = \frac{1}{r_2 - r_1}(-\bar{p}_1 + r_2\bar{p}_2), \tag{6.106}$$

$$\bar{p}_1 = -\frac{(k + a\ell)(E - pG^2)}{\sqrt{\ell(T^2 + \ell G^2)^2}}, \quad \bar{p}_2 = a - qG^2 - \frac{(U + \ell E)(E - pG^2)}{T^2 + \ell G^2}.$$

Inequalities (3.50) now have the following form

$$\Delta_1 = a_{11} > 0, \quad \Delta = \Delta_2 = a_{11}a_{22} - a_{12}^2 > 0. \tag{6.107}$$

Next, we have for (3.63)

$$x_1 = \frac{V}{\Delta}\left[a_{22}\sqrt{a_{11}}\zeta_1 - a_{21}\sqrt{a_{22}}\zeta_2\right],$$

$$x_2 = \frac{V}{\Delta}\left[-a_{12}\sqrt{a_{11}}\zeta_1 + a_{11}\sqrt{a_{22}}\zeta_2\right]. \tag{6.108}$$

From (3.65) we obtain

$$A_{11} = A_{22} = \frac{a_{11}a_{22}}{\Delta}, \quad A_{12} = A_{21} = -\frac{a_{12}\sqrt{a_{11}a_{22}}}{\Delta} \tag{6.109}$$

and consequently, the equation of the surface (3.64) is

$$A_{11}\left(\zeta_1^2 + \zeta_2^2 - \frac{2a_{12}}{\sqrt{a_{11}a_{22}}}\zeta_1\zeta_2\right) + \zeta^2 = 1 \quad . \tag{6.110}$$

It is easy to show that (6.110) is the equation of an ellipsoid. Next, we obtain from (6.104)

$$B_{11} = r_1A_{11}, \quad B_{22} = r_2A_{22}, \quad B_{12} = B_{21} = \frac{r_1 + r_2}{2}A_{12} \quad ,$$

$$Q_1 = -\frac{1}{2\Delta}\sqrt{\frac{a_{11}}{S}}\,[S(\beta_1a_{22} - \beta_2a_{12}) + \Delta] \quad , \tag{6.111}$$

$$Q_2 = -\frac{1}{2\Delta}\sqrt{\frac{a_{22}}{S}}\,[S(\beta_2a_{11} - \beta_1a_{21}) + \Delta] \quad .$$

Using (6.109) and (6.111), we write inequalities (6.105) as

$$B_{11} = \frac{r_1}{\Delta}a_{11}a_{22} > 0 \quad ,$$

$$N = \frac{a_{11}a_{22}}{\Delta^2}\left[r_1r_2a_{11}a_{22} - \left(\frac{r_1 + r_2}{2}\right)^2 a_{12}^2\right] > 0 \quad , . \tag{6.112}$$

$$N(\rho + h) > \frac{a_{11}a_{22}}{\Delta}\left[r_1Q_2^2 + r_2Q_1^2 + (r_1 + r_2)\frac{a_{12}Q_1Q_2}{\sqrt{a_{11}a_{22}}}\right] .$$

Fulfillment of inequalities (6.112) guarantees the stability of the system (3.27). Let us examine these inequalities. Obviously, the first is satisfied everywhere, for any form of F^2 that corresponds to conditions (6.107).

The second inequality imposes a certain limitation on the choice of the coefficients of the form F^2. If $r_1r_2 > 0$, then for real r_1 and r_2 this inequality can be written as

$$\frac{a_{11}a_{22}}{a_{12}^2} > \left(\frac{r_1 + r_2}{2}\right)^2 \frac{1}{r_1r_2} \quad . \tag{6.113}$$

Obviously, this inequality is also not essential.

Finally, we obtain for the third inequality

$$(\bar{\rho} + h)\left[r_1 r_2 a_{11} a_{22} - \left(\frac{r_1 + r_2}{2}\right)^2 a_{12}^2\right] >$$

(6.114)

$$> \Delta\left[r_1 Q_1^2 + r_2 Q_1^2 + \frac{(r_1 + r_2)a_{12}}{\sqrt{a_{11}a_{22}}} Q_1 Q_2\right] .$$

The set B of the parameters p_α of the regulator, determined by inequality (6.114), guarantees the system stability for all disturbances.

In the particular case, when the form F^2 is chosen in accordance with conditions $a_{12} = 0$, inequality (6.114) assumes a simpler form

(6.115) $$r_1 r_2 (\rho + h) > r_1 Q_2^2 + r_2 Q_1^2 .$$

Let us now assume that r_1 and r_2 are complex conjugate numbers, $r_1 = p + qi$ and $r_2 = p - qi$. Let us denote

$$x_1 = u + iv, \quad x_2 = u - iv, \quad \beta_1 = a + ib, \quad \beta_2 = a - ib .$$

(3.27) can then be rewritten as

(6.116)
$$\begin{aligned} \dot{u} &= -pu + qv + \sigma , \\ \dot{v} &= -pv - qu , \\ \dot{\sigma} &= 2au - 2bv - \bar{\rho}\sigma - f(\sigma) . \end{aligned}$$

With respect to the real variables u, v, and σ, we can use, as previously, the real metric V^2, putting

$$V^2 = a_{11}u^2 + 2a_{12}uv + a_{22}v^2 + S\sigma^2$$

and subjecting the real numbers a_{11}, a_{12}, and a_{22} to conditions (6.107). Then, repeating the previous arguments, we find

(6.117)
$$\begin{aligned} B_{11} &= \frac{a_{11}}{\Delta}(pa_{22} + qa_{12}), \quad B_{22} = \frac{a_{22}}{\Delta}(pa_{11} - qa_{12}) , \\ B_{12} &= -\frac{\sqrt{a_{11}a_{22}}}{2\Delta}(2a_{12}p - a_{22}q + a_{11}q) , \\ Q_1 &= -\frac{1}{2\Delta}\sqrt{\frac{a_{11}}{S}}\,[\Delta + 2S(a_{22}a + a_{12}b)] , \\ Q_2 &= \frac{\sqrt{a_{22}S}}{\Delta}(a_{12}a + a_{11}b) . \end{aligned}$$

The stability conditions for the system are written in the form of inequalities

(6.118)
$$pa_{22} + qa_{12} > 0 \ ,$$
$$N = \frac{a_{11}a_{22}}{\Delta^2}(pa_{22} + qa_{12})(pa_{11} - qa_{12}) > \frac{(2a_{12}p+qa_{11}-qa_{12})^2}{4\Delta^2} \ ,$$
$$N(\bar{\rho} + h) > B_{11}Q_2^2 + B_{22}Q_1^2 - 2B_{12}Q_1Q_2 \ .$$

In the particular case, when $a_{12} = 0$ and $a_{11} = a_{22}$, we have only the last inequality

(6.119)
$$p(\bar{\rho} + h) > Q_1^2 + Q_2^2 \ .$$

6. INDIRECT CONTROL IN THE CASE OF NEGATIVE SELF EQUALIZATION

Let us consider once more the stability of an indirectly controlled machine with negative self equalization $(n < 0)$.

To shorten the derivations, we assume that the sensing element of the regulator has no mass, and that $\gamma = 0$. The disturbed motion of the system under consideration is described by the equations

(6.120)
$$\dot{\varphi} = -\frac{n}{T_\alpha}\varphi = \frac{1}{T_\alpha}\xi \ ,$$
$$\dot{\zeta} = -\frac{1}{T_1}\zeta + \beta f(\sigma) \ ,$$
$$\dot{\eta} = -\frac{\delta}{T_k}\eta + \frac{1}{T_k}\varphi, \quad \xi = f(\sigma), \quad \sigma = \eta - \zeta \ .$$

The solution of this problem can be obtained by the Lur'e method developed in Section 3 of Chapter VI. We again turn to the general form of the initial equations (2.160), which will now be written

(6.121)
$$\dot{\eta}_k = \sum_{\alpha=1}^{n} b_{k\alpha}\eta_\alpha + h_k h \sum_{\alpha=1}^{n} p_\alpha\eta_\alpha + h_k\varphi(\sigma) \ ,$$
$$\sigma = \sum_{\alpha=1}^{n} p_\alpha\eta_\alpha \ ,$$

where the function $\varphi(\sigma)$ is defined in (6.13).

We introduce symbols for the coefficients of the linearized system

$$\bar{b}_{k\alpha} = b_{k\alpha} + hh_k p_\alpha \quad , \tag{6.122}$$

which is obtained from (6.121) when $\varphi(\sigma) \equiv 0$.

As in Section 3 of Chapter VI, the basic assumption in the solution of this problem is that the linearized system is stable; we therefore assume that the equation

$$D(\mu) = 0 \tag{6.123}$$

has roots μ_k that satisfy the condition $\operatorname{Re} \mu_k < 0$ $(k = 1, \ldots, n)$. Consequently, the constants of the regulator will be assumed such that (6.123) satisfies the Hurwitz inequalities

$$\Delta_1 > 0, \ldots, \Delta_n > 0 \quad . \tag{6.124}$$

Inequalities (6.124) contain the coefficient h and thereby limit the choice of the subclass of functions $\varphi(\sigma)$ of this problem. The determinant $D(\mu)$ is

$$D(\mu) = \begin{vmatrix} \bar{b}_{11} - \mu & \bar{b}_{12} & \cdots & \bar{b}_{1n} \\ \bar{b}_{21} & \bar{b}_{22} - \mu & \cdots & \bar{b}_{2n} \\ \cdot & \cdot & \cdot & \cdot \\ \bar{b}_{n1} & \bar{b}_{n2} & \cdots & \bar{b}_{nn} - \mu \end{vmatrix} \quad . \tag{6.125}$$

Let us now turn to a study of a control system described by equations

$$\begin{aligned} \dot{\eta}_k &= \sum_{\alpha=1}^{n} \bar{b}_{k\alpha}\eta_\alpha + h_k\varphi(\sigma) \qquad (k = 1, \ldots, n) \quad , \\ \sigma &= \sum_{\alpha=1}^{n} p_\alpha \eta_\alpha \quad . \end{aligned} \tag{6.126}$$

To solve this problem, we would have to reduce (6.126) to the canonical form, and then construct the Lyapunov function. But we can simplify the entire derivation considerably if we note the formal similarity between (2.160) and (6.126). If in the formulas of Section 11, Chapter II we replace all the values $b_{k\alpha}$ by the quantities $\bar{b}_{k\alpha}$, and replace all the numbers $-\rho_k$ by the roots of (6.123), we obtain the following canonical equations for the problem

$$\dot{x}_s = \mu_s x_s + \varphi(\sigma) \qquad (s = 1, \ldots, n) ,$$

(6.127)
$$\sigma = \sum_{k=1}^{n} \gamma_k x_k ,$$

$$\dot{\sigma} = \sum_{k=1}^{n} \beta_k x_k - r\varphi(\sigma) .$$

The variables x_s of the canonical transformation are determined here from the formulas

(6.128)
$$x_s = \frac{1}{H_m(\mu_s)} \sum_{\alpha=1}^{n} D_{\alpha m}(\mu_s)\eta_\alpha \qquad (s = 1, \ldots, n) ,$$

in which $H_m(\mu_s)$ and $D_{\alpha m}(\mu_s)$ are obtained in accordance with the known rules from the determinant (6.125). The inverse transformation is of the form

(6.129)
$$\eta_k = \sum_{\alpha=1}^{n} \frac{H_k(\mu_\alpha)}{D'(\mu_\alpha)} x_\alpha \qquad (k = 1, \ldots, n) .$$

We calculate the constants γ_k, β_k, and r, from the known equations

$$\gamma_k = - \sum_{s=1}^{n} \frac{p_s H_s(\mu_s)}{D'(\mu_k)}, \qquad \beta_k = \mu_k \gamma_k$$

(6.130)

$$r = - \sum_{k=1}^{n} \gamma_k = - \sum_{k=1}^{n} p_k h_k \qquad (k = 1, \ldots, n) .$$

Let us apply these formulas to the problem of (6.120). We put

$$t = T_\alpha \tau, \quad \varphi = \eta_1, \quad \zeta = \eta_2, \quad \eta = \eta_3, \quad \xi = \eta_4, \; b_{11} = -n ,$$

(6.131)
$$b_{14} = -1, \quad b_{22} = -\frac{T_\alpha}{T_i}, \quad b_{23} = \beta h T_\alpha, \quad b_{31} = \frac{T_\alpha}{T_k} ,$$

$$b_{33} = -\frac{\delta T_\alpha}{T_k}, \quad h_2 = \beta T_\alpha, \quad h_4 = T_\alpha$$

and consider the linearized system

$$\begin{aligned}
\dot{\eta}_1 &= \bar{b}_{11}\eta_1 + \bar{b}_{14}\eta_4 \ , \\
\dot{\eta}_2 &= \bar{b}_{22}\eta_2 + hh_2(\eta_3 - \eta_2) \ , \\
\dot{\eta}_3 &= \bar{b}_{31}\eta_1 + \bar{b}_{33}\eta_3 \ , \\
\dot{\eta}_4 &= hh_4(\eta_3 - \eta_2) \ ,
\end{aligned}$$

where

$$\begin{aligned}
&\bar{b}_{11} = b_{11}, \ \bar{b}_{13} = \bar{b}_{12} = 0, \ \bar{b}_{14} = b_{14}, \ \bar{b}_{24} = \bar{b}_{44} = 0 \ , \\
&\bar{b}_{23} = hh_2, \ \ \bar{b}_{22} = b_{22} - hh_2, \ \bar{b}_{41} = 0 \ , \qquad (6.132)\\
&\bar{b}_{42} = -\,hh_4, \ \ \bar{b}_{43} = hh_4 \ .
\end{aligned}$$

Setting up the determinant (6.125), we find

$$D(\mu) = \begin{vmatrix} \bar{b}_{11} - \mu & 0 & 0 & \bar{b}_{14} \\ 0 & \bar{b}_{22} - \mu & \bar{b}_{23} & 0 \\ \bar{b}_{31} & 0 & \bar{b}_{33} - \mu & 0 \\ 0 & \bar{b}_{42} & \bar{b}_{43} & -\mu \end{vmatrix} \ , \qquad (6.133)$$

and, expanding this determinant, we obtain the characteristic equation of the linearized problem

$$\mu^4 - (\bar{b}_{11} + \bar{b}_{22} + \bar{b}_{33})\mu^3 + [\bar{b}_{22}\bar{b}_{33} + \bar{b}_{11}\bar{b}_{22} + \bar{b}_{11}\bar{b}_{33}]\mu^2 - $$
$$ - [\bar{b}_{11}\bar{b}_{22}\bar{b}_{33} + \bar{b}_{14}\bar{b}_{31}\bar{b}_{43}]\mu + \bar{b}_{14}\bar{b}_{31}[\bar{b}_{43}\bar{b}_{22} - \bar{b}_{23}\bar{b}_{42}] = 0 \ .$$

Insertion of (6.131) and (6.132) yields

$$\begin{aligned}
a_0 &= 1 \ , \\
a_1 &= n + \frac{T_\alpha}{T_1} + \beta h T_\alpha + \frac{\delta T_\alpha}{T_k} \ , \\
a_2 &= \frac{\delta T_\alpha}{T_k}\left(\frac{T_\alpha}{T_1} + \beta h T_\alpha\right) + \frac{\delta T_\alpha}{T_k} n + n\left(\frac{T_\alpha}{T_1} + \beta h T_\alpha\right) \ , \qquad (6.134)\\
a_3 &= h\,\frac{T_\alpha^2}{T_1} + \frac{n\delta T_\alpha}{T_k}\left(\frac{T_\alpha}{T_1} + \beta h T_\alpha\right) \ , \\
a_4 &= \frac{hT_\alpha^3}{T_k T_1} \ ,
\end{aligned}$$

where a_i are the coefficients of the characteristic equation.

The Hurwitz inequalities, which guarantee the stability of the linearized system, are of the form

$$a_1 > 0, \quad a_2 > 0, \quad a_3 > 0, \quad a_4 > 0 \ ,$$
(6.135)
$$(a_1 a_2 - a_3) a_3 > a_1^2 a_4 \ .$$

To conclude the transformation of the initial equations into the canonical form, let us calculate the quantities (6.130). Since $p_1 = p_4 = 0$ and $p_3 = - p_2 = 1$, we obtain from (6.133)

$$H_2 = \begin{vmatrix} \bar{b}_{11} - \mu & 0 & 0 & \bar{b}_{14} \\ 0 & h_2 & \bar{b}_{23} & 0 \\ \bar{b}_{31} & 0 & \bar{b}_{33} - \mu & 0 \\ 0 & h_4 & \bar{b}_{43} & -\mu \end{vmatrix} = - \beta T_\alpha \mu (\mu + n) \left(\mu + \frac{\delta T_\alpha}{T_k} \right),$$

$$H_3 = \begin{vmatrix} \bar{b}_{11} - \mu & 0 & 0 & \bar{b}_{14} \\ 0 & \bar{b}_{22} - \mu & h_2 & 0 \\ \bar{b}_{31} & 0 & 0 & 0 \\ 0 & \bar{b}_{42} & h_4 & -\mu \end{vmatrix} = \frac{T_\alpha^2}{T_i} \left(\mu + \frac{T_\alpha}{T_i} \right) .$$

Using formulas (6.130), we calculate β_k and r

(6.136)
$$\beta_1 = - \frac{\mu_1}{D'(\mu_1)} \left[\frac{T_\alpha^2}{T_i} \left(\mu_1 + \frac{T_\alpha}{T_i} \right) - \beta \mu_1 T_\alpha (\mu_1 + n) \left(\mu_1 + \frac{\delta T_\alpha}{T_k} \right) \right] ,$$

$$\beta_2 = - \frac{\mu_2}{D'(\mu_2)} \left[\frac{T_\alpha^2}{T_i} \left(\mu_2 + \frac{T_\alpha}{T_i} \right) - \beta \mu_2 T_\alpha (\mu_2 + n) \left(\mu_2 + \frac{\delta T_\alpha}{T_k} \right) \right] ,$$

$$\beta_3 = - \frac{\mu_3}{D'(\mu_3)} \left[\frac{T_\alpha}{T_i} \left(\mu_3 + \frac{T_\alpha}{T_i} \right) - \beta \mu_3 T_\alpha (\mu_3 + n) \left(\mu_3 + \frac{\delta T_\alpha}{T_k} \right) \right] ,$$

$$\beta_4 = - \frac{\mu_4}{D'(\mu_4)} \left[\frac{T_\alpha}{T_i} \left(\mu_4 + \frac{T_\alpha}{T_i} \right) - \beta \mu_4 T_\alpha (\mu_4 + n) \left(\mu_4 + \frac{\delta T_\alpha}{T_k} \right) \right] ,$$

$$r = - p_2 h_2 - p_3 h_3 = \beta T_\alpha$$

and obtain finally the canonical equations

$$(6.137)\qquad \begin{aligned} \dot{x}_s &= \mu_s x_s + \varphi(\sigma) \qquad (s = 1, \ldots, 4)\ , \\ \sigma &= \sum_{k=1}^{4} \gamma_k x_k\ , \\ \dot{\sigma} &= \sum_{k=1}^{4} \beta_k x_k - \beta T_\alpha \varphi(\sigma)\ . \end{aligned}$$

These equations are investigated in the next section.

7. ANOTHER METHOD OF FORMULATING SIMPLIFIED STABILITY CRITERIA

In the investigation of the stability of control systems, considered in the preceding section [and describable by (6.127)], we could have employed the arguments given in Section 3 of this chapter. Noting the formal similarity between (6.127) and (6.53), we can assume that all the theorems on stability proved in that section remain valid also for the control systems of Section 6. But, as we have already verified, the application of these theorems to any particular problem involves great computational difficulties. Therefore, we outline in this section a method for formulating simplified stability criteria, based on the analysis of equations that contain only two unknowns.

Thus, let us consider (6.127) and assume that the simple roots μ_k include real roots $\mu_1, \ldots, \mu_s$ and complex conjugate pairs $\mu_{s+1}, \ldots, \mu_n$.

We assume that $n = 2q$ is an even number, and take the V-function to be the positive definite function

$$V = -\sum_{i=1}^{n-1} \left(\frac{a_i^2 x_i^2}{2\mu_i} + \frac{a_{i+1}^2 x_{i+1}^2}{2\mu_{i+1}} + \frac{2a_i a_{i+1} x_i x_{i+1}}{\mu_i + \mu_{i+1}} \right) + \int_0^\sigma \varphi(\sigma)d\sigma\ ,$$

where i takes on only odd values $1, 3, \ldots, n-1$.

The total derivative of the function V, calculated with (6.127) in analogy with the method used in Section 7, Chapter IX, is

$$\frac{dV}{dt} = -\sum_{i=1}^{n-1} (a_i x_i + a_{i+1} x_{i+1})^2 - r\varphi^2(\sigma)\ ,$$

or

$$\frac{dV}{dt} = - \sum_{i=1}^{n-1} [a_i x_i + a_{i+1} x_{i+1} + \sqrt{r}\varphi(\sigma)]^2 .$$

Here the constants a_k are determined from

$$(6.138)\qquad \begin{aligned} &\beta_i - \frac{a_i^2}{\mu_i} - \frac{2a_i a_{i+1}}{\mu_i + \mu_{i+1}} = 0 , \\ &\beta_{i+1} - \frac{a_{i+1}^2}{\mu_{i+1}} - \frac{2a_i a_{i+1}}{\mu_i + \mu_{i+1}} = 0 \qquad (i = 1, 3, \ldots, n-1) , \end{aligned}$$

or from

$$(6.139)\qquad \begin{aligned} &\beta_i + 2\sqrt{r}a_i - \frac{a_i^2}{\mu_i} - \frac{2a_i a_{i+1}}{\mu_i + \mu_{i+1}} = 0 , \\ &\beta_{i+1} + 2\sqrt{r}a_{i+1} - \frac{a_{i+1}^2}{\mu_{i+1}} - \frac{2a_i a_{i+1}}{\mu_i + \mu_{i+1}} = 0 , \\ &(i = 1, 3, \ldots, n-1) . \end{aligned}$$

Analogously, taking the V-function to be

$$V = - \sum_{i=1}^{n-1} \left(\frac{a_i^2 x_i^2}{2\mu_i} + \frac{a_{i+1}^2 x_{i+1}^2}{2\mu_{i+1}} + \frac{2a_i a_{i+1} x_i x_{i+1}}{\mu_i + \mu_{i+1}} \right)$$

and making the a_i satisfy the relations

$$(6.140)\qquad \begin{aligned} &\gamma_i - \frac{a_i^2}{\mu_i} - \frac{2a_i a_{i+1}}{\mu_i + \mu_{i+1}} = 0 , \\ &\gamma_{i+1} - \frac{a_{i+1}^2}{\mu_{i+1}} - \frac{2a_i a_{i+1}}{\mu_i + \mu_{i+1}} = 0 \qquad (i = 1, 3, \ldots, n-1) , \end{aligned}$$

we obtain an expression for the total derivative

$$\frac{dV}{dt} = - \sum_{i=1}^{n-1} (a_i x_i + a_{i+1} x_{i+1})^2 - \sigma\varphi(\sigma) .$$

In all three cases the derivative of the V-function is negative definite. Consequently, relations (6.138), as well as relations (6.139) or (6.140), are sufficient conditions for the stability of the system.

We have thus proved the following theorem: The system (6.121) has absolute asymptotic stability if the regulator constants are such that the q pairs of systems of quadratic equations (6.138), [(6.139), (6.140)], {in which $\mu_1, \ldots, \mu_s$ and $\beta_1, \ldots, \beta_s$ [$\gamma_1, \ldots, \gamma_s$] are real, and $\mu_{s+1}, \ldots, \mu_n$ and $\beta_{s+1}, \ldots, \beta_n$ [$\gamma_{s+1}, \ldots, \gamma_n$] are complex conjugate} have at least one solution containing real roots $a_1, \ldots, a_s$ and complex conjugate roots $a_{s+1}, \ldots, a_n$, and if, in addition, inequalities (6.124) are satisfied.

With the aid of this theorem it is also possible to solve the problem of indirect control, considered in the preceding section. For this purpose it is enough to form the inequalities

$$\Gamma_1^2 > 0, \quad D_1^2 > 0 \quad ,$$

compiled with the aid of the quantities β_1, and β_2, and the inequalities

$$\Gamma_2^2 > 0, \quad D_2^2 > 0 \quad ,$$

compiled with the aid of the quantities β_3 and β_4, where β_1, β_2, β_3 and β_4 are determined by (6.136).

Naturally, the problem can be solved with these inequalities only if they are compatible with inequalities (6.135).

CHAPTER VII: PROGRAMMED CONTROL

1. STATEMENT OF THE STABILITY PROBLEM

So far we have considered the operation of an automatic control system, when it was necessary to maintain one specified steady state mode of the regulated object. However, the actual tasks that must be performed by regulators are usually more varied and more extensive. For example, in some cases it is necessary to assign to the regulated object, during the control process, another steady state or an unsteady mode of operation, different from the initial one. In the latter case, this mode changes with time in accordance with a specified law, and is called programmed-control mode.

In this connection, it is always important in practice to specify that the regulator which maintains the initial steady state mode in a system must also maintain any other new steady state mode (or the programmed-control mode) which it is desired to specify for the system.

This chapter is devoted to an investigation of this problem and to a proof of one theorem, which is qualitative in character. Let there be a regulated object, whose state is described by differential equations of the form

$$\dot{\zeta}_k = \psi_k(\zeta_1, \ldots, \zeta_m, \zeta) \qquad (k = 1, \ldots, m) \quad , \tag{7-1}$$

where ζ_k are the coordinates of the object, ζ the coordinate of the regulator, ψ_k the specified functions of the arguments that enter into these coordinates, functions analytic in a certain region N. Since we wish to maintain in the regulated object a certain steady state, characterized by the coordinates $\zeta_1^*, \ldots, \zeta_m^*$, we change the position of the regulator to another position, specified by the quantity ζ^*. These coordinates satisfy the equations $\psi_k(\zeta_1^*, \ldots, \zeta_m^*, \zeta^*) = 0$ $(k = 1, \ldots, m)$. The purpose of connecting the regulator to the regulated object is to maintain the steady state of the object by maintaining the regulator in exactly this position ζ^*. Let us assume for simplicity that the regulator is describable by equations of the form

(7.2)
$$\frac{d}{dt}(\zeta - \zeta^*) = f(\sigma) \ ,$$
$$\sigma = \sum_{\alpha=1}^{m} p_\alpha(\zeta_\alpha - \zeta_\alpha^*) - (\zeta - \zeta^*) \ ,$$

where $f(\sigma)$ is a function of class (A), and p_α are constant parameters of the regulator.

The first fundamental problem in the theory of automatic control consists of a choice of the regulator parameters such that stability of the steady state of the control system (7.1) and (7.2) is guaranteed, this steady state being characterized by the obvious solution

(7.3)
$$\zeta_1 = \zeta_1^*, \ \ldots, \ \zeta_m = \zeta_m^*, \ \zeta = \zeta^* \ .$$

The solution of this problem, as is known, reduces to an investigation of equations of disturbed motion

(7.4)
$$\dot\eta_k = \sum_{\alpha=1}^{m} b_{k\alpha}\eta_\alpha + n_k\xi + \Xi_k(\eta_1, \ \ldots, \ \eta_m, \ \xi) \qquad (k = 1, \ \ldots, \ m)$$
$$\dot\xi = f(\sigma), \ \sigma = \sum_{\alpha=1}^{m} p_\alpha\eta_\alpha - \xi \ ,$$

where

(7.5)
$$\eta_k = \zeta_k - \zeta_k^*, \qquad \xi = \zeta - \zeta^* \ ,$$
$$b_{k\alpha} = \left(\frac{\partial\psi_k}{\partial\zeta_\alpha}\right)^*, \quad n_k = \left(\frac{\partial\psi_k}{\partial\zeta}\right)^*, \qquad (\alpha, \ k = 1, \ \ldots, \ m) \ .$$

The asterisk denotes here that the partial derivatives are taken for values of the arguments, defined by (7.3). The functions Ξ_k denote here the nonlinear terms, formed as a result of a series expansion of ψ_k in the vicinity of the point (7.3).

Proceeding to the solution of the problem of the stability of such a system, we usually discard the term Ξ_k in (7.4), and consider only the first-approximation equations*

* For the validity of such action see Chapter X.

$$\dot{\eta}_k = \sum_{\alpha=1}^{m} b_{k\alpha}\eta_\alpha + n_k\xi \qquad (k = 1, \ldots, m) \quad ,$$

$$\dot{\xi} = f(\sigma), \qquad \sigma = \sum_{\alpha=1}^{m} p_\alpha\eta_\alpha - \xi \quad .$$

Henceforth we shall assume that this problem has already been solved, and that the stability criterion of the solution (7.3) has been formulated in some form or another. For the sake of being definite, we shall assume that the regulator design is based on the same stability criterion as given in Section 1 of Chapter VI.

Let us assume that during the operation of the regulated object it became necessary to specify a new steady state, differing from (7.3), or an unsteady state, which must change with time in accordance with a specified law. In the general case, this state is described by known continuous functions of time

$$\zeta_1 = \zeta_1^{**}(t), \ \ldots, \ \zeta_m = \zeta_m^{**}(t), \ \zeta = \zeta^{**}(t) \quad . \tag{7.6}$$

These functions, $m + 1$ in number, should serve as some particular solution of the initial equations (7.1):

$$\dot{\zeta}_k^{**} \equiv \psi_k(\zeta_1^{**}, \ \ldots, \ \zeta_m^{**}, \ \zeta^{**}), \qquad (k = 1, \ldots, m) \quad .$$

In order to actually realize this specified change in state, we introduce into the regulator an element of programmed control,* without changing either the structure or the constants of the regulator. In this case the equations of the regulator are written in the following manner:

$$\begin{aligned} &\frac{d}{dt}(\zeta - \zeta^*) = f(\sigma) \quad , \\ &\sigma = \sum_{\alpha=1}^{m} p_\alpha(\zeta_\alpha - \zeta_\alpha^*) - (\zeta - \zeta^*) - \sigma_{**}(t) \quad . \end{aligned} \tag{7.7}$$

They differ from (7.2) only in the presence of one new term $\sigma_{**}(t)$. Obviously, the function $\sigma_{**}(t)$ should reflect the structure of the programming element of the regulator, and should satisfy, in accordance with the choice of solution (7.6), the following equation

* V. L. Lossievskiĭ, Avtomaticheskie regulyatory (Automatic Regulators), Oborongiz, 1944.

$$\frac{d}{dt}(\zeta^{**} - \zeta^{*}) = f(\sigma^{**}) \quad ,$$

where

$$\sigma^{**} = \sum_{\alpha=1}^{m} p_\alpha(\zeta_\alpha^{**} - \zeta_\alpha^{*}) - (\zeta^{**} - \zeta^{*}) - \sigma_{**}(t) \quad .$$

In this particular case, when $\sigma_{**} \equiv 0$, solution (7.6) becomes solution (7.3). If σ_{**} is a non-vanishing constant, solution (7.6) represents a trivial solution of the system (7.1) and (7.7), differing from solution (7.3). It will describe a new steady state of the system.

In the general case, when $\sigma_{**}(t)$ is a continuous bounded function, the solution (7.6) describes that desired program variation in the state of the regulated object which we wish to obtain. In connection with the above, we pose the following problem. Let us assume that we have solved the problem of the stability of the steady state (7.3), and have set the regulator in a position at which the realizability of this state is guaranteed. The question is, will this same setting of the regulator lead to the stability of any programmed change in the state (7.6) of the regulated object, a change determined by the specified continuous and bounded function $\sigma_{**}(t)$? If, however, the solution (7.6) is not stable, what are the new methods of setting the regulator in such a way as to guarantee stability of solution (7.6)?

We shall prove below one theorem in the theory of programmed automatic control, which gives an answer to this problem for one particular case.

2. THEOREM ON PROGRAMMED CONTROL

Thus, we deal with the stability of the solution (7.6) of (7.1) and (7.7), in which $\sigma_{**}(t)$ is any specified continuous and bounded function.

Putting

$$\text{(7.8)} \qquad \begin{gathered} \eta_1 = \zeta_1 - \zeta_1^{**}, \ \ldots, \ \eta_m = \zeta_m - \zeta_m^{**}, \ \xi = \zeta - \zeta^{**} , \\ \sigma - \sigma^{**} = \mu \quad , \end{gathered}$$

we formulate, in the usual manner, the equations of the disturbed motion

$$\dot{\zeta}_k^{**} + \dot{\eta}_k = \psi_k^{**} + \sum_{k=1}^{m} \left(\frac{\partial\psi_k}{\partial\zeta_\alpha}\right)^{**} \eta_\alpha + \left(\frac{\partial\psi_k}{\partial\zeta}\right)^{**} \xi + \Xi_k^{**}(\eta_1, \ldots, \eta_n, \xi, t),$$

$$\dot{\zeta}^{**} + \dot{\xi} = f(\sigma^{**}) + \left(\frac{\partial f}{\partial\sigma}\right)^{**} \mu + \ldots \ (k = 1, \ldots, m) \quad .$$

Performing the obvious cancellations and discarding the nonlinear terms, we obtain as the first approximation

$$(7.9)\qquad \begin{aligned} \dot{\eta}_k &= \sum_{\alpha=1}^{m} b_{k\alpha}^{**}\eta_\alpha + n_k^{**}\xi \qquad (k = 1, \ldots, m) \,, \\ \dot{\xi} &= S\mu, \quad \mu = \sum_{\alpha=1}^{m} p_\alpha\eta_\alpha - \xi \,. \end{aligned}$$

Here

$$(7.10)\qquad \begin{aligned} S &= \left(\frac{\partial f}{\partial \sigma}\right)^{**}, \quad b_{k\alpha}^{**} = \left(\frac{\partial \psi_k}{\partial \zeta_\alpha}\right)^{**} \,, \\ n_k^{**} &= \left(\frac{\partial \psi_k}{\partial \zeta}\right)^{**} \qquad (k, \alpha = 1, \ldots, m) \end{aligned}$$

are known functions of time.

As follows from experiment, for all functions $f(\sigma)$ of class (A) encountered in the practice of automatic control, the quantity S will assume only positive values, and in extreme cases zero values. As to the functions $b_{k\alpha}^{**}$ and n_k^{**}, their behavior is determined by the characteristics of the regulated object and by the form of the specified program.

In this book we shall study that particular case of programmed control in which the coefficients $b_{k\alpha}^{**}$ and n_k^{**}, can be treated as certain functions whose variation is sufficiently small for all values $t > 0$ compared with the values of $b_{k\alpha}$ and n_k in (7.5).

Thus, we assume

$$(7.11)\qquad \begin{aligned} b_{k\alpha}^{**} &= b_{k\alpha} + \varphi_{k\alpha}(t), \\ n_k^{**} &= n_k + m_k(t) \qquad (\alpha, k = 1, \ldots, m) \,, \end{aligned}$$

where $\varphi_{k\alpha}$ and m_k are known functions of time, bounded in modulus by a sufficiently small positive value ε for all $t > 0$. Such a case of programmed control will undoubtedly occur by virtue of the analyticity of the function ψ_k in (7.1) if, in addition, $|\sigma_{**}(t)| < \varepsilon$.

Control systems for which the process of programmed control is described by such equations will be denoted by the letter π. In this case, studying the stability of the obvious solution

$$\eta_1^* = 0, \ \ldots, \ \eta_m^* = 0, \quad \xi^* = 0 \tag{7.12}$$

of (7.9), we can employ the method used to investigate the system (7.4) in Chapter VI. Actually, let us introduce a new variable μ [(7.9)], which will be used to eliminate ξ. This will permit rewriting (7.9) as

$$\dot{\eta}_k = \sum_{\alpha=1}^{m} \bar{b}_{k\alpha}\eta_\alpha - n_k\mu + \sum_{\alpha=1}^{m} \varphi^{o}_{k\alpha}(t)\eta_\alpha - m_k\mu \ ,$$

$$\dot{\mu} = \sum_{\alpha=1}^{m} \bar{p}_\alpha\eta_\alpha - (\bar{\rho} + S)\mu + \tag{7.13}$$

$$+ \sum_{\alpha=1}^{m}\left[\sum_{\beta=1}^{m} p_\beta\varphi^{o}_{\beta\alpha}\right]\eta_\alpha - \left(\sum_{\alpha=1}^{m} p_\alpha m_\alpha\right)\mu \ .$$

Here

$$\bar{b}_{k\alpha} = b_{k\alpha} + n_k p_\alpha, \quad \bar{p}_\alpha = \sum_{\beta=1}^{m} p_\beta \bar{b}_{\beta\alpha}, \quad \bar{\rho} = \sum_{\alpha=1}^{m} p_\alpha n_\alpha \ ,$$

(7.14)

$$\varphi^{o}_{k\alpha} = \varphi_{k\alpha} + m_k p_\alpha \qquad (\alpha, \ \beta, \ k = 1, \ \ldots, \ m) \ .$$

Let us introduce a non-singular linear transformation

$$x_s = \sum_{\alpha=1}^{m} C_\alpha^{(s)}\eta_\alpha \qquad (s = 1, \ \ldots, \ m)$$

and choose its coefficients $C_\alpha^{(s)}$ to satisfy

$$- r_s C_\beta^{(s)} = \sum_{\alpha=1}^{m} C_\alpha^{(s)}\bar{b}_{\alpha\beta}, \ - 1 = \sum_{\alpha=1}^{m} C_\alpha^{(s)} n_\alpha \qquad (\alpha, \ \beta = 1, \ \ldots, \ m) \ . \tag{7.15}$$

Let us assume that this is possible; then the inverse transformation becomes

$$\eta_\beta = \sum_{k=1}^{m} D_k^{(\beta)} x_k \qquad (\beta = 1, \ \ldots, \ m) \ . \tag{7.16}$$

In this case, as can be readily verified, the transformed equations can be written

(7.17)
$$\dot{x}_s = -r_s x_s + \mu + \sum_{k=1}^{m} \Phi_k^{(s)}(t) x_k \qquad (s = 1, \ldots, m),$$
$$\dot{\mu} = \sum_{\alpha=1}^{m} \bar{\beta}_\alpha x_\alpha - (\bar{\rho} + S)\mu + \sum_{\alpha=1}^{m} Q_\alpha(t) x_\alpha - R_\mu ,$$

where

$$\Phi_k^{(s)}(t) = \sum_{\alpha=1}^{m} \sum_{\beta=1}^{m} C_\alpha^{(s)} \varphi_{\alpha\beta}^{o}(t) D_k^{(\beta)}, \qquad R(t) = \sum_{\alpha=1}^{m} p_\alpha m_\alpha(t) ,$$

$$Q_k^{(t)} = \sum_{\alpha=1}^{m} \sum_{\beta=1}^{m} \bar{p}_\beta D_k^{(\alpha)} \varphi_{\alpha\beta}^{o}(t), \qquad \beta_\alpha = \sum_{k=1}^{m} \bar{p}_\beta D_\alpha^{(k)} \quad (k, s = 1, \ldots, m).$$

By virtue of the limitations imposed on the quantities $\varphi_{k\alpha}$ and m_α, we have a system of inequalities

(7.18)
$$|\Phi_k^{(s)}(t)| < \varepsilon \sum_{\alpha=1}^{m} \sum_{\beta=1}^{m} |C_\alpha^{(s)} D_k^{(\beta)}| \; |1 + p_\alpha| ,$$
$$|Q_k(t)| < \varepsilon \sum_{\alpha=1}^{m} \sum_{\beta=1}^{m} |\bar{p}_\beta D_k^{(\alpha)}| \; |1 + p_\alpha| ,$$
$$|R(t)| < \varepsilon \sum_{\alpha=1}^{m} |p_\alpha| .$$

The reduction of the initial equations to the form (7.17) is always possible, if the numbers r_s are chosen to be the simple roots of the equation

(7.19)
$$\Delta(r) = |\bar{b}_{\alpha\beta} + r\delta_{\alpha\beta}| = 0 .$$

We shall assume that (7.19) has real roots $r_1, \ldots, r_s$ and complex conjugate roots $r_{s+1}, \ldots, r_m$, having the property $\operatorname{Re} r_k > 0$ $(k = 1, \ldots, m)$. The latter can be obtained by subjecting the choice of the regulator parameters to the Hurwitz inequalities

(7.20)
$$\Delta_1 > 0, \ldots, \Delta_m > 0 ,$$

written for the equation

$$\Delta(-r) = 0 \quad .$$

It can be shown that if ε is sufficiently small, the inequalities (7.20) are the necessary and sufficient conditions for the stability of a system (7.4) having an ideal regulator. Obviously, these must be considered the necessary stability conditions for this problem.

Let us consider the positive definite form

$$V = \sum_{k=1}^{m} \sum_{i=1}^{m} \frac{a_k a_i x_k x_i}{r_k + r_i} + \frac{1}{2} \chi^2 \mu^2 + \Phi \quad ,$$

in which $\chi^2 > 0$, $a_1, \ldots, a_s$ are real, and $a_{s+1}, \ldots, a_m$ are complex-conjugate pairs. Its derivative, calculated in accordance with (7.17), is

$$\dot{V} = -\left(\sum_{k=1}^{m} a_k x_k \right)^2 + 2\mu \sum_{k=1}^{m} \sum_{i=1}^{m} \frac{a_k a_i x_k}{r_k + r_i} +$$

$$+ \chi^2 \mu \sum_{k=1}^{m} \bar{\beta}_k x_k - \chi^2 (\bar{\rho} + S) \mu^2 -$$

$$- \sum_{k=1}^{m} r_k A_k x_k^2 - C_1 (r_{s+1} + r_{s+2}) x_{s+1} x_{s+2} - \cdots$$

$$+ W_1 (x_1, \ldots, x_m, \mu) \ .$$

Adding to the right half of this equation the expression

$$\mu^2 + 2\mu \sum_{k=1}^{m} a_k x_k$$

and then subtracting it, we obtain

$$\dot{V} = -\left[\sum_{k=1}^{m} a_k x_k + \mu \right]^2 - [\chi^2 (\bar{\rho} + S) - 1] \mu^2 - \sum_{k=1}^{s} r_k A_k x_k^2 - C_1 (r_{s+1} + r_{s+2}) x_{s+1} x_{s+2} -$$

$$- \cdots + \sum_{k=1}^{s} \left[A_k + \chi^2 \bar{\beta}_k + 2a_k + 2a_k \sum_{i=1}^{m} \frac{a_i}{r_k + r_i} \right] x_k + \sum_{\alpha=1}^{n-s} \left[C_\alpha + \chi^2 \bar{\beta}_{s+\alpha} + \right.$$

$$\left. 2a_{s+\alpha} + 2a_{s+\alpha} \sum_{i=1}^{m} \frac{a_i}{r_{s+\alpha} + r_i} \right] x_{s+\alpha} - W_1 (x_1, \ldots, x_m) \quad .$$

Subjecting the choice of regulator parameters to the relations

$$A_k + x^2\bar{\beta}_k + 2a_k + 2a_k \sum_{i=1}^{m} \frac{a_i}{r_k + r_i} = 0 \qquad (k = 1, \ldots, s),$$

$$(7.21)\quad C_\alpha + x^2\bar{\beta}_{s+\alpha} + 2a_{s+\alpha} + 2a_{s+\alpha} \sum_{i=1}^{m} \frac{a_i}{r_{s+\alpha} + r_i} = 0$$

$$(\alpha = 1, \ldots, m - s),$$

$$(7.22)\qquad x^2(\bar{\rho} + S) - 1 > 0,$$

we have

$$\dot{V} = -\left[\sum_{k=1}^{m} a_k x_k + \mu\right]^2 - [x^2(\bar{\rho} + S) - 1]\mu^2 - \sum_{k=1}^{s} r_k A_k x_k^2 -$$

$$(7.23)$$

$$- C_1(r_{s+1} + r_{s+2})x_{s+1}x_{s+2} - \ldots + W_1(x_1, \ldots, x_m, \mu).$$

The function W_1 represents a quadratic form in the variables $x_1, \ldots, x_m, \mu$, and if the process of program regulation is such that the number ε is sufficiently small, the quantity W_1 does not affect the sign of $\dot{V}$. The latter follows from the fact that the function

$$W = \left[\sum_{k=1}^{m} a_k x_k + \mu\right]^2 + [x^2(\bar{\rho} + S) - 1]\mu^2 +$$

$$+ \sum_{k=1}^{s} r_k A_k x_k^2 + C_1(r_{s+1} + r_{s+2})x_{s+1}x_{s+2} + \ldots$$

is positive definite. The upper value ε, for which the above statement retains its force, can always be determined by studying the Sylvester conditions formulated for the function $\dot{V}$. If this condition is satisfied, and consequently the particular solution (7.6) of (7.1) is also satisfied, (7.7) will be stable for all disturbances in the region N and for any function $f(\sigma)$ of class (A). Turning to Section 1 of Chapter VI, we see that conditions (7.21) and (7.22) are the stability conditions of solution (7.3) when $\sigma_{**} = 0$.

In some cases it turns out that if S is sufficiently small, condition (7.22) will contradict conditions (7.20) and (7.21). This situation can be corrected in such cases by imposing additional requirements on the function $f(\sigma)$. Let S_m denote the minimum value of S that is greater than zero for the entire range of variation of σ^{**} in this problem. Then we obtain, instead of inequality (7.22), the following inequality:

$$\kappa^2(\bar{\rho} + S_m) - 1 > 0 \quad . \tag{7.24}$$

In this case it is necessary to choose for this problem from among all functions of class (A), that subclass (A") of the functions $f(\sigma)$ which satisfies (7.24). Let us consider the subclass (A') of the functions $f(\sigma)$, considered in Section 1 of Chapter VI, defined by the inequality

$$\kappa^2(\bar{\rho} + h) - 1 > 0 \quad , \tag{7.25}$$

where

$$h = \left[\frac{df}{d\sigma} \right]_{\sigma=0} \quad .$$

Comparing both subclasses of the functions $f(\sigma)$, we can say that if $h < S_m$ and inequality (7.25) is satisfied, then (7.24) is also satisfied, and the stability conditions of the solutions (7.3) and (7.6) are identical; on the other hand, if $h > S_m$, then (7.24) may not be satisfied although inequality (7.25) is, particularly at the same values of σ for which $S < h$.

We have thus proved the following theorem: Let a regulator be set to maintain a steady state process, determined by the obvious solution (7.3), in a control system of class (π). Then, for the same setting, it will maintain any other unsteady control mode that is sufficiently close to the first mode $(|\sigma_{**}(t)| < \varepsilon)$, in accordance with a program that is determined by a specified continuous and bounded function $\sigma_{**}(t)$ — no matter what the disturbances are in the region N, and no matter what the function $f(\sigma)$ of class (A) or subclass (A") may be.

Let us make one remark. Throughout this chapter we have emphasized that the general statement of the problem of programmed control is considered only for those cases when $|\sigma_{**}(t)| < \varepsilon$. Naturally, in the general case one can always find such a function $\sigma_{**}(t)$ for which the theorem proved becomes invalid. But, on the other hand, one can also find cases of programmed control, different from those considered above, in which the formulated theorem retains its force. These include, for example, the case characterized by a bounded function $\sigma_{**}(t)$, which satisfies the following equation:

$$\lim_{t \to \infty} \sigma_{**}(t) = \text{const.}$$

Under this assumption, analogous limiting equations will be obtained also for the coefficients $b_{k\alpha}^{**}$ and n_k^{**} of the system of first-approximation equations.

In this case the stability of the system, calculated from the limiting values of the above coefficients, will also guarantee the stability of the specified programmed control process.

B. S. Razumikhin has pointed out that the theorem remains valid also for the case of slowly-varying functions $\sigma_{**}(t)$, $b_{k\alpha}^{**}$ and n_k^{**}.

The case of programmed control just described can be realized by applying the controlling pulse signal $\sigma_{**}(t)$ directly to the actuator or else to the amplifier.

Another case, however, is also possible in programmed control, when the specified program is realized with the aid of special programmed transducers, installed in the sensing elements of the regulator. In this case, the regulator equation is written as follows:

$$\frac{d}{dt}(\zeta - \zeta^* - \bar{\zeta}) = f(\sigma) ,$$

$$\sigma = \sum_{\alpha=1}^{m} p_\alpha(\zeta_\alpha - \zeta_\alpha^* - \bar{\zeta}_\alpha) - (\zeta - \zeta^* - \bar{\zeta}) .$$

The functions $\bar{\zeta}_\alpha$, $\bar{\zeta}$ are chosen from the condition that the following equations must be satisfied

$$\zeta_\alpha^* + \bar{\zeta}_\alpha = \zeta_\alpha^{**}, \quad \zeta^* + \bar{\zeta} = \zeta^{**} \qquad (\alpha = 1, \ldots, m) .$$

Retaining the symbols (7.8) for the coordinates of the disturbed motion, we obtain the following equations

$$(7.26) \qquad \dot{\eta}_k = \sum_{\alpha=1}^{m} b_{k\alpha}^{**}\eta_\alpha + \dot{n}_k^{**}\xi, \quad \dot{\xi} = f(\sigma), \quad \sigma = \sum_{\alpha=1}^{m} p_\alpha\eta_\alpha - \xi .$$

In their form, (7.26) are the same as the equations of disturbed motion of a system in the case when $\bar{\zeta}_\alpha = \zeta_\alpha^{**} \equiv 0$; essentially, they differ only in the fact that the coefficients $b_{k\alpha}^{**}$ and n_k^{**} are now determined by the form of the program motion and depend explicitly on time.

With respect to (7.26), it is possible to repeat the above

argument and to prove the program-control theorem. We shall not dwell on this.

We shall add still one more remark to the above. In some particular cases of the program-control problem, it becomes meaningful to speak of system stability over a finite time interval.

The statement of the problem of the stability of a system during a finite time interval and one of the methods of solution are given in Chapter XI.

CHAPTER VIII: THE PROBLEM OF CONTROL QUALITY

1. SECOND FUNDAMENTAL PROBLEM OF THE THEORY OF AUTOMATIC CONTROL

In most investigations of control systems, we cannot restrict ourselves merely to an examination of the problem of steady state stability. Very frequently it is necessary to know certain quantities that determine the character of the control process, i.e., the character of the convergence of the disturbed motion to an undisturbed one. These characteristics are customarily called the characteristics of the control quality and the problem of their determination is called the problem of control quality.

Various control systems must satisfy different requirements with respect to the character of the control process.

The first and simplest statement of the problem of control quality belongs to the founder of the theory of automatic control, I. A. Vyshnegradskiĭ (for the case of linear systems). Vyshnegradskiĭ was interested in the conditions that must be satisfied by the regulator parameters in order that the so-called control transient have an aperiodic or oscillatory character. He obtained these conditions for one particular problem.

In Vyshnegradskiĭ's formulation, these conditions reduce to a definite limitation on the roots of the characteristic equation: if all the roots of this equation are real, the transient is aperiodic; if some are complex, the transient becomes oscillatory.

The further development of the problem of control quality followed the path of generalization and refinement of the problem stated by Vyshnegradskiĭ. At first this generalization consisted of constructing the regions of aperiodic or oscillatory stability (instability) of a linear system of order not higher than the fourth, and also the determination of means of damping the disturbed motion and measures for static stability of the system. The set of these problems constitutes the scope of the problem of so-called degree of stability. One definition of the degree of stability of a linear system was given by Ya. Z. Tsypkin and P. V. Bromberg.* By degree of stability of a linear system they understand the quantity

$$\xi = \min |\operatorname{Re} \lambda_k| \qquad (k = 1, \ldots, h) ,$$

where λ_k are the roots of the characteristic equation of the system.

Knowing the degree of stability makes it possible sometimes to calculate the upper value of the time t^* of the conditional damping of the control process. Then the quality of the process of regulation is considered to be the better, the less the value of t^*.

At the present time many investigators sometimes consider it also necessary to calculate another quantity characteristic of control quality ℓ^*, the greatest deviation of the regulated parameter from its steady state.

Many investigators have determined these two characteristics of the quality of regulation (t^* and ℓ^*) for linear systems. The best results have been obtained by Soviet authors, including, in addition to those mentioned above, N. G. Chetaev, A. A. Krasovskiǐ, A. A. Fel'dbaum, and V. V. Solodovnikov.**

Recently a third point of view concerning the problem of quality of regulation appeared consisting in the following approach: One considers linear systems under the influence of disturbing forces, with only a single limitation on modulus, and one calculates the maximum possible deviations of these systems from their steady state. It is assumed that better control is obtained in a system in which the deviations reach the lower values. Such a trend in the solution of the problem of quality was developed principally by B. V. Bulgakov, to whom the basic results are due in this case.

Note should also be made of the work by N. D. Moiseev, devoted to so-called technical stability.

In this book the problem of quality will be considered as the

---

* Ya. Z. Tsypkin and P. V. Bromberg. Degree of Stability of Linear Systems. Trudy NISO, No. 9, Published by BNT Min. of the Aircraft Ind., 1946; P. V. Bromberg, On the Quality of Linear Systems of Control, Trudy, No 48, Oborongiz, 1952.

** N. G. Chetaev, The Duration of the Transient Process in a Linear System, PMM, Vol. XV, No. 3, 1951; A. A. Krasovskiǐ, On the Degree of Stability of Linear Systems, Trudy, VVIA im. N. E. Zhukobskiǐ, No. 281, 1948. A. A. Fel'dbaum, Integral Criteria of the Quality of Regulation, AiT, Vol. IX, No. 1, 1948. V. V. Solodovnikov, Izv. AN SSSR, OTN, No. 12, 1945.

second problem of the theory of automatic control, and will be formulated in the following manner.*

Given a control system of the form (4.1), let us assume that by solving the first problem in the theory of automatic control, we have constructed the region B of the regulator parameters, for which this system is absolutely stable. Obviously, this system has at every point B a least positive characteristic number. Let us denote this by λ^*.

Turning then to (4.1), we see that any disturbed motion of this system can be considered as the motion of the representative point M in phase space of the variables $x_1, \ldots, x_n, \sigma$. Observing this motion it is natural first to examine the distance $\sqrt{2}R$ of the representative point from the origin

$$2R^2 = \sum_{k=1}^{s} x_k^2 + x_{s+1}x_{s+2} + \ldots + x_{n-1}x_n + \sigma^2$$

and the rate of change of this distance $\sqrt{2}\dot{R}$. Let ε be a positive number as small as desired. Let

$$\lambda^* = \lambda_1 + \varepsilon$$

and

$$R^2(t) = \bar{R}^2(t)e^{-2\lambda^* t} \quad .$$

By virtue of the definition of λ^*, the function

$$\bar{R}^2(t)e^{2\varepsilon t} = \bar{R}^2(t)e^{2(\lambda^*+\varepsilon)}$$

tends to increase without limit, while the function

$$\bar{R}^2(t)e^{-2\varepsilon t} = R^2(t)e^{2(\lambda^*-\varepsilon)t}$$

tends to zero. The functions $R^2(t)$ and $\bar{R}^2(t)$ are related by

$$R^2(t) = [\bar{R}^2(t)e^{-2\varepsilon t}]e^{-2\lambda_1 t} \quad . \tag{8.1}$$

Now let t^* be the time, during which the phase distance $\sqrt{2}R(t)$ diminishes by a factor e^a, where a is a specified positive number. We shall call t^* the time of the conditional damping of the control process. To estimate the value of t^*, we have

$$\frac{R^2(t*)}{R^2(o)} = e^{-2\alpha} = \frac{[\bar{R}^2(t^*)e^{-2\varepsilon t^*}]}{R^2(o)} e^{-2\lambda_1 t^*} \quad .$$

---

* A. M. Letov, Boundary Conditions of the Least Characteristic Number of One Class of Control Systems, PMM, Volume XV, No. 5, 1951.

Since the function $\bar{R}^2(t)e^{-2\varepsilon t}$ tends to zero, then at sufficiently large t^* the following inequality

$$\frac{\bar{R}^2(t^*)e^{-2\varepsilon t^*}}{R^2(0)} \leq 1$$

is correct. Therefore, for a sufficiently large a we have

$$t^* < \frac{a}{\lambda_1} = \frac{a}{\lambda^* - \varepsilon} \quad . \tag{8.2}$$

Thus, with accuracy to within ε, the lowest characteristic number λ^* limits from above the conditional measure of damping (for sufficiently large a). It can therefore serve as a measure of the quality of regulation in the system (4.1).

From this point of view, the regulator for which this number has a greater value can be considered to be the better regulator.

In connection with inequality (8.2), it is appropriate to separate the second problem in control theory into two problems, the direct and the inverse problems. The direct problem is formulated as follows: Given a control system and its absolute stability region, it is necessary to determine the conditional damping time everywhere in the region B. The inverse problem has the following formulation: for a system of the form (4.1), find the subregion $B' \epsilon B$, for which the conditional damping time t^* does not exceed a specified limit $\bar{t}^*$.

To solve these problems it is necessary to know the functional dependence of the number λ^* on the parameters of the control system. If this dependence could be established, then the problem of quality formulated here would be solved in the same manner as was done by Ya. Z. Tsypkin and P. V. Bromberg for linear systems. So far, however, there is no effective method for calculating the number λ^*, and therefore, it becomes necessary to estimate it by some method. A known method for this purpose is the general method of Lyapunov for determining the boundaries of the entire set of chracteristic numbers; this method will be developed here in application to the system (4.1). Naturally, this development must not be taken to be the solution of the formulated second problem in the theory of automatic control. However, the Lyapunov method does provide us with a means of investigating the maximum capabilities of regulators in the sense of attaining some specified control quality, and explains at the same time the role of its individual parameters. In this lies the theoretical and practical value of the proposed application of the Lyapunov method to the theory of automatic control, in spite of the fact that the question of how close these maximum capabilities are to the true ones remains open.

In accordance with the remarks, we now formulate the direct and inverse problems.

Direct problem: for a given control system of the type (4.1), and for a given region B, of its absolute stability, determine at each point of B the limits of the lowest characteristic number, λ^*.

Inverse problem: for a given control system of type (4.1), find a subregion $B' \in B$, in which the determined lower limit of the least characteristic number reaches a maximum possible value.

Let λ_1 and λ' be the lower and upper limits of the entire set of characteristic numbers of system (4.1). Obviously we have, accurate to within ε:

$$\frac{a}{\lambda'} \leq t^* \leq \frac{a}{\lambda_1} .$$

We shall develop below two methods solving the second problem of the theory of automatic control.

2. SOLUTION OF THE DIRECT PROBLEM

Let there be an arbitrary control system, defined by equations of the form (3.13), and let all r_k be real numbers. Let us consider the positive definite function

$$V = \frac{1}{2} \sum_{k=1}^{m} x_k^2 + \frac{1}{2} \sigma^2 ,$$

equal to half the square of the distances of the representative point M to the origin. Its total derivative, calculated in accordance with (3.13) is

$$\dot{V} = - \sum_{k=1}^{m} r_k x_k^2 - \bar{\rho}\sigma^2 - r\sigma f(\sigma) + \sum_{k=1}^{m} (1 + \bar{\beta}_k) x_k \sigma$$

or, what is the same,

$$\dot{V} = - \sum_{k=1}^{m} r_k x_k^2 - (\bar{\rho} + hr)\sigma^2 + \sum_{k=1}^{m} (1 + \bar{\beta}_k) x_k \sigma - r\sigma\varphi(\sigma) .$$

Let us consider the discriminant of the form $-[\dot{V} + r\sigma\varphi(\sigma)]$, the determinant

$$
(8.3)\qquad \Delta = \begin{vmatrix} r_1 & 0 & \cdots & 0 & -\frac{1}{2}(1+\bar{\beta}_1) \\ 0 & r_2 & \cdots & 0 & -\frac{1}{2}(1+\bar{\beta}_2) \\ \vdots & \vdots & & \vdots & \vdots \\ 0 & 0 & \cdots & r_n & -\frac{1}{2}(1+\bar{\beta}_n) \\ -\frac{1}{2}(1+\bar{\beta}_1) & -\frac{1}{2}(1+\bar{\beta}_2) & \cdots & -\frac{1}{2}(1+\bar{\beta}_m) & \bar{\rho} + hr \end{vmatrix}
$$

and all its diagonal minors

$$
(8.4)\qquad \Delta_1 = r_1, \ \Delta_2 = r_2\Delta_1, \ \ldots, \ \Delta_m = r_m\Delta_{m-1} \quad ,
$$

and let us find the conditions necessary and sufficient for the function $-\dot{V} - r\sigma\varphi(\sigma)$ to be positive. These are expressed by the inequality

$$
(8.5)\qquad \Delta_1 > 0, \ \ldots, \ \Delta_m > 0, \ \Delta = \Delta_m \left[\bar{\rho} + hr - \frac{1}{4}\sum_{k=1}^{m} \frac{(1+\beta_k)^2}{r_k}\right] > 0 \quad ,
$$

the fulfillment of which guarantees the stability of the system (3.13) for all disturbances, no matter what the function $f(\sigma)$ of class (A) or subclass (A') may be.

Consequently, for any control system of the type (3.13), the fulfillment of conditions (8.5) guarantees the existence of a certain smallest characteristic number $\lambda^* \geq 0$, or, what is the same, the existence of a definite quality of regulation. Let us calculate the limits of the number λ^*. For this purpose, we carry out the λ-transformation of the canonical variables, putting

$$
(8.6)\qquad y_k = x_k e^{\lambda t}, \quad \mu = \sigma e^{\lambda t} \qquad (k = 1, \ldots, m) \quad .
$$

In the new variables, the equations of motion become

$$
(8.7)\qquad \begin{aligned} \dot{y}_k &= -(r_k - \lambda)y_k + \mu \quad , \\ \dot{\mu} &= \sum_{k=1}^{n} \bar{\beta}_k y_k - (\bar{\rho} - \lambda)\mu - F(\mu, t) \end{aligned} \qquad (k = 1, \ldots, m) \quad ,
$$

where

$$F(\mu, t) = re^{\lambda t} f(\mu e^{-\lambda t}) \quad . \tag{8.8}$$

(8.7) have the obvious solution

$$y_1^* = 0, \ \ldots, \ y_m^* = 0, \ \mu^* = 0 \ . \tag{8.9}$$

No matter what $\lambda > 0$, the stability of the solution guarantees the stability of the solution $x_k^* = 0 \ (k = 1, \ldots, m)$, $\sigma^* = 0$. The reverse is true only for those λ that satisfy the condition $0 \leq \lambda < \lambda^*$. Actually, let λ^* be the smallest characteristic number of any group of functions of the canonical variables, transformed in accordance with formulas (8.6). If the transformation parameter λ increases from zero and assumes a set of real positive numbers, then as these values pass through the value λ^*, the solutions (8.9) of the λ-transformed equations become unstable. This phenomenon should be accompanied by simultaneous violation of the stability conditions of solution (8.9). These stability conditions can be obtained in a more simple manner from an examination of the positive definite function

$$W = \frac{1}{2} \sum_{k=1}^{m} y_k^2 + \frac{1}{2} \mu^2 \quad . \tag{8.10}$$

It is easy to see that after carrying out all the derivations on the determination of the derivative with respect to time, we obtain the stability conditions of solution (8.9), which differ from conditions (8.5) only in that the quantities r_k and $\bar{\rho} + hr$ are replaced in the latter quantities $r_k - \lambda$ and $\bar{\rho} + hr - \lambda$. Thus, we obtain the following stability condition

$$\begin{gathered} \Delta_1(\lambda) = r_1 - \lambda > 0, \ \ldots, \ \Delta_m(\lambda) = \Delta_{m-1}(\lambda)(r_m - \lambda) > 0 \ , \\ \Delta(\lambda) = \Delta_m(\lambda) \left[\bar{\rho} + hr - \lambda - \frac{1}{4} \sum_{k=1}^{m} \frac{(1+\bar{\beta}_k)^2}{r_k - \lambda} \right] > 0 \quad . \end{gathered} \tag{8.11}$$

When $\lambda = 0$, conditions (8.11) become the same as conditions (8.5). If λ increases continuously, the violation of any one of the inequalities (8.11) should be preceded by its becoming an equality. In order to estimate that value of λ for which this takes place, we number the roots of equation (3.9), in such a manner* that $r_1 < r_2 < \cdots < r_m < r_{m+1}$, where

---

* The assumption that r_{m+1} is greater than each of the roots of (3.9) is not essential. The arguments retain their force also in that case, when $\bar{\rho} + hr$ occupies any intermediate position along the quantities $r_1, r_2, \ldots$.

$$r_{m+1} = \bar{\rho} + h \quad . \tag{8.12}$$

Let us furthermore denote

$$\varphi(\lambda) = 4(r_1 - \lambda) \ldots (r_m - \lambda)(r_{m+1} - \lambda) \quad ,$$

$$\psi(\lambda) = \sum_{k=1}^{m} (r_1 - \lambda) \ldots (r_{k-1} - \lambda)(1 + \bar{\beta}_k)^2 (r_{k+1} - \lambda) \ldots (r_m - \lambda) \quad , \tag{8.13}$$

and consider the equation

$$\varphi(\lambda) = \psi(\lambda) = 4\Delta(\lambda) = 0 \quad . \tag{8.14}$$

It is necessary, for what follows, to investigate its roots.

Let us prove the following lemma: (8.14) has only real roots, which are positive if conditions (8.5) are fulfilled, the least of these being $\lambda_1 \leq r_1$.

The proof of the lemma follows from examination of the course of the functions $\varphi(\lambda)$ and $\psi(\lambda)$. Let us assume that $1 + \beta_k \neq 0$ for all k. It is seen from (8.13) that the points $\lambda = r_k$, $(k = 1, \ldots, m + 1)$ are the zeros of the function $\varphi(\lambda)$, while the zeros of the function $\psi(\lambda)$ of which there are $(m - 1)$, lie in the intervals (r_1, r_2), ..., (r_{m-1}, r_m). Let us denote the zeros of the function $\psi(\lambda)$ by $\bar{r}_1$, ..., $\bar{r}_{m-1}$. Next, it is clear that the zeros of the functions $\partial\varphi/\partial\lambda$ lie between the zeros of the functions $\varphi(\lambda)$, and the zeros of the function $\partial\psi/\partial\lambda$ lie between the zeros of the function $\psi(\lambda)$. Consequently, the function $\varphi(\lambda)$, in the intervals $(-\infty, r_1)$, $(r_{m+1}, +\infty)$ and the function $\psi(\lambda)$ in the intervals $(-\infty, r_1)$ and $(r_{m-1}, +\infty)$ are monotonic. The roots of (8.14) correspond to the intersections between the curves $y = \varphi(\lambda)$ and $y = \psi(\lambda)$. But $\varphi(0) > \psi(0)$ [by virtue of (8.5)], and therefore these roots, of which there are $(m + 1)$, lie in the intervals $(0, r_1)$, (r_1, r_2), (r_2, r_3), ..., (r_{m-1}, r_m), $(r_{m+1}, +\infty)$, (Figure 9). Thus, the lemma is proved.

It may turn out that the absolute stability of the control system will be guaranteed by satisfying some other sufficient conditions, for example (4.12) or (6.10) and (6.15). However, as can be readily ascertained in this case an analogous lemma is true.

From the proved lemma it follows that the first violation of conditions (8.11) will pass through $\lambda = \lambda_1$, and then one can no longer guarantee the stability of solution (8.9) of (8.7). Consequently, the root λ_1 of (8.14) can serve as the lower boundary of the least characteristic number λ^*.

Analogously, we also determine the upper limit of the number λ^*. As before, considering the function W [(8.10)] and using (8.7), we

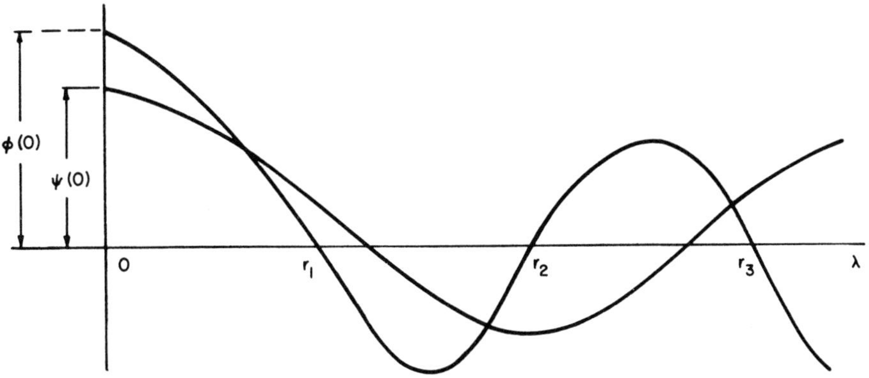

FIGURE 9

obtain its total derivative

$$\dot{W} = \sum_{k=1}^{m} (\lambda - r_k) y_k^2 + (\lambda - \bar{r}_{m+1})\mu^2 + \sum_{k=1}^{m} (1 + \bar{\beta}_k) y_k \mu - \mu\Phi(\mu, t) \quad ,$$

where

$$\bar{r}_{m+1} = \bar{\rho} + H, \quad \Phi(\mu, t) = F(\mu, t) - H\mu \quad . \tag{8.15}$$

By virtue of the definition of the number H (Page 22) we have that $\mu\Phi(\mu) < 0$ for all $\mu \neq 0$.

Let us write up the discriminant $\bar{\Delta}(\lambda)$ of the form $\dot{W} + \mu\Phi(\mu)$, i.e., the determinant

$$\bar{\Delta}(\lambda) = \begin{vmatrix} \lambda - r_1 & 0 & \cdots & 0 & \frac{1}{2}(1 + \bar{\beta}_1) \\ 0 & \lambda - r_2 & \cdots & 0 & \frac{1}{2}(1 + \bar{\beta}_1) \\ \vdots & & & \vdots & \vdots \\ 0 & 0 & \cdots & \lambda - r_m & \frac{1}{2}(1 + \bar{\beta}_m) \\ \frac{1}{2}(1 + \bar{\beta}_1) & \frac{1}{2}(1 + \bar{\beta}_2) & \cdots & \frac{1}{2}(1 + \bar{\beta}_m) & \lambda - \bar{r}_{m+1} \end{vmatrix} \tag{8.16}$$

and its diagonal minors

$$\bar{\Delta}_1 = \lambda - r_1, \ \ldots, \ \bar{\Delta}_m = \bar{\Delta}_{m-1}(\lambda - r_m) \quad . \tag{8.17}$$

The fulfillment of the inequalities

$$\bar{\Delta}_1 > 0, \ \bar{\Delta}_2 > 0, \ \ldots, \ \bar{\Delta}_m > 0, \ \bar{\Delta} > 0 \tag{8.18}$$

serves as the necessary and sufficient condition for the function $\dot{W} + \mu\Phi(\mu)$ to be positive definite, no matter what the function $f(\mu)$ of class (A) or subclass (A').

Let us consider the equation

$$\bar{\Delta}(\lambda) = 0 \quad . \tag{8.19}$$

It differs from (8.14) only in its free term, which contains, instead of r_{m+1} [(8.12)], the quantity $\bar{r}_{m+1}$ [(8.15)]. For (8.19), the following lemma holds: (8.19) has only real roots, and if conditions (8.5) are fulfilled, these roots are positive, and the largest of these is $\bar{\lambda}_{m+1} > \bar{r}_{m+1}$. The proof of this lemma is analogous to that of the preceding lemma.

By virtue of this lemma and of formulas (8.17), all inequalities (8.18) are satisfied when $\lambda \geq \bar{\lambda}_{m+1}$. Consequently, when $\lambda > \bar{\lambda}_{m+1}$, the function $\dot{W}$ becomes positive definite and therefore the phase distance $\sqrt{2W}$ of the representative point M increases continuously. Consequently, $\bar{\lambda}_{m+1}$ can serve as the upper limit of the entire set of characteristic numbers of the given system.

Thus, we have proved the following theorem: For any control system of the type (4.1), the entire set of its characteristic numbers is included in the interval $(\lambda_1, \bar{\lambda}_{m+1})$, where λ_1 is the least root of (8.14), and $\bar{\lambda}_{m+1}$ is the largest root of (8.19).

This theorem answers the question asked in the direct problem. It can be proved when the roots r_k of (3.9) include complex conjugate roots. In this case, instead of λ_1 and $\bar{\lambda}_{m+1}$ it is necessary to consider the quantities $\operatorname{Re} \lambda_1$ and $\operatorname{Re} \bar{\lambda}_{m+1}$.

3. SOLUTION OF THE INVERSE PROBLEM

Proceeding to the solution of the inverse problem, let us turn again to Figure 9. It is easy to understand that λ_1 reaches its upper limit $\bar{\lambda}_1 = r_1$ in two cases: either when $\varphi(0) \longrightarrow \infty$, or when $\psi(0) \longrightarrow 0$.

Let us first examine the first case. With the regulator parameters chosen arbitrarily in B, $\varphi(0) \longrightarrow \infty$ only if $h \longrightarrow \infty$, i.e., if the slope of the characteristic $f(\sigma)$ (the characteristic of the actuator) is increased without limit. The above limit is reached when $h = \infty$, i.e., in the case of an ideal regulator. On the other hand, it is clear that in this case the lower limit of the least characteristic number coincides with the number itself. Actually, the equations of disturbed motion in the case of an ideal regulator are (1.4) and (1.11). But these are linear differential equations with constant coefficients, and (3.9) is their characteristic equation. Consequently, in this case, the inequality $\lambda^* = r_1$ is correct.

In the second case, it is possible to obtain the same result for λ_1, if one specifies that the following relations be satisfied

$$\psi(0) = \sum_{k=1}^{m} r_1 \cdots r_{k-1}(1 + \bar{\beta}_k)^2 r_{k+1} \cdots r_m = 0 \quad ,$$

(8.20)

$$\bar{\rho} + h > r_1 \ .$$

But by virtue of the fact that the real numbers r_k $(k = 1, \ldots, m)$ are positive, the relations (8.20) are equivalent to

$$(8.21) \qquad 1 + \beta_1 = 0, \ \ldots, \ 1 + \bar{\beta}_m = 0, \ \bar{\rho} + h > r_1 \ .$$

If these are fulfilled, all the conditions (8.11) are fulfilled for $0 \leq \lambda < r_1$; if $\lambda \geq r_1$, conditions (8.11) are violated. Consequently, $\lambda = r_1$ is the lower limit of the least characteristic number.

Thus, the solution of the inverse problem is given by the two following theorems.

THEOREM 1. The lower limit λ_1 of the least characteristic number λ^* of system (3.13) reaches its upper limiting value (the number r_1 itself), only in the case of an ideal regulator [no matter what the parameters p_α $(\alpha = 1, \ldots, m)$ of the regulator, as defined in B, may be].

THEOREM 2. The lower limit λ_1 of the least characteristic number λ^* of the system (3.13) reaches its upper limit r_1, if the parameters of the regulator are defined in B in accordance with relations (8.12) [no matter what the function $f(\sigma)$ of class (A) may be, whose slope is bounded from below by the number h, satisfying inequality (8.21)]. In this case, the inequality $\lambda^* \geq r_1$ is in force.

The last theorem calls for one comment. In some particular problems, the requirement that (8.21) be satisfied leads to a repeated examination of the inverse problem for the case in which all the roots r_k are the same. In these cases, (8.21) become approximate, and the degree of approximation can be made as high as desired.

4. FIRST AND SECOND BULGAKOV PROBLEMS

Using the theory developed here, let us see what are the maximum capabilities of the regulator in the two Bulgakov problems. Let us start with the equation for r. We have

$$r^2 - \frac{U + \ell E}{\sqrt{(T^2+\ell G^2)}} r + \frac{k + a\ell}{\ell} = 0 \quad . \tag{8.22}$$

It is first necessary to satisfy the conditions $\operatorname{Re} r_k > 0$ $(k = 1, 2, \ldots)$. For this it is necessary and sufficient to satisfy the inequality

$$k + a\ell > 0 \quad , \tag{8.23}$$

for $U = \ell E$ is always greater than zero. No matter what the value of k, this inequality can always be fulfilled by choosing a sufficiently large product $A\ell$ of the two constants of the regulator.

Let us consider first the case of an ideal regulator. Inequality (8.23) determines the region B of absolute stability of the control system (2.81). From Theorem 1, the upper limit of the least characteristic number is

$$\lambda_1 = \operatorname{Re} r_1 \quad ,$$

where

$$r_1 = \frac{1}{2} \frac{U + \ell E}{\sqrt{\ell (T^2+\ell G^2)}} - \sqrt{\frac{1}{4} \frac{(U+\ell E)^2}{\ell (T^2+\ell G^2)} - \frac{k + a\ell}{\ell}} \quad .$$

The limiting value of λ^* is reached when the single condition

$$(U + \ell E)^2 \leq 4(k + a\ell)(T^2 + \ell G^2) \quad , \tag{8.24}$$

is satisfied in this case we obtain

$$\lambda^* = \operatorname{Re} r_1 = \frac{1}{2} \frac{U + \ell E}{\sqrt{\ell (T^2+\ell G^2)}} \quad ,$$

and the conditional damping time will be limited by the inequality

$$t^* \sqrt{r} < \frac{2a^* \sqrt{\ell (T^2+\ell G^2)}}{U + \ell E} .$$

If r is replaced by its value in accordance with formula (2.82), we obtain finally

$$t^* \leq \frac{2a^* (T^2+\ell G^2)}{U + \ell E} .$$

For a regulator containing no proportional feedback $(\ell = \infty)$,

(8.25) $$t^* \leq \frac{2a^* G^2}{E} .$$

In this case, the upper boundary of the time of conditional damping can be made sufficiently small by correspondingly decreasing the ratio G^2/E.

If, however, the regulator contains no second-derivative control signal $(G^2 = 0)$, we obtain

(8.26) $$t^* \leq \frac{2a^* T^2}{U + \ell E} .$$

In this case the upper limit of the time of conditional attentuation depends on the inertia of the regulated object, and also on the magnitude of the total damping $U + \ell E$, which depends both on the natural damping constant U and on the artificial damping constant ℓE.

Comparing inequalities (8.25) and (8.26), we conclude that from the point of view of control quality, the better regulator is the one without derivative feedback.

Let us now assume that the regulator is not ideal, and consider the problem of the number λ^* reaching its upper limit. Conditions (8.21), in the case of real r_1 and r_2, become

(8.27) $$1 + \bar{\beta}_1 = 0, \quad 1 + \bar{\beta}_2 = 0, \qquad h > r_1 + \sqrt{r}\,(E - pG^2) .$$

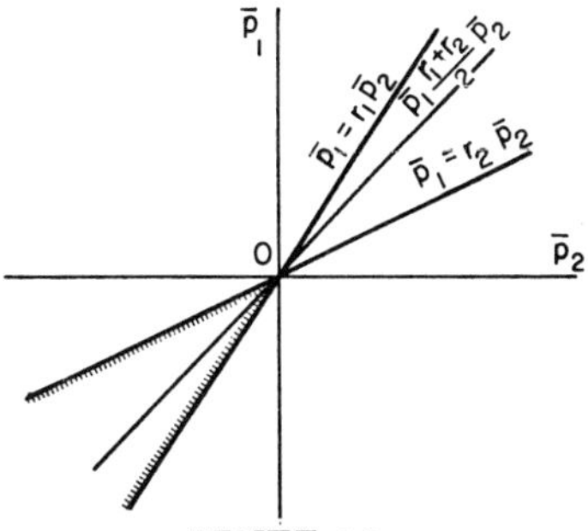

FIGURE 10

Let us first examine the equations of (8.27). From the formulas (3.26) it is seen that these equations can be definitely satisfied if it is possible to satisfy the relations

$$\bar{p}_1 - r_1\bar{p}_2 < 0, \; -\bar{p}_1 + r_2\bar{p}_2 < 0, \; 2\bar{p}_1 = (r_1 + r_2)\bar{p}_2 \quad . \tag{8.28}$$

Actually, if (8.28) is satisfied, we have $\bar{\beta}_1 < 0$, $\bar{\beta}_2 < 0$, and $\bar{\beta}_1 = \bar{\beta}_2$, and the equations of (8.27) are satisfied unconditionally. Let us now consider the plane of the variables $\bar{p}_1$, $\bar{p}_2$ (Figure 10), and plot the region of those points $\bar{p}_1$, $\bar{p}_2$, where relations (8.28) are satisfied. The existence of this region is guaranteed by the fulfillment of only two inequalities $\bar{p}_1 < 0$ and $\bar{p}_2 < 0$. In expanded form these inequalities are

$$-\frac{(k+a\ell)(E-pG^2)}{\sqrt{\ell(T^2+\ell G^2)}} < 0, \; a - qG^2 - \frac{(U+\ell E)(E-pG^2)}{T^2 + \ell G^2} < 0 \quad . \tag{8.29}$$

By virtue of inequality (8.23), the first of these simplifies to

$$E - pG^2 < 0 \quad .$$

Thus, for a regulator with an actuator of finite velocity, the subregion B' of region B of absolute stability of the control system is determined by the inequalities

$$k + a\ell > 0, \; E - pG^2 > 0, \; h > r_1 + \sqrt{r}\,(E - pG^2) \quad , \tag{8.30}$$

$$\frac{(U+\ell E)(E-pG^2)}{T^2 + \ell G^2} > a - qG^2 \quad . \tag{8.31}$$

It is easy to see that this region contains an infinite set of parameters a, ℓ, E, G^2 and satisfying inequalities (8.30) and (8.31). Let us consider, for example, a regulator that contains no proportional feedback ($\ell = \infty$), and let us plot the curves

$$x = py \quad , \tag{8.32}$$

$$x^2 + qy^2 - pxy = a \quad , \tag{8.33}$$

where $x = E/G$ and $y = G$. Since $q < 0$, the region B occupies that portion of the first quadrant of the plane xy which lies inside the hyperbola (8.33) (Figure 11) and is bounded by the x axis.

The geometric locus of the points where λ_1 reaches its upper limiting value is determined by (8.28). In this case it gives a third-order curve

$$qxy^2 + p(2a - x^2)y + x(x^2 - 3a) = 0 \quad . \tag{8.34}$$

Using (8.33) and (8.34), it is easy to establish that the curve (8.34) passes nowhere in the region B when $q < 0$. Consequently, in the case of a regulator containing no proportional feedback ($\ell = \infty$, $G^2 \neq 0$), there does not exist in the region B a single point in which the lower limit of the least characteristic number reaches its maximum value.

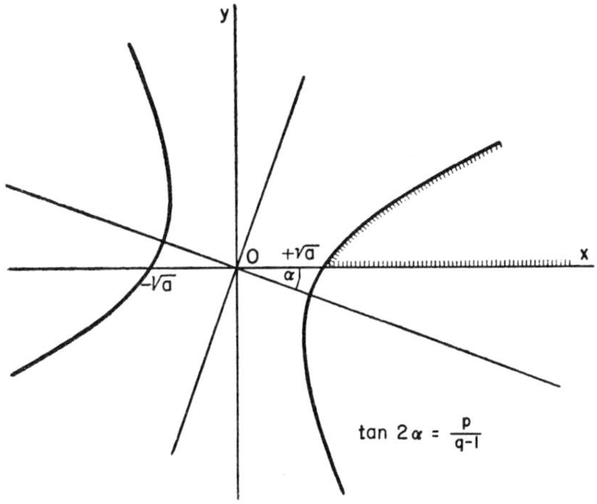

FIGURE 11

On the other hand, in the case of a regulator with proportional feedback ($0 < \ell < \infty$, $G^2 = 0$), the region B is determined by the conditions

$$k + a\ell > 0, \quad E > 0, \quad \frac{(U+\ell E)E}{T^2} > a \quad . \tag{8.35}$$

Considering k as a parameter and denoting $a\ell = x$, $\ell E/T = y$, we plot the curve

$$y\left(y + \frac{U}{T}\right) = x \quad , \tag{8.36}$$

$$k + x = 0 \quad , \tag{8.37}$$

$$y\left(y + \frac{U}{T}\right)^2 - 2ky - \left(3y + \frac{U}{T}\right)x = 0 \quad . \tag{8.38}$$

The parabola (8.36) and the line (8.37) bound the region B (cross-hatched region in Figure 12).

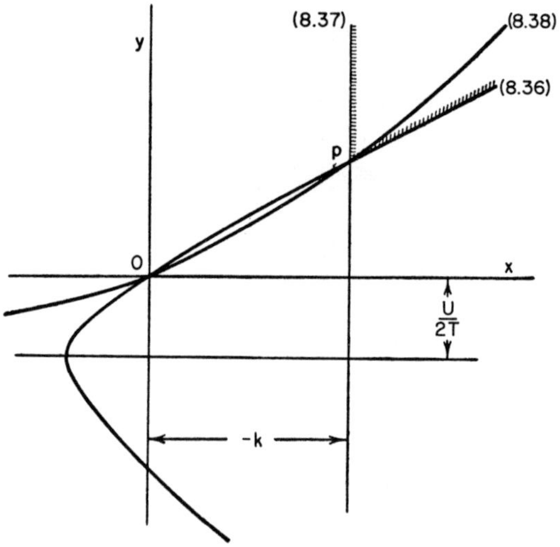

FIGURE 12

It remains now to establish whether the curve (8.38) passes through this region. To ascertain this, let us find the points of intersection of the curve (8.38) and the parabola (8.36). For this purpose, let us insert into (8.38) the value of x determined by (8.36). After having done this, we see that the first point of intersection of these curves is the origin, and for the other points we have

$$\left(y + \frac{U}{T}\right)^2 - 2k - \left(3y + \frac{U}{T}\right)\left(y + \frac{U}{T}\right) = 0 ,$$

or, after simplification

$$k + y\left(y + \frac{U}{T}\right) = 0 . \tag{8.39}$$

The roots of (8.39) answer the problem posed. From (8.39) we find

$$y_1, y_2 = -\frac{U}{T2} \pm \sqrt{\left(\frac{U}{2T}\right)^2 - k} .$$

These roots are real and one is positive. It is easy to establish that this corresponds to the intersection of the straight line (8.37) and the parabola (8.36).

Next, we note that the curve (8.38) has an asymptote, whose equation is

$$y = -\frac{U}{3T} .$$

Turned towards this asymptote is that end of the curve which corresponds to the values $x < 0$. We next find the slopes of the tangents to the parabola (8.36) and to the curve (8.38). They are

$$\left(\frac{dy}{dx}\right)_I = \frac{1}{2y + \frac{U}{T}} ,$$

$$\left(\frac{dy}{dx}\right)_{II} = \frac{3y + \frac{U}{T}}{\left(y + \frac{U}{T}\right)^2 + 2y\left(y + \frac{U}{T}\right) - 2k - 3x} .$$

In the vicinity of the origin, where x and y are sufficiently small, we have

$$\left(\frac{dy}{dx}\right)_I = \frac{T}{U} , \quad \left(\frac{dy}{dx}\right)_{II} = \frac{UT}{U^2 - 2kT^2} ,$$

and, consequently, in the interval $0 < x < -k$ the curve (8.38) lies below the parabola (8.36). At the point P we have

$$\left(\frac{dy}{dx}\right)_I = \frac{T}{\sqrt{U^2 - 4kT^2}} , \quad \left(\frac{dy}{dx}\right)_{II} = \frac{\left[-U + 3\sqrt{U^2 - 4kT^2}\,\right]T}{U^2 - 4kT^2 + 2U\sqrt{U^2 - 4kT^2}} .$$

Comparison shows that at this point

$$\left(\frac{dy}{dx}\right)_{II} > \left(\frac{dy}{dx}\right)_I ,$$

and consequently, in the interval $-k < x < \infty$, the curve (8.38) lies above the parabola (8.36). This analysis establishes accurately the course of the curve (8.38).

Thus, the set of points x, y, located on the curve (8.38) in the region B, comprises that sought subregion $B' \in B$, in which the lower limit of the least characteristic number reaches its upper limiting values; this is correct for all characteristics $f(\sigma)$ of the actuator, whose slope satisfy the inequality

(8.40) $$h > r_1 + \sqrt{rE} \quad .$$

5. SECOND METHOD OF SOLVING THE QUALITY PROBLEM*

We describe below a second method of solving the problem of quality, based on the use of the third canonical form of the initial equations. The basic feature of this form is that it permits solving the stability and quality problems for control systems by studying only (3.57) and (3.61), which determine the phase-space metric of these systems.

As follows from the form of the above equations, the problem reduces to a study of only the function W.

We considered before the possible solution of the problem of stability, based on this method. Turning now to the problem of quality, we again turn to the first equation (3.61). We shall assume first that the numbers $r_1, \ldots, r_m$ are real. Then the function W is real, and to estimate the quality of the transient process we can consider only the abbreviated equation

(8.41) $$\dot{\bar{R}} = - W\bar{R} \quad .$$

From this we find

(8.42) $$R < R(0)e^{-\int W\, dt} \quad ,$$

where W is a function that is defined on the surface (3.64).

The stability of the obvious solution corresponds to any sign-definite positive function W.*

We write the Sylvester inequalities for the function W. We have

(8.43)
$$\begin{vmatrix} B_{11} & \cdots & B_{1k} \\ \cdot & \cdots & \cdot \\ B_{k1} & \cdots & B_{kk} \end{vmatrix} > 0, \quad (k = 1, \ldots, m), \quad \begin{vmatrix} B_{11} & \cdots & B_{1m} & Q_1 \\ \cdot & \cdots & \cdot & \cdot \\ B_{m1} & \cdots & B_{mm} & Q_m \\ Q_1 & \cdots & Q_m & \rho + h \end{vmatrix} > 0 \quad ,$$

---

* A. M. Letov, Concerning the Theory of Quality of Nonlinear Control Systems. Avtomatika i Telemekhanika, Volume XIV, No. 5, 1953.

** N. P. Erugin noted that for the stability of the obvious solution it is sufficient that W be a positive definite function.

where

$$B_{rs} = \frac{\sqrt{a_{rr}a_{ss}}}{\Delta^2} \sum_{\alpha=1}^{m} \sum_{\beta=1}^{m} \frac{r_\alpha + r_\beta}{2} a_{\alpha\beta}\Delta_{s\alpha}\Delta_{r\beta} \quad ,$$

(8.44)

$$Q_k = - \frac{\sqrt{a_{kk}}}{2\Delta\sqrt{S}} \sum_{\alpha=1}^{m} \left(S\beta_\alpha + \sum_{\beta=1}^{m} a_{\alpha\beta} \right) \Delta_{k\alpha}$$

$$r,\ s,\ k = 1,\ \ldots,\ m) \quad .$$

Thus, if the parameters of the regulator are chosen in accordance with inequalities (8.43), the control system, characterized by simple and real $r_1, \ldots, r_m$, is absolutely stable. Inequalities (8.43) define the region of stability B.

Proceeding now to the problem of quality of the control system, let us consider inequality (8.42). Obviously, the damping rate of the transient in the fulfillment of inequality (8.43) is determined by the values of the function W, which it assumes on the closed surface (3.64). Consequently, in analogy with the method used by N. G. Chetaev* for linear systems, the solution of the problem of the quality of control systems of the given class reduces to finding the extremal values of the function W [(3.62)] on the surface (3.64). As is known**, these are equal to the values of the roots of equation

(8.45)

$$\begin{vmatrix} B_{11} - \lambda A_{11} & \cdots & B_{1m} - \lambda A_{1m} & Q_1 \\ \cdot & \cdot & \cdot & \cdot \\ B_{m1} - \lambda A_{m1} & \cdots & B_{mm} - \lambda A_{mm} & Q_m \\ Q_1 & \cdots & Q_m & \bar{\rho} + h - \lambda \end{vmatrix} = 0 \quad ,$$

and if inequalities (8.43) are satisfied, these roots are real and positive. Let these roots be $\lambda_1, \ldots, \lambda_{m+1}$ and arranged in ascending order. Then the transient damps out not slower than the function $e^{-\lambda_1 t}$, and not faster than the function $e^{-\lambda_{m+1} t}$, i.e.,

$$R(0)e^{-\lambda_{m+1} t} < R(t) < R(0)e^{-\lambda_1 t} \quad .$$

* N. G. Chetaev, The Duration of the Transient Process in a Linear System, PMM, Volume XV, No. 3, 1951. P. V. Bromberg, On the Quality of Linear Control Systems, Trudy, No. 48, Oborongiz, 1952.

** V. I. Smirnov, Kurs vysshei̮ matematiki, (Course of Higher Mathematics), Vol. III, Part I, Gostekhizdat, 1951.

Let us expand equation (8.45)

$$B_0\lambda^{m+1} + B_1\lambda^m + \dots + B_m\lambda + B_{m+1} = 0 \quad . \tag{8.46}$$

Its coefficients depend on the parameters of the control system and on coefficients of the form (3.61). Let us assume that these coefficients have been chosen and fixed in some manner. Then the sought subset B' is determined by those values of the parameters p_α of the regulator, from among the set B, for which the least real and positive root λ_1 of (8.46) is the largest.

Thus, we have proved the following theorem: the parameters p_α of the regulator, for which the control system has the best quality, form a subset B', in which the least real root of (8.46) has a maximum.

All the above can be repeated also when the numbers $r_1, \dots, r_m$ include complex conjugate ones. For this purpose it is necessary to employ the Hermitian metric of the unitary space of variables, determining the distance of the representative point M from the origin. However, with the example given below, it will be shown that it is possible to avoid these considerations and to reduce the problem to the case considered above.

Let us consider a system, described by (3.19). In Chapter VI, we gave the inequalities (8.43) for this system. To reduce the calculations, we put $a_{11} = a_{22}$, and $a_{12} = 0$. (8.46) is of the form

$$\lambda^3 - \left(\bar{\rho} + h + \frac{r_1 + r_2}{\Delta^2}\right)\lambda^2 + \left[\frac{r_1 r_2}{\Delta^2} + (\bar{\rho} + h)\frac{r_1 + r_2}{\Delta^2} - \frac{Q_1^2}{A_{11}} - \frac{Q_2^2}{A_{22}}\right]\lambda - r_1 r_2\frac{(\bar{\rho}+h)}{\Delta^4} + r_1^2 Q_2^2 + r_2 Q_1^2 = 0 \quad . \tag{8.47}$$

If r_1 and r_2 are complex conjugate numbers, then this equation breaks up into two:

$$\lambda = p$$

and

$$(p - \lambda)(\bar{\rho} + h - \lambda) = Q_1^2 + Q_2^2 \quad . \tag{8.48}$$

These can be readily analyzed.

CHAPTER IX: STABILITY OF CONTROL SYSTEMS WITH TWO ACTUATORS

1. STATEMENT OF PROBLEM*

Present day technology affords many examples of various control systems containing two or more actuators. Therefore, the desire to extend the Lur'e method and other methods of determining the Lyapunov function to include systems of this type is quite well founded.

For simplicity, we shall consider regulated objects describable by the system (1.1) and assume that these are subjected to the action of regulators with two actuators.

Let μ_1 and μ_2 be the coordinates, and n_{k1} and n_{k2} be the constant parameters of the corresponding regulator.

To simplify the arguments, we shall disregard in the equations of disturbed motion the constants of the inertia and of the connected load of the actuators.

If in some particular problem we find it necessary to take some of these factors into account, we can employ the same methods of normal transformations developed in the first chapter. Therefore, putting $m = n$, we write the initial equations in the following form

$$\dot{\eta}_k = \sum_{\alpha=1}^{n} b_{k\alpha}\eta_\alpha + n_{k1}\xi_1 + n_{k2}\xi_2 \quad (k = 1, \ldots, n) \quad ,$$

$$(9.1) \qquad \dot{\xi}_1 = f_1(\sigma_1), \quad \sigma_1 = \sum_{\alpha=1}^{n} p_{1\alpha}\eta_\alpha - r_{11}\xi_1 - r_{12}\xi_2 \quad ,$$

$$\dot{\xi}_2 = f_2(\sigma_2), \quad \sigma_2 = \sum_{\alpha=1}^{n} p_{2\alpha}\eta_\alpha - r_{21}\xi_1 - r_{22}\xi_2 \quad .$$

---

* A. M. Letov, Stability of Control Systems with Two Actuators, PMM, Volume XVII, No. 4, 1953.

We shall assume that the functions $f_1(\sigma_1)$ and $f_2(\sigma_2)$ belong to class (A) or to subclass (A'). As to the controlling pulse signals σ_1 and σ_2, each of these describes the contol sequence chosen for the corresponding actuator. These sequences are characterized by constants $p_{1\alpha}$, $p_{2\alpha}$, r_{11}, ..., r_{22}. The numbers r_{11}, ..., r_{22} will be called the cross-connection constants.

The steady state of the system is determined by the relations

$$\sum_{\alpha=1}^{n} b_{k\alpha}\eta_\alpha + n_{k1}\xi_1 + n_{k2}\xi_2 = 0 \qquad (k = 1, \ldots, n) , \tag{9.2}$$

$$\left| \sum_{\alpha=1}^{n} p_{1\alpha}\eta_\alpha - r_{11}\xi_1 - r_{12}\xi_2 \right| \leq \sigma_{1*} , \tag{9.3}$$

$$\left| \sum_{\alpha=1}^{n} p_{2\alpha}\eta_\alpha - r_{21}\xi_1 - r_{22}\xi_2 \right| \leq \sigma_{2*} , \tag{9.4}$$

where σ_{1*}, σ_{2*} are the fixed non-negative numbers, characterizing the dead zones of the actuator. Various cases of solvability of relations (9.2) are possible. Let us consider one of these. Let us assume that the determinant

$$\begin{vmatrix} b_{11} & \cdots & b_{1n} \\ \cdot & \cdots & \cdot \\ b_{n1} & \cdots & b_{nn} \end{vmatrix} \tag{9.5}$$

does not vanish. Then (9.2) are solvable with respect to η_α, and we find

$$\eta_\alpha = \sum_{\rho=1}^{2} A_{\alpha\rho}\xi_\rho \qquad (\alpha = 1, \ldots, n) , \tag{9.6}$$

where $A_{\alpha\rho}$ are certain constants.

Inserting the values of η_α into inequalities (9.3) and (9.4) we obtain a system of two inequalities, containing the unknowns ξ_1 and ξ_2

$$\begin{aligned} |A_1\xi_1 + B_1\xi_2| &\leq \sigma_{1*} , \\ |A_2\xi_1 + B_2\xi_2| &\leq \sigma_{2*} , \end{aligned} \tag{9.7}$$

where $A_1, \ldots, B_2$ are certain constants. These inequalities define a certain parallelogram in the $\xi_1\xi_2$ plane.

Thus, it follows from formulas (9.4) and inequalities (9.5), that the relations (9.2) - (9.4) have a continuum of solutions η_α^*, ξ_1^*, and ξ_2^*, where ξ_1^* and ξ_2^* are the coordinates of the points belonging to the above parallelogram, and

$$\eta_\alpha^* = \sum_{\rho=1}^{2} A_{\alpha\rho}\xi_\rho^* \qquad (\alpha = 1, \ldots, n) \ . \tag{9.8}$$

This continuum defines the region of these values of the variables η_α, ξ_1 and ξ_2, at which the regulator remains insensitive to disturbed motions of the regulated object.

In the particular case when $\sigma_1^* = \sigma_2^* = 0$, the insensitivity region degenerates into a point

$$\eta_\alpha^* = 0 \quad (\alpha = 1, \ldots, n), \quad \xi_1^* = \xi_2^* = 0 \ . \tag{9.9}$$

Hereinafter, when we speak of the stability of any steady state of a system (9.1), we shall have in mind the state described by solution (9.9), or by any other solution belonging to the above continuum. We shall denote this solution in the following manner

$$\eta_k = \eta_k^* \ (k = 1, \ldots, n), \quad \xi_1 = \xi_1^*, \quad \xi_2 = \xi_2^* \ . \tag{9.10}$$

The initial equations (9.1) have a normal form.

2. CANONICAL FORM OF EQUATIONS OF CONTROL SYSTEMS

Let us assume that the control system is inherently stable, and consequently, the roots of the equation

$$D(\rho) = |b_{\alpha\beta} + \rho\delta_{\alpha\beta}| = 0 \tag{9.11}$$

satisfy the condition $\mathrm{Re}\ \rho_k > 0 \ (k = 1, \ldots, n)$.

Let us define the canonical variables x_s by the equations

$$x_s = \sum_{\alpha=1}^{n} c_\alpha^{(s)}\eta_\alpha + u_1^{(s)}\xi_1 + u_2^{(s)}\xi_2 \qquad (s = 1, \ldots, n) \tag{9.12}$$

and let us choose the transformation constants in such a way as to satisfy the relations

$$(9.13) \qquad -\rho_s C_\beta^{(s)} = \sum_{\alpha=1}^{n} b_{\alpha\beta} C_\alpha^{(s)} \qquad (s, \beta = 1, \ldots, n),$$

$$(9.14) \qquad \begin{aligned} -\rho_s u_1^{(s)} &= \sum_{\alpha=1}^{n} n_{\alpha 1} C_\alpha^{(s)} \\ -\rho_s u_2^{(s)} &= \sum_{\alpha=1}^{n} n_{\alpha 2} C_\alpha^{(s)} . \end{aligned}$$

Then the equations of the disturbed motion of the system will have the following form in the new variables

$$(9.15) \qquad \dot{x}_s = -\rho_s x_s + u_1^{(s)} f_1(\sigma_1) + u_2^{(s)} f_2(\sigma_2) \qquad (s = 1, \ldots, n) .$$

Obviously, the n linear homogeneous equations (9.13) with respect to $C_\alpha^{(s)}$ give non-zero solutions, if the constants ρ_s are simple roots of (9.11). In this case, for each value $s = 1, \ldots, n$, (9.13) are solvable with accuracy to within one arbitrary constant. We shall assume that these have been solved and that the transformation (9.12) has been effected. If the roots ρ_k are simple, the transformation will be non-singular and can be resolved relative to η_α. We have

$$(9.16) \qquad \eta_\alpha = \sum_{k=1}^{n} D_k^{(\alpha)} x_k + \sum_{\rho=1}^{2} G_{\alpha\rho} \xi_\rho \qquad (\alpha = 1, \ldots, n),$$

where $D_k^{(\alpha)}$, $G_{\alpha\rho}$ are constants, which will be determined later.

Let us now derive equations for $\dot{\sigma}_1$ and $\dot{\sigma}_2$. For this purpose, let us turn to the equations

$$(9.17) \qquad \begin{aligned} \sigma_1 &= \sum_{\alpha=1}^{n} p_{1\alpha}\eta_\alpha - r_{11}\xi_1 - r_{12}\xi_2 , \\ \sigma_2 &= \sum_{\alpha=1}^{n} p_{2\alpha}\eta_\alpha - r_{21}\xi_1 - r_{22}\xi_2 \end{aligned}$$

and differentiate them. This leads to

$$\dot{\sigma}_\rho = \sum_{\alpha=1}^{n} p_{\rho\alpha}\dot{\eta}_\alpha - r_{\rho 1}f_1(\sigma_1) - r_{\rho 2}f_2(\sigma_2) \qquad (\rho = 1, 2) \quad . \tag{9.18}$$

Differentiating next (9.12) and using (9.15), we obtain

$$\sum_{\alpha=1}^{n} C_\alpha^{(s)}\dot{\eta}_\alpha = -\rho_s x_s \qquad (s = 1, \ldots, n) \quad . \tag{9.19}$$

In the case of the non-singular transformation (9.12), (9.19) can be solved with respect to $\dot{\eta}_\alpha$, and we obtain

$$\dot{\eta}_\alpha = \sum_{k=1}^{n} \Gamma_k^{(\alpha)} x_k \qquad (\alpha = 1, \ldots, n) \quad , \tag{9.20}$$

where $\Gamma_k^{(\alpha)}$ are certain constants. Consequently, in accordance with (9.18) and (9.20), the sought canonical equations assume the following final form

$$\begin{aligned} \dot{x}_s &= -\rho_s x_s + u_1^{(s)}f_1(\sigma_1) + u_2^{(s)}f_2(\sigma_2) \qquad (s = 1, \ldots, n) \,, \\ \dot{\sigma}_1 &= \sum_{\alpha=1}^{n} \beta_{1\alpha}x_\alpha - r_{11}f_1(\sigma_1) - r_{12}f_2(\sigma_2) \,, \\ \dot{\sigma}_2 &= \sum_{\alpha=1}^{n} \beta_{2\alpha}x_\alpha - r_{21}f_1(\sigma_1) - r_{22}f_2(\sigma_2) \,. \end{aligned} \tag{9.21}$$

Here the constants $\beta_{1\alpha}$ depend on the parameters $p_{1\alpha}$, and the constants $\beta_{2\alpha}$ depend on the parameters $p_{2\alpha}$; the calculation of these constants is given below.

3. FORMULAS FOR THE COEFFICIENTS OF THE CANONICAL TRANSFORMATION

For practical application of the canonical equations to the solution of particular problems, it is very important to have formulas that permit immediate calculations of the constants $\beta_{s\alpha}$ in terms of the initial data. In many systems with one regulating device, such formulas, as noted earlier, were obtained by A. I. Lur'e.

In order to obtain the required formulas, let us turn to (9.13).

In the case of simple ρ_s the solutions of these equations can be represented in the form

$$C_k^{(s)} = A_i^{(s)} D_{ik}(\rho_s) \qquad (k, s = 1, \ldots, n) , \tag{9.22}$$

where D_{ik} stand for the cofactors of the elements of the i'th row of determinant (9.11), and $A_i^{(s)}$ are the proportionality factors, to be determined in the following manner. Adding term by term the relations (9.14) and replacing $C_k^{(s)}$ in these by means of expressions (9.22), we obtain

$$- \rho_s (u_1^{(s)} + u_2^{(s)}) = A_i^{(s)} \left[\sum_{\alpha=1}^{n} (n_{\alpha 1} + n_{\alpha 2}) D_{i\alpha}(\rho_s) \right]$$

$$(s = 1, \ldots, n) .$$

Let us now require the fulfillment of the following conditions

$$u_1^{(s)} + u_2^{(s)} = 1 \qquad (s = 1, \ldots, n) . \tag{9.23}$$

We then obtain

$$A_i^{(s)} = \frac{1}{H_i(\rho_s)} \qquad (s = 1, \ldots, n) , \tag{9.24}$$

where

$$H_i(\rho_s) = - \frac{1}{\rho_s} \sum_{\alpha=1}^{n} (n_{\alpha 1} + n_{\alpha 2}) D_{i\alpha}(\rho_s) \qquad (s = 1, \ldots, n). \tag{9.25}$$

Thus, the constants of the transformation are determined by the following formulas

$$C_k^{(s)} = \frac{D_{ik}(\rho_s)}{H_i(\rho_s)} \qquad (s = 1, \ldots, n) , \tag{9.26}$$

(9.27

$$u_1^{(s)} = - \frac{1}{\rho_s} \sum_{\alpha=1}^{n} n_{\alpha 1} \frac{D_{i\alpha}(\rho_s)}{H_i(\rho_s)} ,$$

$$u_2^{(s)} = - \frac{1}{\rho_s} \sum_{\alpha=1}^{n} n_{\alpha 2} \frac{D_{i\alpha}(\rho_s)}{} \qquad (s = 1, \ldots, n) .$$

Now (9.19) can be rewritten as

$$\sum_{k=1}^{n} D_{ik}(\rho_s)\dot{\eta}_k = -\rho_s H_i(\rho_s)x_s \qquad (s = 1, \ldots, n) .$$

It is easy to verify that they differ from (2.24) only in their right halves. On the basis of this, we can dispense with all the intermediate considerations connected with the search for their solutions, and use formulas (2.33) directly. For this purpose it is enough to replace the expressions $H_k(\rho_j)(x_j - \xi)(\rho_{n+1} - \rho_j)$ in formulas (2.32) by means of the expressions $-\rho_k H_k(\rho_j)x_j)$. Having done so, we obtain

$$(9.28) \qquad \dot{\eta}_k = -\sum_{j=1}^{n} \frac{\rho_j H_k(\rho_j)}{D'(\rho_j)} x_j \qquad (k = 1, \ldots, n) .$$

In order to obtain the sought expression $\beta_{k\alpha}$, it remains for us to insert the values of (9.28) into (9.18). This yields

$$(9.29) \qquad \begin{aligned} \beta_{1j} &= -\sum_{\alpha=1}^{n} \frac{\rho_j H_\alpha(\rho_j)}{D'(\rho_j)} p_{1\alpha} , \\ \beta_{2j} &= -\sum_{\alpha=1}^{n} \frac{\rho_j H_\alpha(\rho_j)}{D'(\rho_j)} p_{2\alpha} \end{aligned} \qquad (j = 1, \ldots, n) .$$

4. CONSTRUCTION OF THE LYAPUNOV FUNCTION

Everything done above with respect to the construction of the Lyapunov function for systems with one regulator can be repeated also for control systems of the present type. However, we shall not repeat this in its entirety.

In the present chapter we shall only extend the Lur'e and Malkin methods to include systems with two regulators. This is quite sufficient to show, on the one hand, the methods used for such an extension, and on the other, to indicate the difficulty of obtaining stability criteria.

Let us assume that (9.11) has s real roots $\rho_1, \ldots, \rho_s$ and $n - s$ complex conjugate roots $\rho_{s+1}, \ldots, \rho_n$; all the roots will be assumed to be simple.

As the Lyapunov function for an inherently stable system, we take the positive definite function

$$V = F + \int_0^{\sigma_1} f_1(\sigma_1)d\sigma_1 + \int_0^{\sigma_2} f_2(\sigma_2)d\sigma_2 \ , \tag{9.30}$$

where

$$F = \sum_{k=1}^{n} \sum_{i=1}^{n} \frac{a_k a_i x_k x_i}{\rho_k + \rho_i} \ . \tag{9.31}$$

Here $a_1, \ldots, a_s$ are any real numbers, and $a_{s+1}, \ldots, a_n$ are arbitrary complex conjugate numbers. The total derivative of the function V will be, according to (9.21), of the following form

$$\begin{aligned} \dot{V} = &- \left(\sum_{k=1}^{n} a_k x_k \right)^2 - \\ &- r_{11}f_1^2(\sigma_1) - r_{22}f_2^2(\sigma_2) - (r_{12} + r_{21})f_1(\sigma_1)f_2(\sigma_2) + \\ &+ f_1(\sigma_1) \sum_{k=1}^{n} \left[\beta_{1k} + 2a_k \sum_{i=1}^{n} \frac{a_i u_1^{(i)}}{\rho_k + \rho_i} \right] x_k + \\ &+ f_2(\sigma_2) \sum_{k=1}^{n} \left[\beta_{2k} + 2a_k \sum_{i=1}^{n} \frac{a_i u_2^{(i)}}{\rho_k + \rho_i} \right] x_k \ . \end{aligned} \tag{9.32}$$

We could obtain a criterion for the system stability from formulas (9.32). However, before doing this, we shall use the elementary and already known transformation to reduce the expression of $\dot{V}$ into one of the following two forms

$$\begin{aligned} \dot{V} = &- \left[\sum_{k=1}^{n} a_k x_k + f_1(\sigma_1) + f_2(\sigma_2) \right]^2 - \\ &-(r_{11}-1)f_1^2(\sigma_1) - (r_{22}-1)f_2^2(\sigma_2) - (r_{12}+r_{21}-2)f_1(\sigma_1)f_2(\sigma_2) + \\ &+ \sum_{k=1}^{n} \left[\beta_{1k} + 2a_k + 2a_k \sum_{i=1}^{n} \frac{a_i u_1^{(i)}}{\rho_k + \rho_i} f_1(\sigma_1)x_k + \right. \\ &+ \sum_{k=1}^{n} \left[\beta_{2k} + 2a_k + 2a_k \sum_{i=1}^{n} \frac{a_i u_2^{(i)}}{\rho_k + \rho_i} \right] f_2(\sigma_2)x_k \end{aligned} \tag{9.33}$$

or

$$\dot{V} = -\left[\sum_{k=1}^{n} a_k x_k + \sqrt{r_{11}} f_1(\sigma_1) + \sqrt{r_{22}} f_2(\sigma_2)\right]^2 -$$

$$- (r_{12} + r_{21} - 2\sqrt{r_{11} r_{22}}) f_1(\sigma_1) f_2(\sigma_2) +$$

$$+ \sum_{k=1}^{n}\left[\beta_{1k} + 2\sqrt{r_{11}} a_k + 2a_k \sum_{i=1}^{n} \frac{a_i u_1^{(i)}}{\rho_k + \rho_i}\right] x_k f_1(\sigma_1) + \tag{9.34}$$

$$+ \sum_{k=1}^{n}\left[\beta_{2k} + 2\sqrt{r_{22}} a_k + 2a_k \sum_{i=1}^{n} \frac{a_i u_2^{(i)}}{\rho_k + \rho_i}\right] x_k f_2(\sigma_2) \quad .$$

In the latter expression it is assumed that $r_{11} > 0$ and $r_{22} > 0$.

In accordance with forms (9.32), (9.33), and (9.34) and the fundamental theorem of the Lyapunov direct method, the stability conditions of the investigated systems can be reduced in final analysis to the fulfillment of the relations

$$\begin{aligned} &\beta_{1k} + 2a_k \sum_{i=1}^{n} \frac{a_i u_1^{(i)}}{\rho_k + \rho_i} = 0 \quad , \\ &\beta_{2k} + 2a_k \sum_{i=1}^{n} \frac{a_i u_2^{(i)}}{\rho_k + \rho_i} = 0 \end{aligned} \qquad (k = 1, \ldots, n) \quad , \tag{9.35}$$

$$r_{11} > 0, \; 4r_{11}r_{22} > (r_{12} + r_{21})^2 \tag{9.36}$$

or of the relations

$$\begin{aligned} &\beta_{1k} + 2\dot{a}_k + 2a_k \sum_{i=1}^{n} \frac{a_i u_1^{(i)}}{\rho_k + \rho_i} = 0 \quad , \\ &\beta_{2k} + 2a_k + 2a_k \sum_{i=1}^{n} \frac{a_i u_2^{(i)}}{\rho_k + \rho_i} = 0 \end{aligned} \qquad (k = 1, \ldots, n) \; , \tag{9.37}$$

$$r_{11} > 1, \; 4(r_{11} - 1)(r_{22} - 1) > (r_{12} + r_{21} - 2)^2 \tag{9.38}$$

or, finally, of the relations

(9.39)
$$\beta_{1k} + 2\sqrt{r_{11}}a_k + 2a_k \sum_{i=1}^{n} \frac{a_i u_1^{(i)}}{\rho_k + \rho_i} = 0 \quad ,$$
$$(k = 1, \ldots, n) \quad ,$$
$$\beta_{2k} + 2\sqrt{r_{22}}a_k + 2a_k \sum_{i=1}^{n} \frac{a_i u_2^{(i)}}{\rho_k + \rho_i} = 0$$

(9.40)
$$r_{12} + r_{21} = 2\sqrt{r_{11}r_{22}} \quad .$$

If conditions (9.35) and (9.36) [or (9.37) and (9.38), or (9.39) and (9.40)] are fulfilled, the total derivative of the function V will be negative definite, thanks to which the fulfillment of the above relations actually guarantees the stability of the system. With this, relations (9.35), (9.37), and (9.39) narrow down the choice of the parameters $p_{1\alpha}$ and $p_{2\alpha}$, while relations (9.36), (9.38), and (9.40) narrow down the choice of the cross-connection constants r_{11}, r_{12}, r_{21}, and r_{22}.

Relations (9.39), unlike the analogous relations (9.37) and (9.35), contain the coefficients r_{11} and r_{22}. This circumstance can, in many cases, facilitate substantially the choice of the parameters p_1 and p_2, for r_{11} and r_{22} can always be considered as arbitrary positive numbers.

Thus, we have proved the following theorem: The steady state (9.10) of the control system (9.1) has absolute asymptotic stability if, for given real $\rho_1, \ldots, \rho_s$ and complex conjugate $\rho_{s+1}, \ldots, \rho_n$, for which $\operatorname{Re} \rho_k > 0$ $(k = 1, \ldots, n)$, it is possible to choose constants $p_{1\alpha}$ and $p_{2\alpha}$ in such a way that there exists a single system s of real numbers $a_1, \ldots, a_s$ and $n - s$ complex conjugate pairs $a_{s+1}, \ldots, a_n$, satisfying relations (9.35) and (9.37) or else relations (9.39), and, in addition, the coefficients of the cross-connections satisfy the relation (9.36) [or (9.38), (9.40)].

A remark is in order here. Relations (9.35), (9.37), and (9.39) are determined with an accuracy to within the choice of n constants of the canonical transformation and have the property that their first halves contain only $p_{1\alpha}$, and the second only $p_{2\alpha}$. Therefore, for an actual investigation of the problem of the solvability of the relations with respect to the unknowns $a_1, \ldots, a_n$, it is sometimes convenient to employ other relations which are equivalent to the first ones. To obtain these, we proceed in the following manner. We introduce the notation

(9.41) $u_1^{(s)} + u_2^{(s)} = k_s, \; u_1^{(s)} - u_2^{(s)} = m_s \qquad (s = 1, \ldots, n)$

and perform term-by-term addition, and also term-by-term subtraction of relations (9.35) [or correspondingly (9.37), and (9.39)]. This leads to the following stability criteria

(9.42)
$$\beta_{1k} + \beta_{2k} + 2a_k \sum_{i=1}^{n} \frac{a_i k_i}{\rho_k + \rho_i} = 0 , \qquad (k = 1, \ldots, n) ,$$
$$\beta_{1k} - \beta_{2k} + 2a_k \sum_{i=1}^{n} \frac{a_i m_i}{\rho_k + \rho_i} = 0$$

(9.43)
$$r_{11} > 0, \; 4r_{11}r_{22} > (r_{12} + r_{21})^2$$

or

(9.44)
$$\beta_{1k} + \beta_{2k} + 4a_k + 2a_k \sum_{i=1}^{n} \frac{a_i k_i}{\rho_k + \rho_i} = 0 , \qquad (k = 1, \ldots, n) ,$$
$$\beta_{1k} - \beta_{2k} + 2a_k \sum_{i=1}^{n} \frac{a_i m_i}{\rho_k + \rho_i} = 0$$

(9.45)
$$r_{11} > 1, \; 4(r_{11} - 1)(r_{22} - 1) > (r_{12} + r_{21} - 2)^2$$

or, finally,

(9.46)
$$\beta_{1k} + \beta_{2k} + 2(\sqrt{r_{11}} + \sqrt{r_{22}})a_k + 2a_k \sum_{i=1}^{n} \frac{a_i k_i}{\rho_k + \rho_i} = 0 , \qquad (k = 1, \ldots, n),$$
$$\beta_{1k} - \beta_{2k} + 2(\sqrt{r_{11}} - \sqrt{r_{22}})a_k + 2a_k \sum_{i=1}^{n} \frac{a_i m_i}{\rho_k + \rho_i} = 0$$

(9.47)
$$r_{12} + r_{21} = 2\sqrt{r_{11}r_{22}} .$$

5. EXAMPLE

By way of an example of the application of the above method, let us consider the case of the control system for which $n = 2$. Here we shall assume, for simplicity, that the numbers ρ_1 and ρ_2 are real.

Turning to (9.39), we write

$$\beta_{11} + 2\sqrt{r_{11}}a_1 + \frac{u_1^{(1)}}{\rho_1} a_1^2 + \frac{2u_1^{(2)}a_1 a_2}{\rho_1 + \rho_2} = 0 \quad , \tag{9.48}$$

$$\beta_{12} + 2\sqrt{r_{11}}a_2 + + \frac{u_1^{(2)}}{\rho_2} a_2^2 + \frac{2u_1^{(1)}a_1 a_2}{\rho_1 + \rho_2} = 0 \quad , \tag{9.49}$$

$$\beta_{21} + 2\sqrt{r_{22}}a_1 + \frac{u_2^{(1)}}{\rho_1} a_1^2 + \frac{2u_2^{(2)}a_1 a_2}{\rho_1 + \rho_2} = 0 \quad , \tag{9.50}$$

$$\beta_{22} + 2\sqrt{r_{22}}a_2 + \frac{u_2^{(2)}}{\rho_2} a_2^2 + \frac{2u_2^{(1)}a_1 a_2}{\rho_1 + \rho_2} = 0 \tag{9.51}$$

and attempt to satisfy (9.48) - (9.41) by means of numbers a_1 and a_2.

We multiply (9.48) and (9.49) by $\sqrt{r_{22}}$, and (9.50) and (9.51) by $-\sqrt{r_{11}}$. Then, after term-by-term addition of (9.48) to (9.50), and (9.49) to (9.51), we obtain

$$\begin{aligned} \frac{a_1^2}{\rho_1} U_1 + \frac{2a_1 a_2}{\rho_1 + \rho_2} U_2 &= B_1 \quad , \\ \frac{a_2^2}{\rho_2} U_2 + \frac{2a_1 a_2}{\rho_1 + \rho_2} U_1 &= B_2 \quad , \end{aligned} \tag{9.52}$$

where

$$U_s = u_1^{(s)}\sqrt{r_{22}} - u_2^{(s)}\sqrt{r_{11}}, \quad B_s = \beta_{2s}\sqrt{r_{11}} - \beta_{1s}\sqrt{r_{22}} \quad . \tag{9.53}$$

Next, we multiply the first of these equations by U/ρ_1, and the second by U_2/ρ_2 and add these. In addition, we multiply the first equation by $U_1\rho_1$, and the second by $U_2\rho_2$, and also add them. After these operations we obtain

(9.54)
$$\left[\frac{a_1U_1}{\rho_1} + \frac{a_2U_2}{\rho_2}\right]^2 = \Gamma^2 ,$$
$$[a_1U_1 + a_2U_2]^2 = D^2 ,$$

where

(9.55)
$$\Gamma^2 = \frac{B_1U_1}{\rho_1} + \frac{B_2U_2}{\rho_2}, \qquad D^2 = \rho_1B_1U_1 + \rho_2B_2U_2 .$$

If the regulator constants are such that the inequalities

(9.56)
$$\Gamma^2 > 0, \quad D^2 > 0$$

are fulfilled, then (9.54) can be solved, and we obtain

(9.57)
$$a_1 = \frac{C_1}{U_1}, \quad a_2 = \frac{C_2}{U_2} ,$$

where C_1 and C_2 are certain functions of the regulator parameters, equal to

(9.58)
$$C_1 = \frac{\pm \rho_1}{\rho_2 - \rho_1}(\rho_2\Gamma - D), \qquad C_2 = \frac{\pm \rho_2}{\rho_2 - \rho_1}(D - \rho_1\Gamma) .$$

Next, we multiply (9.48) by β_{21}, and (9.50) by $-\beta_{11}$, and add; we then multiply (9.49) by β_{22}, and add it to (9.51) which is multiplied by $-\beta_{12}$. Then, noting that $a_1 \neq 0$ and $a_2 \neq 0$, we find

(9.59)
$$2B_1 + \frac{a_1R_{11}}{\rho_1 + \rho_1} + \frac{a_2R_{12}}{\rho_1 + \rho_2} = 0 ,$$
$$2B_2 + \frac{a_1R_{21}}{\rho_1 + \rho_2} + \frac{a_2R_{22}}{\rho_2 + \rho_2} = 0 ,$$

where

(9.60)
$$R_{1s} = \beta_{21}u_1^{(s)} - \beta_{11}u_2^{(s)}$$
$$R_{2s} = \beta_{22}u_1^{(s)} - \beta_{12}u_2^{(s)} \qquad (s = 1, 2) .$$

If we now turn to (9.13) and (9.14), and take the quantities $C_2^{(s)}$

as the arbitrary constants of the canonical transformation, we get

$$
(9.61)\qquad
\begin{aligned}
&u_1^{(s)} = r_1^{(s)} C_2^{(s)}, \qquad u_2^{(s)} = r_2^{(s)} C_2^{(s)} \ , \\
&C_1^{(s)} = - \frac{\rho_s + b_{22}}{b_{12}} C_2^{(s)}, \qquad (s = 1,\ 2) \ ,
\end{aligned}
$$

$$
(9.62)\qquad
\begin{aligned}
&r_1^{(s)} = - \frac{1}{\rho_s} \left[n_{21} - \frac{n_{11}(\rho_s + b_{22})}{b_{12}} \right] , \\
& \qquad\qquad (s = 1,\ 2) \ . \\
&r_2^{(s)} = - \frac{1}{\rho_s} \left[n_{22} - \frac{n_{12}(\rho_s + b_{22})}{b_{12}} \right]
\end{aligned}
$$

On the other hand, solving (9.19) we obtain

$$
(9.63)\qquad
\begin{aligned}
&\dot{\eta}_1 = \frac{1}{\Delta} (- \rho_1 C_2^{(2)} x_1 + \rho_2 C_2^{(1)} x_2) \ , \\
&\dot{\eta}_2 = \frac{1}{\Delta} (- \rho_2 C_1^{(1)} x_2 + \rho_1 C_1^{(2)} x_1) \ ,
\end{aligned}
$$

where

$$
\Delta = \frac{\rho_2 - \rho_1}{b_{12}} C_2^{(1)} C_2^{(2)} \ .
$$

Then, according to (9.63), the expressions for $\beta_{k\alpha}$ become

$$
(9.64)\qquad
\begin{aligned}
&\beta_{11} = - \frac{\rho_1}{(\rho_2 - \rho_1) C_2^{(1)}} [b_{12} p_{11} + (\rho_2 + b_{22}) p_{12}] = \frac{\gamma_{11}}{C_2^{(1)}} \ , \\
&\beta_{12} = \frac{\rho_2}{(\rho_2 - \rho_1) C_2^{(2)}} [b_{12} p_{11} + (\rho_1 + b_{22}) p_{12}] = \frac{\gamma_{12}}{C_2^{(2)}} \ , \\
&\beta_{21} = - \frac{\rho_1 b_{12}}{(\rho_2 - \rho_1) C_2^{(1)}} [b_{12} p_{21} + (\rho_2 + b_{22}) p_{22}] = \frac{\gamma_{21}}{C_2^{(1)}} \ , \\
&\beta_{22} = \frac{\rho_2 b_{12}}{(\rho_2 - \rho_1) C_2^{(2)}} [b_{12} p_{21} + (\rho_1 + b_{22}) p_{22}] = \frac{\gamma_{22}}{C_2^{(2)}} \ .
\end{aligned}
$$

Let us now return to (9.59). By the conditions of the theorem.

we should satisfy these equations by means of solutions (9.57). When these solutions are inserted into (9.59), we find

$$2B_1 + \frac{C_1R_{11}}{2\rho_1 U_1} + \frac{C_2R_{12}}{(\rho_1+\rho_2)U_2} = 0 \quad ,$$

(9.65)

$$2B_2 + \frac{C_1R_{21}}{(\rho_1+\rho_2)U_1} + \frac{C_2R_{22}}{2\rho_2 U_2} = 0 \quad .$$

Next, according to formulas (9.53), (9.62) and (9.64), we obtain one group of equations

$$U_s = (r_1^{(s)}\sqrt{r_{22}} - r_2^{(s)}\sqrt{r_{11}})C_2^{(s)} = U_s^{(1)}C_2^{(s)} \quad ,$$

(9.66) $\qquad (s = 1,\ 2)$,

$$B_s = \frac{\gamma_{2s}\sqrt{r_{11}} - \gamma_{1s}\sqrt{r_{22}}}{C^{(s)}} = \frac{B_s'}{C_2^{(s)}}$$

and then, from formulas (9.60), (9.62), and (9.64), the second group of equations

$$R_{1s} = [\gamma_{21}r_1^{(s)} - \gamma_{11}r_2^{(s)}]\,\frac{C_2^{(s)}}{C_2^{(1)}} = \frac{R_{1s}'C_2^{(s)}}{C_2^{(1)}} \quad ,$$

(9.67) $\qquad (s = 1,\ 2)$.

$$R_{2s} = [\gamma_{22}r_1^{(s)} - \gamma_{12}r_2^{(s)}]\,\frac{C_2^{(s)}}{C_2^{(2)}} = \frac{R_{2s}'C_2^{(s)}}{C_2^{(2)}}$$

Now we insert the values of (9.66) and (9.67) into (9.65). Taking into account their structure with respect to the arbitrary quantities $C_2^{(1)}$ and $C_2^{(2)}$, we obtain, after obvious simplification,

$$2B_1' + \frac{C_1R_{11}'}{2\rho_1 U_1'} + \frac{C_2R_{12}'}{(\rho_1+\rho_2)U_2'} = 0 \quad ,$$

(9.68)

$$2B_2' + \frac{C_1R_{21}'}{(\rho_1+\rho_2)U_1'} + \frac{C_2R_{22}'}{2\rho_2 U_2'} = 0 \quad .$$

Equations (9.68) contain six parameters of the regulator: $p_{11}, p_{12}, p_{21}, p_{22}$,

r_{11}, and r_{22}. Solving these equations with respect to any pair of these parameters, for example, p_{12} and p_{21}, we obtain

$$p_{12} = p_{12}(p_{11}, p_{22}, r_{11}, r_{22}), \quad p_{21} = p_{21}(p_{11}, p_{22}, r_{11}, r_{22}) \,. \tag{9.69}$$

Thus, if the formulas (9.69) have been found, the stability region of the system in the space of the parameters p_{11} and p_{12}, $r_{11} > 0$, and $r_{22} > 0$, is separated by one two inequalities (9.56).

In this particular case, when $r_{11} = r_{22} = 1$, these inequalities become

$$\begin{aligned} r^2 &= (p_{11} - p_{21})[(n_{21} - n_{22})b_{12} + (n_{12} - n_{11})b_{22}] + \\ &\quad + (p_{22} - p_{12})[(n_{21} - n_{22})b_{11} + (n_{12} - n_{11})b_{21}] > 0 \,, \\ D^2 &= (p_{21} - p_{11})[(n_{21} - n_{22})b_{12} - (n_{12} - n_{11})b_{11}] + \\ &\quad + (p_{22} - p_{12})[(n_{21} - n_{22})b_{22} - (n_{12} - n_{11})b_{21}] > 0 \,. \end{aligned} \tag{9.70}$$

This example shows that the stability criterion for a control system, obtained on the basis of the above method, can be difficult to realize in view of the necessity of maintaining relation (9.69).

There is no doubt whatever about the fulfillment of condition (9.40).

6. CONSTRUCTION OF SIMPLIFIED STABILITY CRITERIA

In this section we shall detail the Malkin method for constructing simplified criteria of stability as applied to control systems with two regulators.*

For this purpose, we again return to (9.1), and to simplify the discussion, and also to make the discussions general to a certain extent, we introduce the notation

$$\begin{gathered} \xi_1 \quad \eta_{n+1}, \; \xi_2 = \eta_{n+2}, \; \varphi_1(\sigma_1) = f_1(\sigma_1) - h_1\sigma_1, \; \varphi_2(\sigma_2) = f_2(\sigma_2) - h_2\,\sigma_2, \\ b_{n+1,s} = b_{n+2,s} = 0 \quad (s = 1, \ldots, n + 2) \,, \\ - r_{11} = p_{1,n+1}, \quad - r_{12} = p_{1,n+2} \,, \\ - r_{21} = p_{2,n+1}, \quad - r_{22} = p_{2,n+2} \end{gathered} \tag{9.71}$$

---

* A. M. Letov and A. P. Duvakin, Stability of Control Systems With Two Control Organs. PMM, Volume XVIII, No. 2, 1954.

and assume that $f_1(\sigma_1)$ and $f_2(\sigma_2)$ are functions of subclass (A').

Denoting henceforth the number $n + 2$ by n, and separating in the functions $f_1(\sigma_1)$ and $f_2(\sigma_2)$ the linear terms $h_1(\sigma_1)$ and $h_2(\sigma_2)$, we write the initial equations in the following general form

$$\dot{\eta}_k = \sum_{\alpha=1}^{n} b_{k\alpha}\eta_\alpha + \sum_{s=1}^{2} u_{sk}\varphi_s(\sigma_s) \qquad (k = 1, \ldots, n) \;,$$

(9.72)

$$\sigma_s = \sum_{\alpha=1}^{n} p_{s\alpha}\eta_\alpha \qquad (s = 1, 2) \;.$$

We use $b_{k\alpha}$ and u_{sk} for constant quantities, where the number $b_{k\alpha}$ generally speaking includes those that depend on the regulator parameters; $\varphi_s(\sigma_s)$ are functions of class (A), satisfying the conditions

$$\varphi_s(\sigma_s) = 0, \quad \text{if} \quad \sigma_s = 0 \;,$$

$$\sigma_s\varphi(\sigma_s) > 0, \quad \text{if} \quad \sigma_s = 0 \;.$$

We shall assume that the equilibrium position

(9.73) $$\overset{*}{\eta}_1 = \overset{*}{\eta}_2 = \ldots = \overset{*}{\eta}_n = 0$$

is unique over the entire space of the variables. We shall also assume that all the roots of the equation

(9.74) $$|b_{k\alpha} + \delta_{k\alpha}\rho| = 0$$

are simple and satisfy the condition

(9.75) $$\text{Re } \rho_k > 0 \qquad (k = 1, \ldots, n).$$

Our problem is to establish sufficient conditions for asymptotic stability of the trivial solution (9.73) for all initial disturbances.

Let us consider (9.72), assuming for the time being all $u_{sk} = 0$, and let us specify an arbitrary, negative definite quadratic form $W(\eta_1, \ldots, \eta_n)$. Then the form

(9.76) $$-W = \sum_{\alpha=1}^{n}\sum_{\beta=1}^{n} A_{\alpha\beta}\eta_\alpha\eta_\beta$$

will be positive definite, and its coefficients can be arbitrary real numbers, satisfying the Sylvester inequalities

$$\begin{vmatrix} A_{11} & \cdots & A_{1k} \\ \cdot & \cdots & \cdot \\ A_{k1} & \cdots & A_{kk} \end{vmatrix} > 0 \qquad (k = 1, \ldots, n) \quad . \tag{9.77}$$

As was shown by Lyapunov, who investigated equations of the form (9.72) for $u_{sk} = 0$, there exists a unique quadratic form

$$F = \sum_{\alpha=1}^{n} \sum_{\beta=1}^{n} B_{\alpha\beta}\eta_\alpha\eta_\beta \qquad (B_{\alpha\beta} = B_{\beta\alpha}) \quad , \tag{9.78}$$

satisfying the equation

$$\sum_{i=1}^{n} \frac{\partial F}{\partial \eta_i} \left[\sum_{\alpha=1}^{n} b_{i\alpha}\eta_\alpha \right] = W \quad , \tag{9.79}$$

which, under conditions (9.75), must be positive definite. The coefficients of the form (9.78) are determined by comparison of the coefficients of like terms in the left and right half of (9.79).

Now let $u_{sk} \neq 0$. For the Lyapunov function V we take then the positive definite function

$$V = F + 2\int_0^{\sigma_1} \varphi_1(\sigma_1)d\sigma_1 + 2\int_0^{\sigma_2} \varphi_2(\sigma_2)d\sigma_2 \quad . \tag{9.80}$$

This function is suitable for the investigation of stability for all finite disturbances, since it is unbounded, owing to the presence of the quadratic form F in all the variables $\eta_1, \ldots, \eta_n$.

The total derivative of the function (9.80) with respect to time, written down by virtue of (9.72) and taken with the opposite sign, is given by the equation

$$-\dot{V} = -W + 2\sum_{k=1}^{2} \sum_{\alpha=1}^{m} p_{\alpha k}\eta_\alpha\varphi_k(\sigma_k) + 2\sum_{k=1}^{2} \sum_{s=1}^{2} r_{sk}\varphi_s(\sigma_s)\varphi_k(\sigma_k) \quad , \tag{9.81}$$

where

$$P_{\alpha k} = - \sum_{i=1}^{n} (b_{i\alpha}p_{ki} + B_{i\alpha}u_{ki}) \qquad (\alpha, k = 1, \ldots, n) ,$$

(9.82)

$$- r_{sk} = \sum_{i=1}^{n} p_{ki}u_{si} \qquad (s = 1, 2) .$$

The trivial solution (9.72) will be asymptotically stable if the expression (9.81) is a positive definite function of the arguments $\eta_1, \ldots, \eta_n, \varphi_1$ and φ_2. Considering the expression (9.81) as a quadratic form of $n + 2$ arguments $\eta_1, \ldots, \eta_n, \varphi_1$, and φ_2, and using the Sylvester theorem, we conclude that the necessary and sufficient condition for this form to be positive is that all the principal minors of its discriminant D_{n+2} be positive, where

$$D_{n+2} = \begin{vmatrix} A_{11} & \cdots & A_{1n} & P_{11} & P_{22} \\ \cdot & \cdot & \cdot & \cdot & \cdot \\ A_{n1} & \cdots & A_{nn} & P_{n1} & P_{n2} \\ P_{11} & \cdots & P_{n1} & 2r_{11} & r_{12} + r_{21} \\ P_{12} & \cdots & P_{n2} & r_{12} + r_{21} & 2r_{22} \end{vmatrix} . \tag{9.83}$$

But, as can be seen from (9.77), the first n minors of this determinant are positive by virtue of the fact that $- W$ is positive definite. Therefore, there remain two essential conditions

$$D_{n+1} > 0 \quad \text{and} \quad D_{n+2} > 0 . \tag{9.84}$$

From this we obtain the following theorem: The undisturbed motion (9.73) of the system (9.72) is asymptotically stable for all initial disturbances, if conditions (9.75) and (9.84) are satisfied simultaneously.

By way of an example of a system with two regulators, let us consider the simplest system, describable by equations

$$\begin{aligned} \dot{x}_1 &= - \rho_1 x_1 + u_{11}\varphi_1(\sigma_1) + u_{21}\varphi_2(\sigma_2) , \\ \dot{x}_2 &= - \rho_2 x_2 + u_{12}\varphi_1(\sigma_1) + u_{22}\varphi_2(\sigma_2) , \\ \dot{\sigma}_1 &= p_{11}x_1 + p_{12}x_2 - r_{11}\varphi_1(\sigma_1) , \\ \dot{\sigma}_2 &= p_{21}x_1 + p_{22}x_2 - r_{22}\varphi_2(\sigma_2) . \end{aligned} \tag{9.85}$$

We studied such a system in Section 5 of this chapter.

Let us find the sufficient condition for the absolute stability of the steady state of the system, describable by equations

$$x_1^* = x_2^* = \sigma_1^* = \sigma_2^* = 0 \quad .$$

With this, the numbers ρ_k will be considered to satisfy condition (9.75).

Let the Lyapunov function be of the form

$$V = F + 2\int_0^{\sigma_1} \varphi_1(\sigma_1)d\sigma_1 + 2\int_0^{\sigma_2} \varphi_2(\sigma_2)d\sigma_2 \quad ,$$

where

$$F = B_{11}x_1^2 + 2B_{12}x_1x_2 + B_{22}x_2^2$$

is an arbitrary positive definite function, and the second and third terms satisfy the condition

$$\int_0^{\infty} \varphi_1(\sigma_1)d\sigma_1 = \infty, \quad \int_0^{\infty} \varphi_2(\sigma_2)d\sigma_2 = \infty \quad .$$

The derivative dV/dt, obtained from (9.85), is determined by the equation

$$\begin{aligned} -\dot{V} = {} & A_{11}x_1^2 + 2A_{12}x_1x_2 + A_{22}x_2^2 + 2P_{11}x_1\varphi_1(\sigma_1) + \\ & + 2P_{21}x_2\varphi_1(\sigma_1) + 2P_{12}x_1\varphi_2(\sigma_2) + 2P_{22}x_2\varphi_2(\sigma_2) + \\ & + 2r_{11}\varphi_1^2(\sigma_1) + 2r_{22}\varphi_2^2(\sigma_2) \quad , \end{aligned}$$

where

$$A_{11} = 2\rho_1 B_{11}, \quad A_{12} = (\rho_1 + \rho_2)B_{12}, \quad A_{22} = 2\rho_2 B_{22},$$

$$P_{11} = -B_{11}u_{11} - B_{12}u_{12} - p_{11} \quad ,$$

$$P_{12} = -B_{11}u_{21} - B_{12}u_{22} - p_{21} \quad ,$$

$$P_{21} = -B_{12}u_{11} - B_{22}u_{12} - p_{12} \quad ,$$

$$P_{22} = -B_{12}u_{21} - B_{22}u_{22} - p_{22} \quad .$$

Let us denote

$$D_{31} = \begin{vmatrix} A_{11} & A_{12} & P_{11} \\ A_{12} & A_{22} & P_{21} \\ P_{11} & P_{21} & 2r_{11} \end{vmatrix} , \qquad D_{32} = \begin{vmatrix} A_{11} & A_{12} & P_{12} \\ A_{12} & A_{22} & P_{22} \\ P_{12} & P_{22} & 2r_{22} \end{vmatrix} ,$$

$$D_4 = \begin{vmatrix} A_{11} & A_{12} & P_{11} & P_{12} \\ A_{12} & A_{22} & P_{21} & P_{22} \\ P_{11} & P_{21} & 2r_{11} & 0 \\ P_{12} & P_{22} & 0 & 2r_{22} \end{vmatrix} .$$

Let the second regulator be disconnected, i.e., $u_{21} = u_{22} = p_{21} = p_{22} = r_{22} = 0$. Then the sufficient condition for the stability of the obvious solution of the system (9.85) is

$$D_{31} > 0 \quad . \tag{9.86}$$

If the first regulator is disconnected $(u_{11} = u_{12} = p_{11} = p_{12} = r_{11} = 0)$, the stability condition has the same form

$$D_{32} > 0 \quad . \tag{9.87}$$

If both regulators are disconnected, then the undisturbed motion of the system (9.85) will be stable, if, simultaneously with one of the conditions (9.86) or (9.87), there is satisfied the condition

$$D_4 > 0 \quad . \tag{9.88}$$

If $B_{11} = 1$, $B_{22} = 1$, and $B_{12} = 0$, these conditions become

$$\begin{aligned} D_{31} &= 8r_{11}\rho_1\rho_2 - 2\rho_1(u_{12} + p_{12})^2 - 2\rho_2(u_{11} + p_{11})^2 > 0 \quad , \\ D_{32} &= 8r_{22}\rho_1\rho_2 - 2\rho_1(u_{22} + p_{22})^2 - 2\rho_2(u_{21} + p_{21})^2 > 0 \quad , \\ D_4 &= 2r_{22}D_{31} + 2r_{11}D_{32} - 16r_{11}r_{22}\rho_1\rho_2 + \\ &\quad + [(u_{11} + p_{11})(u_{22} + p_{22}) - (u_{12} + p_{12})(u_{21} + p_{21})]^2 > 0 \quad . \end{aligned} \tag{9.89}$$

The stability conditions (9.89) are broader compared with the conditions of Section 5, for they do not contain equations that express additional relations between the regulator parameters.

CHAPTER X: TWO SPECIAL PROBLEMS IN THE THEORY OF STABILITY OF CONTROL SYSTEMS

1. STABILITY IN THE FIRST APPROXIMATION

All our considerations of the absolute stability of control systems have been based on the assumption that (1.1), which describe the disturbed motion of the regulated object, are linear. In the general case, however, these equations must be looked upon as being first-approximation equations. The actual equations of disturbed motion contain nonlinear terms, characterizing the regulated object. They are of the form

$$\dot{\eta}_k = \sum_{k=1}^{n} b_{k\alpha}\eta_\alpha + h_k\xi + \Xi_k(\eta_1 \quad \ldots, \eta_n, \xi) \qquad (k = 1, \ldots, n) ,$$

(10.1)

$$\dot{\xi} = - \rho_{n+1}\xi + f(\sigma), \qquad \sigma = \sum_{\alpha=1}^{n} p_\alpha\eta_\alpha - \xi \quad ,$$

where Ξ_k are certain functions of the arguments $\eta_1, \ldots, \eta_n$ and ξ, which assume zero values when $\eta_1 = \cdots = \eta_n = \xi = 0$. In connection with this, the question arises of finding strict conditions whose fulfillment guarantees the correctness of all the arguments developed above concerning the absolute stability of the system (10.1), made on the basis of an investigation of the first-approximation equations.

The principal purpose of this exposition reduces to indicating such cases, in which conclusions concerning the stability of control system (10.1), reached through an investigation of the first-approximation equations, remain valid, at least for all disturbances η_{ko} and ξ_o that have a sufficiently small modulus.

Thus, let us consider (10.1). We assume that the functions Ξ_k

satisfy the conditions*

$$(10.2) \qquad |\Xi_k(\eta_1, \ldots, \eta_n, \xi)| < \alpha\{|\eta_1| + |\eta_2| + \ldots + |\eta_n| + |\xi|\} ,$$

where α is a sufficiently small positive number. Wishing to solve this problem, we again employ transformation (2.1) to canonical variables. Repeating the arguments of Section 1 of Chapter II, we obtain instead of (4.1) (editor's note: the author has omitted the nonlinear terms in the equation below for $\dot{\sigma}$)

$$\dot{x}_k = -\rho_k x_k + f(\sigma) + X_k(x_1, \ldots, x_n, x_{n+1}, \sigma) ,$$

$$(10.3) \qquad \sigma = \sum_{k=1}^{n+1} \gamma_k x_k , \qquad (k = 1, \ldots, n+1) .$$

$$\dot{\sigma} = \sum_{k=1}^{n+1} \beta_k x_k - rf(\sigma)$$

Here the constants ρ_k are the roots of (2.5), of which $\rho_1, \ldots, \rho_s$ are real and $\rho_{s+1}, \ldots, \rho_{n+1}$ are complex conjugate pairs. All the roots ρ_k are considered to be simple and satisfying the condition

$$(10.4) \qquad \operatorname{Re} \rho_k \geq 0 \qquad (k = 1, \ldots, n+1) .$$

The values of β_k and γ_k are determined from the known formulas (2.10).

Functions X_k are found from the functions Ξ_k after eliminating from the latter the non-canonical variables $\eta_1, \ldots, \eta_n$ and ξ. By virtue of assumption (10.2), the functions X_k satisfy the following conditions, which are correct within a certain region G of the origin

$$(10.5) \qquad |X_k(x_1, \ldots, x_{n+1}, \sigma)| < \varepsilon\{|x_1| + \ldots + |x_{n+1}|\} ,$$

where ε is a sufficiently small positive number. Let us note that the property (10.5) is inherent in all the functions X_k, which are expandable in the vicinity of G in converging power series, starting at least with second-order terms.

Proceeding to a study of the stability of control systems, let us consider a V-function of the form

---

* I. G. Malkin, Teoriya ustoĭchivosti dvizheniya (Theory of Stability of Motion), Gostekhizdat, 1952.

(10.6) $$V = \Phi + F + \int_0^\sigma f(\sigma)d\sigma \quad .$$

Let us assume that none of the numbers ρ_k vanishes. Repeating the arguments given in Section 3 of Chapter IV, and demanding that the relations (4.12) be satisfied, we obtain an expression for the total derivative V:

$$\dot{V} = - W_1 + W_2$$

where

$$W_1 = \sum_{k=1}^{s} \rho_k A_k x_k^2 + C_1(\rho_{s+1} + \rho_{s+2})x_{s+1}x_{s+2} + \cdots$$

$$\cdots + C_{n-s}(\rho_n + \rho_{n+1})x_n x_{n+1} + \left[\sum_{k=1}^{n+1} a_k x_k + f(\sigma) \right]^2 ,$$

(10.7) $$W_2 = \sum_{k=1}^{s} A_k x_k X_k + C_1(x_{s+1}X_{s+2} + x_{s+2}X_{s+1}) +$$

$$+ C_3(x_{s+3}X_{s+4} + x_{s+4}X_{s+3}) + \cdots$$

$$\cdots + \sum_{i=1}^{n+1} \sum_{k=1}^{n+1} \frac{a_i a_k}{\rho_k + \rho_i} (x_k X_i + x_i X_k) \quad .$$

Under the existing limitations that we impose on the constants A_k, C_α and a_k, and on the function X_k, we cannot gain any definite idea concerning the sign of the function W_2. However, for the region G near the origin, we have:

$$|W_2| < \varepsilon \left\{ \sum_{k=1}^{s} A_k|x_k| + C_1|x_{s+1}| + C_1|x_{s+2}| + \cdots \right.$$

$$\left. \cdots + C_{n-s}|x_n| + C_{n-s}|x_{n+1}| + 2 \sum_{k=1}^{n+1} \sum_{i=1}^{n+1} \left| \frac{a_k a_i}{\rho + k\rho_i} \right| |x_k| \right\} \times$$

$$\times \{|x_1| + \cdots + |x_{n+1}|\} \quad .$$

Let us assume

$$x_s = Ry_s, \quad f(\sigma) = R\zeta, \quad R = +\sqrt{\sum_{k=1}^{n+1} x_k^2 + f^2(\sigma)} \; .$$

(10.8)

$$(s = 1, \ldots, n + 1) \quad ,$$

where y_s are new variables, satisfying the equation

(10.9) $$y_1^2 + \ldots + y_{n+1}^2 + \zeta^2 = 1 \quad .$$

Let M be the largest of the numbers

$$A_k + 2 \sum_{i=1}^{n+1} \left| \frac{a_k a_i}{\rho_k + \rho_i} \right| \qquad (k = 1, \ldots, s) \quad ,$$

$$C_\alpha + 2 \sum_{i=1}^{n+1} \left| \frac{a_{s+\alpha} a_i}{\rho_{s+\alpha} + \rho_i} \right| \qquad (\alpha = 1, \ldots, n - s) \quad .$$

Then

$$|W_2| < \varepsilon M R^2 \{ |y_1| + \ldots + |y_{n+1}| \}^2 = \varepsilon R^2 \bar{W} \quad ,$$

where

$$\bar{W}_2 = M\{ |y_1| + \ldots + |y_{n+1}| \}^2 \quad .$$

On the sphere (10.9), the function $\bar{W}_2$ is bounded from above. On the other hand, we see that the sign-definite function W_1 is

$$W_1 = R^2 \left\{ \sum_{k=1}^{s} \rho_k A_k y_k^2 + C_1(\rho_{s+1} + \rho_{s+2}) y_{s+1} y_{s+2} + \ldots \right.$$

$$\left. \ldots + C_{n-s}(\rho_n + \rho_{n+1}) y_n y_{n+1} + \left[\sum_{k=1}^{n+1} a_k y_k + \zeta \right]^2 \right\} = R^2 \bar{W}_1 \quad ,$$

whereby the function $\bar{W}_1$ assumes positive values W_1 on the sphere (10.9), which are bounded from below. It follows from this, that at sufficiently small values of ε, the sign of the function $\dot{V}$ is determined by the sign of the function $-W_1$, and if conditions (4.12) are fulfilled, the stability of the initial system (10.1) is guaranteed.

The dimensions of the region G are determined by the number ε, for which we can write

$$\min \bar{W}_1 - \varepsilon \max \bar{W}_2 > 0 \quad . \tag{10.10}$$

Let us now assume that among the numbers ρ_k there exists at least one, for example ρ_1, which equals zero, and therefore the control system is neutral with respect to the coordinate η_1. Let us call such a case a critical case. Then, instead of function (10.6), we assume

$$V = \frac{1}{2} A_1 x_1^2 + \sum_{k=2}^{n+1} \sum_{i=2}^{n+1} \frac{a_k a_i x_k x_i}{\rho_k + \rho_i} \int_0^\sigma f(\sigma) d\sigma + \Phi \quad ,$$

and the function Φ will no longer contain the term $(1/2)A_1 x_1^2$.

Calculation of the total derivative of the function V leads us to the equation

$$\dot{V} = - W_1 + W_2 + A_1 x_1 X_1 \quad ,$$

where

$$W_1 = \sum_{k=2}^{s} \rho_k A_k x_k^2 + C_1 (\rho_{s+1} + \rho_{s+2}) x_{s+1} x_{s+2} + \dots$$

$$\dots + \left[\sum_{k=2}^{n+1} a_k x_k + f(\sigma) \right]^2 \quad ,$$

$$W_2 = \sum_{k=2}^{s} A_k x_k X_k + C_1 (x_{s+1} X_{s+2} + x_{s+2} X_{s+1}) + \dots$$

$$\dots + \sum_{k=2}^{n+1} \sum_{i=2}^{n+1} \frac{a_k a_i}{\rho_k + \rho_i} (x_k X_i + x_i X_k) \quad .$$

Here the constants $A_1, \dots, A_s, C_1, \dots, C_{n-s}$ and a_k are subject to relations (4.22).

Obviously, the function W_1 is only a sign-constant function in the variables $x_1, \dots, x_{n+1}$, which assumes either negative or zero values. Therefore, no matter how small the positive number ε, we cannot state, without imposing additional limitations, that the sign of the function $\dot{V}$

is determined everywhere in the vicinity G by the sign of the function $-W_1$. In this case, everything depends on the form of the nonlinear terms X_k, and particularly on X_1. Apparently, it is possible to have here both stable and unstable cases. Consequently, in the presence of one zero root of (2.5), and if only conditions (4.22) are satisfied, it is impossible to guarantee the stability of the system (10.1).

In this critical case, the solution of the first fundamental problem in the theory of automatic control requires a study of the initial equations (10.1), similar to that required in analogous cases in the theory of stability.

We have thus proved the following theorem: In all non-critical cases of a solution of the first fundamental problem in the theory of automatic control, the stability of the initial system (10.1) is guaranteed by the fulfillment of the stability conditions of the first-approximation system (1.31), conditions formed on the basis of the Lur'e theorem. This stability occurs for all functions Ξ_k, satisfying inequalities (10.1), if at the same time the disturbances, superimposed on the system (10.1), are restricted to a sufficiently small region about the origin.

All critical cases, characterized by the vanishing of one or several ρ_k, can be studied on the basis of the second or third canonical forms of the initial equations. In fact, let the equations of the initial system be of the form

$$\dot{\eta}_k = \sum_{\alpha=1}^{n} \bar{b}_{k\alpha}\eta_\alpha + \bar{n}_k\sigma + \Xi_k(\eta_1, \ldots, \eta_n, \sigma) ,$$

$$(10.11) \qquad \dot{\sigma} = \sum_{\alpha=1}^{n} \bar{p}_\alpha\eta_\alpha - \bar{\rho}\sigma - rf(\sigma) + \bar{\bar{\sigma}} ,$$

$$\bar{\bar{\sigma}} = \sum_{\alpha=1}^{n} \bar{p}_\alpha\Xi_\alpha(\eta_1, \ldots, \eta_n, \sigma) .$$

These are obtained from (10.1) by substituting (3.1) and eliminating the variable ξ. All the constants entering here have values that are determined from formulas (3.2) and (3.4).

Let us assume that we have carried out the transformation of (10.11) to the canonical variables (3.6). Then, instead of (3.14) we have

$$\dot{x}_k = - r_k x_k + \sigma + X_k(x_1, \ldots, x_n, \sigma) ,$$

(10.12)

$$\dot{\sigma} = \sum_{k=1}^{n} \bar{\beta}_k x_k - \bar{\rho}\sigma - rf(\sigma) + \bar{\sigma} ,$$

where r_k are simple roots of (3.9), and the numbers $\bar{\beta}_k$ and $\bar{\rho}$ are determined by the formulas (3.13) and (3.11). We shall assume that in any of the above-mentioned critical cases the constants of the regulator satisfy inequalities (3.10); that the numbers $r_1, \ldots, r_s$ are real, while $r_{s+1}, \ldots, r_n$ are complex conjugate pairs; and that each r_k is not equal to zero. The latter limitation is not essential, for cases when the roots of (3.9) vanish are exceptional.

Next, let us assume that the function $f(\sigma)$ belongs to subclass (A').

Let us consider the positive definite function V, defined by (6.4). Its total derivative, calculated in accordance with (10.12) is

$$\dot{V} = - W_1 + W_2 ,$$

where

$$W_1 = \sum_{k=1}^{s} r_k A_k x_k^2 + C_1(r_{s+1} + r_{s+2})x_{s+1}x_{s+2} + \cdots$$

$$\cdots + C_{n-s}(r_n + r_{n-1})x_n x_{n-1} + \left[\sum_{k=1}^{n} a_k x_k + \sigma \right]^2 +$$

$$+ \left[\chi^2(\bar{\rho} + h) - 1 \right]\sigma^2 + \chi^2 \sigma\varphi(\sigma) ,$$

$$W_2 = \sum_{k=1}^{s} A_k x_k X_k + C_1(x_{s+1}X_{s+2} + x_{s+2}X_{s+1}) + \cdots$$

$$\cdots + \sum_{k=1}^{n} \sum_{i=1}^{n} \frac{a_k a_i}{r_k + r_i}(x_k X_i + x_i X_k) + \chi^2 \sigma\bar{\sigma} .$$

In this calculation it is assumed that the numbers $A_1, \ldots, A_s, C_1, \ldots, C_{n-s}, a_1, \ldots, a_n$ and h obey relations (6.10) and (6.15).

Subject to these assumptions regarding the constants $r_1, \ldots, r_n$ and $A_1, \ldots, C_{n-s}$, the function W_1 is sign-definite and assumes everywhere positive values.

With respect to the functions W_1 and W_2, we can repeat all the above considerations and find the ε-characteristic of that region G near the origin at which the sign of the derivative $\dot{V}$ is determined only by the sign of the function $-W_1$. This characteristic will be determined by an inequality analogous to inequality (10.10). In the vicinity G, the function $\dot{V}$ is negative definite, thanks to which relations (6.10) and (6.15) guarantee the stability of the control system (10.1).

Thus, we have proved the following theorem: In all cases (critical and non-critical), the stability of the initial system (10.11) is guaranteed by the fulfillment of the stability conditions of the first-approximation system (3.5), conditions based on the theorem of Section 1 of Chapter VI. This stability occurs for all functions Ξ_k, satisfying inequalities (10.2), provided that in this case the distrubances superimposed on the system (10.11) are confined to a sufficiently small region G near the origin.

2. STABILITY IN THE CASE OF CONSTANT DISTURBING FORCES

In certain investigations of control systems, it becomes of interest to take into account continuously applied disturbing forces. In this section we consider only that case in which the disturbing forces are random or fully determined functions of time, but are sufficiently small in modulus. Thus, let us assume that the equations of the control system are of the form

$$\dot{\eta}_k = \sum_{\alpha=1}^{n} b_{k\alpha}\eta_\alpha + h_k\xi + \varepsilon_k(t) \qquad (k = 1, \ldots, n) ,$$

(10.13)

$$\dot{\xi} = -\rho_{n+1}\xi + f(\sigma), \qquad \sigma = \sum_{\alpha=1}^{n} p_\alpha\rho_\alpha - r\xi ,$$

where $\varepsilon_k(t)$ are disturbing forces, and where all the constants entering into the equations have the same values as before.

The statement of the problem of the stability of the system under constantly acting disturbing forces is contained in the introduction. Since we wish to resolve this problem for the case of control systems of the form (10.13), let us transform two canonical variables in accordance with the formulas

$$x_s = \sum_{\alpha=1}^{n} C_\alpha^{(s)}\eta_\alpha + \xi + y_s \qquad (s = 1, \ldots, n) . \tag{10.14}$$

Differentiating (10.14) and using (10.13), we have

$$\dot{x}_s = \sum_{\alpha=1}^{n} c_\alpha^{(s)} \left[\sum_{\beta=1}^{n} b_{\alpha\beta}\eta_\beta + h_\alpha \xi + \varepsilon_\alpha(t) \right] + (-\rho_{n+1}\xi + f(\sigma)) + \dot{y}_s$$

$$(s = 1, \ldots, n) \quad .$$

If the constants $c_\alpha^{(s)}$ are determined as before with the aid of formulas (2.3) and (2.4), and if the variables y_s are determined as being certain solutions of the differential equations

$$\dot{y}_s = -\rho_s y_s - \sum_{\alpha=1}^{n} c_\alpha^{(s)} \varepsilon_\alpha(t) \qquad (s = 1, \ldots, n) \quad , \tag{10.15}$$

then the first equations assume the canonical form

$$\dot{x}_s = -\rho_s x_s + f(\sigma) \qquad (s = 1, \ldots, n) \quad . \tag{10.16}$$

In formulas (2.3) and (2.4) and equations (10.15) and (10.16), the numbers ρ_k are the roots of (2.5), and we assume, as before, that they are simple and satisfy condition (2.6). Subject to these limitations, the transformation (10.14) will be non-singular and can be resolved with respect to the initial variables $\eta_1, \ldots, \eta_n$. Thus, we find

$$\eta_\alpha = \sum_{k=1}^{n} D_k^{(\alpha)}(x_k - y_k) + G_\alpha \xi \qquad (\alpha = 1, \ldots, n) \quad , \tag{10.17}$$

where the constants $D_k^{(\alpha)}$, and G_α are determined by the formulas (2.36).

To form the last equation of the canonical form, let us introduce a new variable

$$\sigma = \sum_{\alpha=1}^{n} p_\alpha \eta_\alpha - r\xi \quad . \tag{10.18}$$

Using the symbols of (2.10) and the formulas (10.17), we find

$$\sigma = \sum_{k=1}^{n} \gamma_k (x_k - y_k) + \gamma_{n+1} x_{n+1} \quad . \tag{10.19}$$

Differentiating (10.19) and eliminating the quantities $\dot{x}_k$ with the aid

of (10.16), we obtain the equations of the canonical form for the constant disturbing forces

$$\dot{x}_s = -\rho_s x_s + f(\sigma)$$

(10.20)

$$\dot{\sigma} = \sum_{k=1}^{n} \beta_k x_k - rf(\sigma) + \dot{\psi}(t), \quad \sigma = \sum_{k=1}^{n+1} \gamma_k x_k + \psi(t) .$$

Here ψ denotes a disturbing function, of the form

$$\psi = -\sum_{k=1}^{n} \gamma_k y_k . \tag{10.21}$$

Before we proceed to an examination of the stability of the system (10.20), let us write the function ψ in explicit form, expressing it in terms of the solutions of (10.15). The latter are linear equations with constant coefficients. The general solution of each of these is given by the formula

$$y_s = N_s e^{-\rho_s t} + e^{-\rho_s t} \int_0^t e^{\rho_s \tau} \sum_{\alpha=1}^{n} C_\alpha^{(s)} \varepsilon_\alpha(\tau) d\tau \quad (s = 1, \ldots, n), \tag{10.22}$$

where N_s is an integration constant. To perform the canonical transformation, it is sufficient for us to have any one particular solution (10.22). By way of such a solution we take

$$y_s = e^{-\rho_s t} \int_0^t e^{\rho_s \tau} \sum_{\alpha=1}^{n} C_\alpha^{(s)} \varepsilon_\alpha(\tau) d\tau \qquad (s = 1, \ldots, n) . \tag{10.23}$$

In this case the disturbing force will be expressed in terms of the constants of the control system and of the disturbing forces ε_k, and will have the form

$$\psi = -\sum_{k=1}^{n} \gamma_k e^{-\rho_k t} \int_0^t e^{\rho_k \tau} \sum_{\alpha=1}^{n} C_\alpha^{(k)} \varepsilon_\alpha(\tau) d\tau . \tag{10.24}$$

Let us assume that the disturbing forces are described by either random or specified functions $\varepsilon_\alpha(t)$, which are small in modulus. Then for all $t > 0$ we have

(10.25) $$|\varepsilon_\alpha(t)| < \varepsilon \quad ,$$

where ε is a sufficiently small positive number. In this case, to estimate the modulus of the disturbing function ψ, we obtain the inequality

(10.26) $$|\psi| < \varepsilon \sum_{k=1}^{n} \sum_{\alpha=1}^{n} |\gamma_k| \, |C_\alpha^{(k)}| \, \frac{(1-e^{-\mathrm{Re}\,\rho_k t})}{\mathrm{Re}\,\rho_k} \quad .$$

Proceeding now to the problem of the stability of the system (10.20), subjected to the action of disturbing forces of limited modulus, let us consider the V-function, defined by (4.8). Here we shall consider only the ordinary case, i.e., where $\mathrm{Re}\,\rho_k \neq 0$. Calculating $\dot{V}$ on the basis of (10.20), and subjecting the choice of the constant of the regulator to relation (4.12), we then obtain the following final expression for the sought derivative

(10.27) $$\dot{V} = -\sum_{k=1}^{s} \rho_k A_k x_k^2 - C_1(\rho_{s+1} + \rho_{s+2})x_{s+1}x_{s+2} - \cdots$$
$$\cdots - C_{n-s}(\rho_n + \rho_{n-1})x_n x_{n-1} - \left[\sum_{k=1}^{n} a_k x_k + f(\sigma)\right]^2 + f(\sigma)\psi(t)$$

It remains now to estimate the sign of the derivative. For this purpose, we employ the new variables (10.8), and write the expression (10.27) in the form

(10.28) $$\dot{V} = -R^2 W + f(\sigma)\psi(t) \quad ,$$

where

(10.29) $$W = \sum_{k=1}^{s} \rho_k A_k y_k^2 + C_1(\rho_{s+1} + \rho_{s+2})y_{s+1}y_{s+2} + \cdots$$
$$\cdots + \left[\sum_{k=1}^{n} a_k y_k + \zeta\right]^2 \quad .$$

The function W is of definite sign and assumes positive and bounded values on the sphere (10.9).

Let $\min W$ be the smallest of these values, and

$$\max |\psi| = \varepsilon \sum_{k=1}^{n} \sum_{\alpha=1}^{n} \frac{|\gamma_k| \; |C_\alpha^{(k)}|}{\text{Re } \rho_k} .$$

Then there exists an R-vincinity near the origin, having the property that everywhere outside this region the function $\dot{V}$ is negative definite. To estimate the value of R, we make use of the inequality

$$R^2 > \frac{\max |\psi|}{\min W} \max |f(\sigma)| . \tag{10.30}$$

If the constant ε is sufficiently small, then for all finite values of the constants of the control system, the quantity $\max |\psi|$ is also sufficiently small. In this case, the R-vicinity just constructed can be made sufficiently small. This vicinity can be inscribed in the A-vicinity, which figures in the determination of stability in the case of constantly-acting disturbing forces.

Having fixed A, we obtain an A-vicinity of the origin of the coordinates, outside of which the value of the derivative (10.28) assumes everywhere only negative values. Consequently, any trajectory of the representative point M will intersect all the surfaces $V = C$ that include the A-vicinity $(C \geq A)$ of the origin, from the outside in. Thanks to this, after a lapse of a sufficiently long time $t = T$, the point M must fall into the A-region and will remain there for all $t > T$. But the latter indeed signifies stability of the undisturbed motion under continuously applied disturbances.

Let us note that this stability occurs for unlimited disturbances x_{ko} $(k = 1, \ldots, n)$ and for any function $f(\sigma)$ of class (A). Bearing in mind this limitation, we shall henceforth call this stability absolute stability, as we did earlier in the absence of disturbing forces. Thus, we have proved the following theorem: In all non-critical cases of solution of the first fundamental problem of the theory of automatic control, absolute stability of the initial control system (1.31), under continuously applied disturbing forces, is guaranteed by the fulfillment of the criteria derived on the basis of the Lur'e theorem, provided only that the disturbing forces are bounded in modulus to sufficiently positive values.

All the critical cases, characterized by the vanishing of one or several of the numbers ρ_k, require a different examination within the framework of this statement of the problem. We shall carry out this examination on the basis of the canonical equations of the second form. The initial equations are written as follows:

(10.31)
$$\dot{\eta}_k = \sum_{\alpha=1}^{n} \bar{b}_{k\alpha}\eta_\alpha + \bar{n}_\alpha\sigma + \varepsilon_k(t) \ ,$$
$$\dot{\sigma} = \sum_{\alpha=1}^{n} \bar{p}_\alpha\eta_\alpha - \bar{\rho}\sigma - f(\sigma) \qquad (k = 1, \ldots, n) \ .$$

Let the formulas

(10.32)
$$x_s = \sum_{\alpha=1}^{n} C_\alpha^{(s)}\eta_\alpha + y_s \qquad (s = 1, \ldots, n)$$

determine the canonical transformations. Here the constants $C_\alpha^{(s)}$ are chosen in accordance with formulas (3.7) and (3.8), while the functions y_s are determined as the solutions of the differential equations

(10.33)
$$\dot{y}_s = - r_s y_s - \sum_{\alpha=1}^{n} C_\alpha^{(s)}\varepsilon_\alpha(t) \ .$$

Then the equations can be rewritten as follows in the new variables:

(10.34)
$$\dot{x}_s = - r_s x_s + \sigma \ .$$

In formulas (3.7) and (3.8) and in equations (10.33) and (10.34), the quantities r_s are the roots of (3.9). It is assumed that these roots are simple, different from zero, and satisfy the condition $\operatorname{Re} r_k > 0$ $(k = 1, \ldots, n)$. The latter is realized because of fulfillment of conditions (3.10).

Under these assumptions, the transformation (10.32) can be solved with respect to the old variables η_k

(10.35)
$$\dot{\eta}_k = \sum_{k=1}^{n} \bar{D}_\alpha^{(k)}(x_\alpha - y_\alpha) \ ,$$

where $\bar{D}_\alpha^{(k)}$ have the previous values (3.18). Then, using the symbols of (3.13), we obtain finally

(10.36)
$$\dot{x}_s = - r_s x_s + \sigma, \quad \dot{\sigma} = \sum_{\alpha=1}^{n} \bar{\beta}_\alpha x_\alpha - \bar{\rho}\sigma - f(\sigma) + \dot{\psi}(t) \ ,$$

where the function $\psi(t)$ is determined by the equality

$$\psi = -\sum_{\alpha=1}^{n} \bar{\beta}_\alpha y_\alpha \quad . \tag{10.37}$$

If we take by way of the functions y_α the particular solution of (10.32) for zero initial conditions, we obtain for the disturbing function $\psi(t)$ the following expression:

$$\psi = -\sum_{k=1}^{n} \bar{\beta}_k e^{-r_k t} \int_0 e^{-r_k \tau} \sum_{\alpha=1}^{n} C_\alpha^{(k)} \varepsilon_\alpha(\tau) d\tau \quad . \tag{10.38}$$

Subject to the assumption (10.25) we can estimate the modulus of ψ:

$$|\psi| < \varepsilon \sum_{k=1}^{n} \sum_{\alpha=1}^{n} |\bar{\beta}_k| \, |C_\alpha^{(k)}| \, \frac{(1-e^{-\mathrm{Re}\, r_k t})}{\mathrm{Re}\, r_k} \quad . \tag{10.39}$$

If ε is sufficiently small, $|\psi|$ is also sufficiently small.

Let us consider the V-function (6.4), retaining in it the previous assumption concerning the quantities $A_1, \ldots, A_s, C_1, \ldots, C_{n-s}, X,$ and a_k and let us calculate its total derivative $\dot{V}$, using (10.36). As before, we subject the choice of the regulator parameters to relations (6.10) and (6.15). Under these conditions, the derivative $\dot{V}$ is determined by

$$\dot{V} = -\left[\sum_{k=1}^{n} a_k x_k + \sigma\right]^2 - [X^2(\bar{\rho} + h) - 1]\sigma^2 - X^2\sigma\varphi(\sigma) -$$

$$- \sum_{k=1}^{s} r_k A_k x_k^2 - C_1(r_{s+1} + r_{s+2})x_{s+1}x_{s+2} - \cdots + \sigma\psi(t) \quad .$$

Repeating the above arguments, we construct the A-vicinity of the origin, beyond which $\dot{V}$ assumes only negative values.

Consequently, any surface $V = C$ $(C \geq A)$ that contains this vicinity, will be intersected by a trajectory of the representative point from the outside in, so that after the lapse of a sufficiently long time $t = T$, this point will enter into the A-vicinity and will remain there for all $r > T$. But this indeed signifies stability under constantly-acting disturbances.

Thus, we have proved the following theorem: In all non-critical

as well as in all critical cases, absolute stability of the initial system (10.31) under constantly acting disturbing forces is guaranteed by fulfillment of the conditions formed on the basis of the theorem of Section 1 of Chapter VI, provided only the disturbing forces are restricted to a modulus of sufficiently small value.

3. CONCLUSION

The arguments presented in this chapter make possible one general conclusion.

By proving the theorem of the first approximation, we could treat the nonlinear term Ξ_k in (10.1) as a continuously applied disturbing force, restricted only by inequalities (10.2). Then we can readily observe the similarity between the theorems of the two first sections of this chapter. It is seen from these theorems that, no matter what the limitations on the modulus of the constantly-acting force that disturbs the system (1.31), an estimate of its stability, obtained on the basis of the Lur'e theorem, remains correct if this system is stable with the regulator disconnected.

Thus, however, if the control system is neutral with respect to one or several of the coordinates, then in order for the system to be stable under constantly-acting disturbing forces, it is sufficient to choose the regulator parameters in accordance with the criteria obtained on the basis of the theorem of Section 1, Chapter VI.

CHAPTER XI: STABILITY OF UNSTEADY MOTION

1. EQUATIONS OF DISTURBED MOTION

In many investigations of control systems, the problem arises of the stability of their unsteady motions. The latter, unlike the steady motions, are described by particular solutions of the initial differential equations, which in general represent known functions of time. One example of unsteady motion of a control system is the motion in response to a given program of regulation $\sigma_{**}(t)$. Unsteady motion of control systems can also occur as a result of the time variation of the parameters of the regulated object. In either case, the equations of the disturbed motion of the control system will differ substantially from those considered earlier.

Let us assume that the initial equations, describing the state of the regulated object, are of the form (7.1). Let (7.6) determine specified functions, corresponding to unsteady motion of the system. Repeating the arguments presented earlier, we obtain equations for the disturbed motion of the system, in the form

$$\dot{\eta}_k = \sum_{\alpha=1}^{m} b_{k\alpha}\eta_\alpha + n_k\xi \qquad (k = 1, \ldots, m) ,$$

$$(11.1) \qquad \dot{\xi} = f(\sigma) ,$$

$$\sigma = \sum_{\alpha=1}^{m} p_\alpha\eta_\alpha - \xi .$$

Unlike the equations considered before, we shall assume here that all the $b_{k\alpha}$, n_k $(\alpha, k = 1, \ldots, m)$ are specified functions of time, defined for $0 \leq t < T$; the regulator parameters p_α will also be considered

to be functions of time, to be determined. The quantity T is understood to mean any positive number (perhaps also $+\infty$); the subinterval $[0, T)$ is denoted by the letter N.

(11.1) are general. To obtain from the program-control equations (7.9) it is sufficient to put here $f(\sigma) = h\sigma$, and to determine the quantity h with the aid of equation $h = S(t)$.

Let us assume, finally, that to each system of initial values $\eta_k(0)$ and $\sigma(0)$, belonging to a certain bounded region G, there corresponds in N a unique solution of (11.1).

2. STATEMENT OF THE PROBLEM OF STABILITY OF UNSTEADY MOTION

Based on (11.1), let us formulate the problem of the stability of unsteady motion. This formulation will have one essential feature, due to the assumption of a possible limitation on the subinterval of the time variation by means of the finite number T.

The unsteady motion (7.6) of the control system corresponds to the obvious solution

$$(11.2) \qquad \eta_1^* = 0, \ \ldots, \ \eta_m^* = 0, \ \xi^* = 0$$

of the equations of the disturbed motion. Let us transform these equations to the third canonical form.

Returning to Section 5 of Chapter III, we note that the transformation considered there remains fully valid also in the case of variables $b_{k\alpha}$ and n_k. The only circumstance that must be borne in mind in this case is that in the formation of (11.1), the formulas (3.4) for the quantities $\bar{p}_\beta$ must be written in the following form:

$$(11.3) \qquad \bar{p}_\beta = \dot{p}_\beta + \sum_{\alpha=1}^{m} p_\alpha \bar{b}_{\alpha\beta} \ .$$

As before, the canonical equations of the problem will be (3.57), whose first integral is (3.58).

A feature of these equations, for the case of unsteady motion of control systems, is the dependence of the coefficient B_{rs}, Q_s of the form W (3.55) on the time. The time t enters into the coefficients of the form exclusively through the parameters $b_{k\alpha}$, n_k, and p_α of the control system.

Let us formulate the problem.

As follows from the above, the behavior of the representative point M $(\zeta_1, \ldots, \zeta_m, \zeta)$ in any vicinity of the investigated unsteady motion, is described by the radius vector R, whose modulus serves as a solution of the system of differential equations (3.57).

We shall call the control system (11.1) stable in N if, for any arbitrarily specified positive number A, no matter how small, it is possible to choose a positive number λ, such that for arbitrary disturbances $\eta_k(0)$ and $\sigma(0)$, belonging to G and satisfying the condition

(11.4) $$R(0) \leq \lambda ,$$

the following inequality will be fulfilled

(11.5) $$R(t) < A$$

for all $t > 0$, belonging to N.

The problem consists of determining the functions $p_\alpha(t)$ $(\alpha = 1, \ldots, m)$ $[t \in N]$, for which there is guaranteed the stability of the undisturbed unsteady motion (11.2) of system (11.1), no matter what disturbances in G there are, and no matter what the function $f(\sigma)$ of class (A) or subclass (A').

As before, the quantity R can be assumed to be in the form of a Euclidean metric of the space of variables $\eta_1, \ldots, \eta_m$ and σ. Examination of the behavior of the representative point M in phase space makes it possible to draw conclusions concerning the stability of the system both with constants as well as with variable coefficients at any interval N, using at the same time only the elementary tools of mathematical analysis.

In the work by G. V. Kamenkov* and in the work by A. A. Lebedev**, the very same method is used to investigate the quadratic metric of Euclidean space.

3. STABILITY OF UNSTEADY MOTIONS

Proceeding to the solution of the problem, let us note that the equation

$$\dot{\bar{R}} = - W\bar{R}$$

---

* G. V. Kamenkov, Stability of Motion Over a Finite Time Interval, PMM, Volume XVII, No. 5, 1953.

** A. A. Lebedev, Concerning the Problem of Stability of Motion Over a Finite Time Interval, PMM, Volume XVIII, No. 1, 1954.

has a solution

$$\bar{R}(t) = \bar{R}(0)e^{-\int W dt} = R(0)e^{-\int W dt} \quad .$$

Consequently, any function R, determined by differential equations (3.57), will satisfy, for all initial conditions, the following inequality

$$R(t) \leq R(0)e^{-\int W dt} \quad ,$$

if the function $\varphi(\sigma)$ belongs to class (A). It follows, therefore, that for the control system to be stable, it is sufficient that the function W assume positive values on the surface (3.58), no matter what the value of the function F^2, which is positive definite, or the value of the function $\varphi(\sigma)$ of class (A).

Thus, the question of the stability of the given control system reduces to formulating the conditions for the function W to be positive definite. These conditions can be formulated in two forms. For example, let us set up the determinant

$$(11.6) \qquad \begin{vmatrix} B_{11} & \cdots & B_{1m} & Q_1 \\ \cdot & \cdot & \cdot & \cdot \\ B_{m1} & \cdots & B_{mm} & Q_m \\ Q_1 & \cdots & Q_m & \bar{\rho} + h \end{vmatrix} , \quad \begin{vmatrix} B_{11} & \cdots & B_{1k} \\ \cdot & \cdot & \cdot \\ B_{k1} & \cdots & B_{kk} \end{vmatrix} \qquad (k = 1, \ldots, m)$$

and write the Sylvester inequalities. The left halves of these inequalities contain the time. On the other hand, there always exists a linear transformation to new variables, which reduces the real quadratic form W to a canonical form

$$W = \sum_{k=1}^{m+1} \lambda_k x_k^2 \quad ,$$

with the eigenvalues of the form being the positive roots of the equation

$$(11.7) \qquad \Delta(\lambda) = \begin{vmatrix} B_{11} - \lambda & \cdots & B_{1m} & Q_1 \\ \cdot & \cdot & \cdot & \cdot \\ B_{m1} & \cdots & B_{mm} - \lambda & Q_m \\ Q_1 & \cdots & Q_m & \bar{\rho} + h - \lambda \end{vmatrix} = 0 \quad .$$

Let us consider the equation $\Delta(-\lambda) = 0$ and write for this

equation the Hurwitz inequalities

$$\Delta_0 > 0,\ \Delta_1 > 0,\ \ldots,\ \Delta_{m+1} > 0 \quad . \tag{11.8}$$

The left halves of the inequalities contain the time. Consequently, fulfillment of the Sylvester inequalities or of inequalities (11.8) serves as the necessary and sufficient condition for the form W to be positive.

This leads to the following theorem: If the parameters of the regulator are so chosen that the Sylvester inequalities or the inequalities (11.8) are fulfilled for each value $t \in N$, then the control system (11.1) is stable for all disturbances and for all functions $f(\sigma)$ of class (A) or subclass (A').

4. SECOND BULGAKOV PROBLEM IN THE CASE OF VARIABLE COEFFICIENTS

To illustrate the above, let us consider the Bulgakov second problem. The equations of the disturbed motion of the system (2.81) have been studied previously in the case of constant k, U, and T.

To simplify the following arguments, let us transform these equations to new variables, determined with the aid of the equations

$$\sigma = (a - qG^2)\psi + (E - pG^2)\psi - \frac{1}{i}\eta \quad ,$$

$$\zeta = m\psi + n\dot{\psi} \quad ,$$

where m and n are constants, so far arbitrary. In the new variables, (2.81) can be written in the following manner:

$$\begin{aligned} \dot{\eta}_1 &= \bar{b}_{11}\eta_1 + \bar{b}_{12}\eta_2 \quad , \\ \dot{\eta}_2 &= \bar{b}_{21}\eta_1 + \bar{b}_{22}\eta_2 + \bar{n}_2\sigma \quad , \\ \dot{\sigma} &= \bar{p}_1\eta_1 + \bar{p}_2\eta_2 - \bar{\rho}\sigma - \frac{1}{i}f(\sigma) \quad , \end{aligned} \tag{11.9}$$

where

$$\eta_1 = \psi, \ \eta_2 = \zeta, \ \bar{b}_{11} = -\frac{m}{n}, \ \bar{b}_{12} = \frac{1}{n}, \ i = \frac{\ell T^2}{T^2 + \ell G^2},$$

$$\bar{b}_{21} = -m\left[\frac{m}{n} - \frac{U + \ell E}{T^2 + \ell G^2}\right] - \frac{n(k+a\ell)}{T^2 + \ell G^2},$$

$$\bar{b}_{22} = -\left[\frac{U + \ell E}{T^2 + \ell G^2} - \frac{m}{n}\right],$$

$$\bar{p}_1 = \frac{1}{n}\left[\dot{p}_2 + p_2\frac{1}{i}\frac{di}{dt} + p_1 - \frac{p_2(U+\ell E)}{T^2 + \ell G^2}\right], \tag{11.10}$$

$$\bar{p}_2 = \dot{p}_1 - \frac{m}{n}\dot{p}_2 + \frac{p_1}{t}\frac{di}{dt} - \frac{mp_2}{ni}\frac{di}{dt} - \frac{p_2(k+a\ell)}{T^2 + \ell G^2} - \frac{mp_1}{n} + \frac{m}{n}\frac{U + \ell E}{T^2 + \ell G^2}p_2,$$

$$\bar{\rho} = \frac{1}{i}\frac{di}{dt} - \frac{\ell p_2}{T^2 + \ell G^2}.$$

To set up the form W, let us consider the transformation (3.52).

From formulas (3.55) we obtain the following expressions for coefficients of the form:

$$B_{11} = \frac{a_{12}\bar{b}_{12} - a_{22}\bar{b}_{11}}{\Delta_2}a_{11},$$

$$B_{22} = \frac{a_{21}\bar{b}_{21} - a_{11}\bar{b}_{22}}{\Delta_2}a_{22},$$

$$B_{12} = \frac{a_{12}(\bar{b}_{11}+\bar{b}_{22}) - (a_{11}\bar{b}_{12} + a_{22}\bar{b}_{21})}{2\Delta_2}\sqrt{a_{11}a_{22}}, \tag{11.11}$$

$$Q_1 = \frac{S\sqrt{a_{11}}}{2\Delta_2}(a_{12}\bar{p}_2 - a_{22}\bar{p}_1),$$

$$Q_2 = \frac{\sqrt{a_{22}}}{2\Delta_2 S}[-\Delta_2\bar{n}_2 + S^2(a_{21}\bar{p}_1 - a_{11}\bar{p}_2)],$$

$$\Delta_2 = a_{11}a_{22} - a_{12}^2.$$

Let us consider the particular case of the choice of the form F^2, in which $a_{12} = a_{21} = 0$. We then have

(11.12)
$$B_{11} = -\bar{b}_{11}, \quad B_{22} = -\bar{b}_{22},$$
$$B_{12} = -\frac{a_{11}\bar{b}_{12} + a_{22}\bar{b}_{21}}{2\sqrt{a_{11}a_{22}}},$$
$$Q_1 = -\frac{S\bar{p}_1}{2\sqrt{a_{11}}}, \quad Q_2 = -\frac{\sqrt{a_{22}}}{2S}\left[n_2 + \frac{S^2\bar{p}_2}{a_{22}}\right],$$

The first inequality for (11.6) is of the form $m/n > 0$. It is not essential and can always be satisfied. For the left half of the second of these we have

$$B_{11}B_{22} - B_{12}^2 = \bar{b}_{11}\bar{b}_{22} - \frac{(a_{11}\bar{b}_{12}+a_{22}\bar{b}_{21})^2}{4a_{11}a_{22}}.$$

Using formulas (11.12), we obtain

$$4a_{11}a_{22}\frac{m}{n}\left[\frac{U+\ell E}{T^2+\ell G} - \frac{m}{n}\right] > \frac{a_{11}^2}{n^2} + m^2a_{22}\left[\frac{U+\ell E}{T^2+\ell G^2} - \frac{m}{n}\right]^2 +$$
$$+ \frac{n^2a_{22}^2(k+a\ell)^2}{T^2+\ell G^2} - \frac{2mna_{22}^2(k+a\ell)}{T^2+\ell G^2}\left[\frac{U+\ell E}{T^2+\ell G^2} - \frac{m}{n}\right] +$$
$$+ \frac{2a_{11}a_{22}}{n}\left[m\left(\frac{U+\ell E}{T^2+\ell G^2} - \frac{m}{n}\right) - \frac{n(k+a\ell)}{T^2+\ell G^2}\right]$$

Putting

(11.13)
$$\frac{U+\ell E}{T^2+\ell G^2} - \frac{m}{n} = y, \quad \frac{k+a\ell}{T^2+\ell G^2} = x$$

and carrying out elementary transformation on this inequality, we obtain the boundary for the stability region

(11.14)
$$F(xy) = n^2a_{22}^2x^2 + m^2a_{22}^2y^2 - 2mna_{22}^2xy -$$
$$- 2a_{11}a_{22}x - 2\frac{m}{n}a_{11}a_{22}y + \frac{a_{11}^2}{n^2} = 0.$$

It is easy to verify that the curve is a parabola, tangent to the y axis at a point with a positive ordinate a_{11}/mna_{22}, and tangent to the x axis at a point with a positive abscissa a_{11}/n^2a_{22}. The parabola is located in the first quadrant.

The internal points of this parabola correspond to the values of the parameters of the regulator, for which

$$B_{22} = y > 0; \quad F(xy) < 0 \quad . \tag{11.15}$$

It now remains to consider the last condition. It is of the form

$$B_{11}Q_1^2 + B_{22}Q_2^2 < (\rho + h)(B_{11}B_{22} - B_{12}^2) \quad . \tag{11.16}$$

In order for this condition to be fulfilled, it is necessary to have

$$\rho + h > 0 \quad . \tag{11.17}$$

Thus, for the system (2.81) to be stable in the case of unsteady motion, it is enough to choose the regulator parameters in accordance with inequalities (11.15), (11.16), and (11.17). In the case of variable p_α, inequalities (11.16) and (11.17) are differential ones.

CHAPTER XII: CONTROL SYSTEMS CONTAINING TACHOMETRIC FEEDBACK*

1. CASE OF INHERENTLY STABLE CONTROL SYSTEM

Many designs of modern automatic regulators frequently incorporate some tachometric device, which measures the rate of change in the position of the regulating organ, and generates a control signal proportional to this velocity. This signal is used to form the general law of regulation of a particular object. As far as any particular feedback loop goes, the same problems of automatic control arise as with respect to all other control circuits.

In this case, we proceed to study the stability of the system, in the same sense as defined in Chapter I.

The equations of the disturbed motion of the control systems, containing velocity feedback can be written as

$$\dot{\eta}_k = \sum_{\alpha=1}^{m} b_{k\alpha}\eta_\alpha + n_k\xi \qquad (k = 1, \ldots, m) \ ,$$

(12.1)

$$\dot{\xi} = f(\sigma), \qquad \sigma = \sum_{\alpha=1}^{m} p_\alpha\eta_\alpha - r\xi - N\dot{\xi} \ .$$

Unlike the preceding cases, the argument σ of the function $f(\sigma)$ contains the term $N\dot{\xi}$, characterizing the presence in the network of tachometric feedback that generates a control signal proportional to the velocity of the actuator.

In the case when the velcoity of the actuator is sufficiently large, the action of the proportional feedback reduces to a limitation of

---

* Tachometric feedback is sometimes called velocity feedback.

the modulus of the argument σ; there appears therefore a certain possibility, in the case of a simplified treatment of the problem, of assuming the actuator of the regulator to be ideal and of using in the solutions the equation $\sigma = 0$, i.e., the equation

$$N\dot{\xi} + r\xi = \sum_{\alpha=1}^{m} p_\alpha \eta_\alpha \quad . \tag{12.2}$$

Naturally, with such an approach, the stability problem is resolved by using the Hurwitz inequalities.

If, however, we wish to consider the fact that $f(\sigma)$ is restricted and is defined accurately only as belonging to a certain class of functions, and to consider also the known possibility of appearance of considerably large σ, at least in the initial phase of the transient, the Hurwitz inequalities may not be sufficient.

We therefore propose a different approach to the stability of the control system (12.1), based on the direct Lyapunov method. Subject to the assumptions made, it is necessary to find regulator parameters that guarantee the stability of the undisturbed motion of the control system (12.1), corresponding to the trivial solution

$$\eta_1^* = 0, \ \ldots, \ \eta_m^* = 0, \ \xi^* = 0 \tag{12.3}$$

for all disturbances $\eta_n(0)$ and $\xi(0)$, and all functions $f(\sigma)$ of class (A).

Let us initially consider control systems, for which the equation

$$\Delta\ (\rho) = |b_{k\alpha} + \delta_{k\alpha}\rho| = 0 \tag{12.4}$$

has $\rho_1, \ldots, \rho_s$ real roots and $\rho_{s+1}, \ldots, \rho_n$ complex conjugate, simple roots having the property

$$\text{Re}\ \rho_k \geq 0 \qquad (k = 1, \ldots, m) \ . \tag{12.5}$$

In this case, as we have seen in Chapter II, it is possible, when $N = 0$, to reduce equations (12.1) with the aid of a linear non-singular transformation (2.1), to the canonical form

$$\begin{aligned} \dot{x}_k &= -\rho_k x_k + f(\sigma) \qquad (k = 1, \ldots, m) \ , \\ \dot{\sigma} &= \sum_{k=1}^{m} \beta_k x_k - rf(\sigma) \ . \end{aligned} \tag{12.6}$$

Here β_k are determined from the known formulas

$$\beta_k = \frac{1}{Д'(\rho_k)} \sum_{\alpha=1}^{m} p_\alpha N_\alpha(\rho_k) \quad . \tag{12.7}$$

For (12.6) we know the Lur'e method for constructing the Lyapunov functions that solve this stability problem.

If we repeat the process of reducing (12.1) to a canonical form in the case $N \neq 0$, then, as can be readily verified, we obtain equations of the form

$$\begin{aligned} \dot{x}_k &= -\rho_k x_k + f(\sigma) \;, \\ \chi\dot{\sigma} &= \sum_{k=1}^{m} \beta_k x_k - rf(\sigma) \;, \end{aligned} \tag{12.8}$$

in which

$$\chi = 1 + N\frac{df}{d\sigma} \tag{12.9}$$

while the quantities r, ρ_k, and β_k have the same meaning as before.

As can be seen, canonical equations (12.8) of a control system with tachometric feedback differ from the corresponding (12.6) of a control system without feedback only in the presence of a factor χ. For the function $f(\beta)$ of class (A) and of arbitrary $N > 0$, the factor χ is generally speaking a certain bounded, everywhere-positive function of σ (Figure 13). Let there be a form

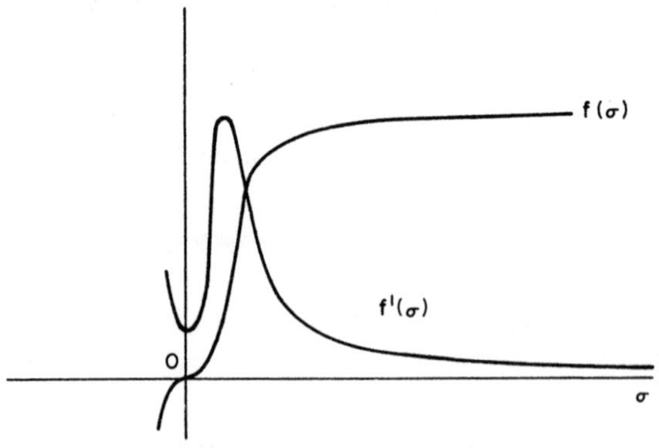

FIGURE 13. CHARACTERISTIC OF THE ACTUATOR

$$F = \sum_{k=1}^{m} \sum_{l=1}^{m} \frac{a_k a_l}{\rho_k + \rho_l} x_k x_l$$

in which $a_1, \ldots, a_s$ are real and $a_{s+1}, \ldots, a_m$ are complex conjugate numbers. As we have seen in Chapter IV, the function F is positive definite everywhere and vanishes at $x_k = 0$. To solve the problem of stability, let us consider the function

$$V = F + \int_0^\sigma xf(\sigma)d\sigma \quad . \tag{12.10}$$

It will be positive definite everywhere. When $N = 0$, the function (12.10) coincides with the Lyapunov function, proposed by Lur'e for the solution of the problem in the case of absence of velocity feedback. Its total derivative is calculated in accordance with (12.8) and will be a negative definite function, provided the following relations are satisfied:

$$\beta_k + 2a_k\sqrt{r} + 2a_k \sum_{l=1}^{m} \frac{a_l}{\rho_k + \rho_l} = 0 \qquad (k = 1, \ldots, m) \quad . \tag{12.11}$$

This derivative will be

$$\dot{V} = - \left[\sum_{k=1}^{m} a_k x_k + \sqrt{r}\ f(\sigma) \right]^2 \quad . \tag{12.12}$$

Relations (12.12) can be considered as equations that determine the unknown a_k. An analysis of these relations leads to the theorem by Lur'e (Chapter IV, Section 3) for the given problem in the case when $N = 0$: the set S of regulator parameters, on which (12.10) have solutions, and which contain $a_1, \ldots, a_s$ real and $a_{s+1}, \ldots, a_m$ complex conjugate numbers, guarantees the absolute stability of the system (12.1). This leads to the corollary: inasmuch as S is independent of N, the system (12.1) is absolutely stable on S for any tachometric feedback constant $N > 0$.

2. CASE OF INHERENTLY UNSTABLE CONTROL SYSTEM

The solution of the problem becomes considerably more complicated in the case of control systems that are unstable when the regulator is turned off.

It is easy to note, that in this case the earlier statement of

the stability problem for any characteristic $f(\sigma)$ of the actuator, belonging to functions of class (A), is no longer valid. In fact, since the system is unstable when the regulator is disconnected, it is clear that it is impossible to attain stability no matter how slow the servomotor. We shall therefore assume henceforth that $f(\sigma)$ belongs to subclass (A').

Let h be a certain positive number, to be chosen later on. Separation of the subclass (A') of the functions $f(\sigma)$, which will be considered later, will be made subject to the following requirements:

$$\text{(12.13)} \qquad \begin{aligned} &1) \quad \left[\frac{df}{d\sigma}\right]_{\sigma=0} \geq h > 0 \\ &2) \quad \sigma\varphi(\sigma) > 0, \; \varphi(\sigma) = f(\sigma) - h\sigma \quad \text{for} \quad |\sigma| < \bar{\sigma} \; . \end{aligned}$$

The first requirement of (12.13) places in class (A) all servomotors that respond to a change in the output signal at a rate not less than the quantity h; the second requirement means that the function $\varphi(\sigma)$ belongs to class (A) of functions, at least in a control range $2\bar{\sigma}$ that is less than or equal to infinity. Here $\bar{\sigma} > 0$ is the abscissa of the point where $f(\sigma)$ intersects the line $h\sigma$. Under these assumptions, let us re-examine (12.1). We introduce a new variable

$$\sigma = \sum_{\alpha=1}^{m} p_\alpha \eta_\alpha - r\xi - Nf(\sigma) \; . \tag{12.14}$$

Elimination of the old variable σ yields new equations for the problem, in the form

$$\text{(12.15)} \qquad \begin{aligned} \dot{\eta}_k &= \sum_{\alpha=1}^{m} \bar{b}_{k\alpha}\eta_\alpha - \frac{n_k}{r}(\sigma + Nf(\sigma)) \; , \\ \chi\dot{\sigma} &= \sum_{\alpha=1}^{m} \bar{p}_\alpha \eta_\alpha - \rho\sigma - (r + \rho N)F(\sigma) \; . \end{aligned}$$

Here we denote

(12.16)
$$\bar{b}_{k\alpha} = b_{k\alpha} + \frac{1}{r} p_\alpha n_k, \quad \bar{p}\beta = \sum_{k=1}^{m} p_k \bar{b}_{k\alpha} ,$$
$$\rho = \frac{1}{r} \sum_{\alpha=1}^{m} p_\alpha n_\alpha, \quad \chi = 1 + N \frac{df}{d\sigma} , \quad (\alpha, \beta, k = 1, \ldots, m) .$$

The linear non-singular transformation

(12.17)
$$- x_s = \sum_{\alpha=1}^{m} C_\alpha^{(s)} \eta_\alpha \qquad (s = 1, \ldots, m)$$

which we used in Chapter III, makes it possible to convert (12.15) to the canonical form

(12.18)
$$\dot{x}_s = r_s x_s + M\sigma + N\varphi(\sigma) ,$$
$$\chi\dot{\sigma} = \sum_{\alpha=1}^{m} \bar{\beta}_\alpha x_\alpha - S\sigma - R\varphi(\sigma) .$$

Here the numbers r_k are the simple roots of the equation

(12.19)
$$\bar{D}(r) = |\bar{b}_{k\alpha} + \delta_{k\alpha} r| = 0$$

while $\bar{\beta}_k$, M, S, R are determined from the formulas

(12.20)
$$M = 1 + hN, \quad S = \rho + hR, \quad R = r + \rho N$$
$$\bar{\beta}_k = \frac{1}{r\bar{D}'(r_k)} \sum_{\alpha=1}^{m} \bar{p}_\alpha \bar{N}_\alpha(r_k) ,$$

in which D' is the derivative of the determinant $D(r)$, and $N_\alpha(r_k)$ are quantities that are determined by formulas (3.17).

Let us examine equation $\bar{D}(-\lambda) = 0$ for which we set up the Hurwitz inequalities

(12.21)
$$\bar{\Delta}_1 > 0, \ldots, \bar{\Delta}_n > 0 .$$

Fulfillment of these inequalities signifies that

(12.22)
$$\operatorname{Re} r_k > 0 \qquad (k = 1, \ldots, m) .$$

Let us assume that all $r_1, \ldots, r_s$ are real, and $r_{s+1}, \ldots, r_m$ are complex conjugate numbers.

Let us examine the positive definite function

$$V = \sum_{k=1}^{m} \sum_{i=1}^{m} \frac{a_k a_i}{r_k + r_i} x_k x_i + \int_0^{\sigma} \chi[M\sigma + N\varphi(\sigma)]d\sigma \quad . \tag{12.23}$$

Its total derivative, calculated in accordance with (12.18) is

$$\begin{aligned} \dot{V} = &- \left(\sum_{\alpha=1}^{m} a_\alpha x_\alpha \right)^2 - [M(\sigma) + N\varphi(\sigma)] \, [S\sigma + R\varphi(\sigma)] + \\ &+ [M\sigma + N\varphi(\sigma)] \sum_{k=1}^{m} \left[\beta_k + 2 \sum_{i=1}^{m} \frac{a_k a_i}{r_k + r_i} \right] x_k \quad . \end{aligned} \tag{12.24}$$

If we add to the right half of the function $\dot{V}$ the quantity

$$[M(\sigma) + N\varphi(\sigma)]^2 + 2[M\sigma + N\varphi(\sigma)] \sum_{\alpha=1}^{m} a_\alpha x_\alpha$$

and then subtract it, we can rewrite this function in the following form:

$$\begin{aligned} \dot{V} = &- \left[\sum_{\alpha=1}^{m} a_\alpha x_\alpha + M\sigma + N\varphi(\sigma) \right]^2 + [M\sigma + N\varphi(\sigma)] \, [(S - M)\sigma + (R - N)\varphi(\sigma)] + \\ &+ [M\sigma + N\varphi(\sigma)] \sum_{k=1}^{m} \left[\bar{\beta}_k + 2a_k + \right. \\ &\cdot \left. + 2 \sum_{i=1}^{m} \frac{a_k a_i}{r_k + r_i} \right] x_k \quad . \end{aligned} \tag{12.25}$$

Using both forms (12.24) and (12.25), it is possible to obtain a stability criterion in two versions. We shall consider only the form (12.25). Like in this case, fulfillment of the conditions

$$\bar{\beta}_k + 2a_k + 2a_k \sum_{i=1}^{m} \frac{a_i}{r_k + r_i} = 0 \qquad (k = 1, \ldots, m) \quad , \tag{12.26}$$

$$S > M \qquad R > N \tag{12.27}$$

makes it possible to obtain a negative definite function $\dot{V}$ that guarantees, as before, absolute stability on the basis of the known Lyapunov theorem.

Relations (12.26) must be considered as equations that determine the unknown numbers a_k.

We have thus proven the following theorem: The set W of regulator parameters, on which (12.26) have solutions containing $a_1, \ldots, a_s$ real and $a_{s+1}, \ldots, a_m$ complex conjugate numbers, and on which there are simultaneously fulfilled inequalities (12.21) and (12.27), guarantees absolute stability of the system (12.1).

The formulated premise generalizes the theorem on the stability of regulated systems of the form (12.1), proved in Chapter VI, Section 1, for the case $N = 0$. It must be noted, that the formulation of the theorem can be simplified by including the inequality $R > N$.

In fact, let us introduce

$$M(\sigma) + N\varphi(\sigma) = Q\zeta , \qquad Q > 0 . \tag{12.28}$$

It is obvious that $\zeta(\sigma)$ is a function of class (A). According to the known theorems on implicit functions*, we can state that (12.28) can be resolved with respect to σ, and that this solution

$$\sigma = \psi(\zeta) \tag{12.29}$$

will represent a function of class (A).

We then have

$$Z = [M(\sigma) + N\varphi(\sigma)]\,[(S - M)\sigma + (R - N)\varphi(\sigma)] = \frac{Q^2}{M}(S - M)\zeta^2 +$$

$$+ Q\left(R - N - \frac{S - M}{M}N\right)\zeta\varphi(\sigma) \quad .$$

But the expression of the factor in front of the second term transforms to the form

$$R - N - \frac{S - M}{M}N = \frac{RM - SN}{M} = \frac{r}{M}$$

thanks to which we obtain

$$Z = \frac{Q^2}{M}(S - M)\zeta^2 + \frac{Qr}{M}\zeta\varphi[\psi(\zeta)] \quad ,$$

---

* E. Goursat, Course of Mathematical Analysis, Volume I, GTTI, 1933.

thus proving our statement.

3. BULGAKOV'S SECOND PROBLEM

By way of an example, let us consider the second problem of Bulgakov.

$$T\ddot{\psi} + U\dot{\psi} + k\psi + \mu = 0$$

(12.28)

$$\dot{\mu} = f^*(\sigma), \quad \sigma = a\psi + E\dot{\psi} + G^2\ddot{\psi} - \frac{1}{\ell}\mu - N\dot{\mu} \quad .$$

All the constants here, with the exception of k, are positive, while k can assume any real value.

The physical meaning of the constants is clear from the written form of the equations. Let us use the symbols of (2.82), and also let

$$N^* = i\sqrt{r}N \quad , \tag{12.29}$$

and let us reduce (12.28) to normal form, containing only dimensionless quantities.

In the case $k \geq 0$, we obtain the already known solutions of the problem, and also the condition $N > 0$.

In the case when $k < 0$, (12.28) are reduced, by the device described above, to the following form:

$$\dot{\eta}_1 = \eta_2, \quad \dot{\eta}_2 = b_{21}\eta_1 + b_{22}\eta_2 + \eta_2\xi \quad ,$$

(12.30)

$$\dot{\xi} = f(\sigma), \quad \sigma = p_1\eta_1 + p_2\eta_2 - \xi - N^*f(\sigma) \quad .$$

Omitting the procedure for reducing this equation to the canonical form, previously described in Section 3, Chapter III, we write down directly relations (12.26), which represent a system of two quadratic equations

$$\bar{\beta}_1 + 2a_1 + \frac{a_1^2}{r_1} + \frac{2a_1a_2}{r_1 + r_2} = 0 \quad ,$$

(12.31)

$$\bar{\beta}_2 + 2a_2 + \frac{a_2^2}{r_2} + \frac{2a_1a_2}{r_1 + r_2} = 0 \quad .$$

Here the symbols represent quantities already known to us

$$\bar{\beta}_1 = \frac{1}{r_2 - r_1} (\bar{p}_1 - r_2\bar{p}_2), \quad \bar{\beta}_2 = \frac{1}{r_2 - r_1} (- \bar{p}_1 + r_2\bar{p}_2)$$

(12.32) $$r_1 + r_2 = \frac{U + \ell E}{\sqrt{\ell(T^2+\ell G^2)}} = - \bar{b}_{22}, \quad \frac{k + a\ell}{\ell} = - \bar{b}_{21}$$

$$\bar{p}_1 = p_2\bar{b}_{21}, \quad \bar{p}_2 = p_1 + p_2\bar{b}_{22}, \quad \rho = - p_2 \ .$$

Conditions (12.21) reduce to only a single inequality

(12.33) $$k + a\ell > 0 \ .$$

It is easy to verify that (12.31) can certainly be solved with respect to the unknowns a_1, a_2 for example, in the case $G^2 = 0$, provided the following inequalities are satisfied:

$$\Gamma^2 = 1 + \frac{E\sqrt{e}}{T}$$

(12.34) $$D^2 = \left(\frac{U + \ell E}{T\sqrt{\ell}} \right)^2 \Gamma^2 - \frac{a(U+\ell E)}{T\sqrt{\ell}} - \frac{k + a\ell}{\ell} (1 + \Gamma)^2 > 0 \ .$$

Of these, the first inequality is trivial, and the second is certainly satisfied if the constants $U + \ell E$ (the sum of the natural damping U and of the artificial damping ℓE) is sufficiently large.

The inequalities (3.8) assume the following form

(12.35) $$1 > \Gamma^2 N^*$$

(12.36) $$h > \frac{\Gamma^2}{1 - N^*\Gamma^2} \ .$$

The first of these connects the choice of the coefficients ℓ, and N of the feedback loops, and the second imposes a limitation from below on the choice of the number h.

The method described in Chapter VIII, Section 5, is used also to solve the quality problem.

4. STABILITY OF AN AUTOMATICALLY CONTROLLED BICYCLE, ROLLING OVER A HORIZONTAL PLANE

By way of the second example of the application of the above method, let us consider an automatically controlled bicycle, rolling over a horizontal plane.

Let θ be the declination of the frame of the bicycle from the vertical and ψ the declination of the controlled front wheel from a line joining the points M_1 and M_2, at which the wheels are tangent to the plane of rolling.

Let us assume that this plane is absolutely rough. We employ the known* equation for the angular momentum about the axis M_1M_2

$$\ddot{\theta} = -\frac{bv}{cd}\dot{\psi} + \frac{g}{d}\left(\theta - \frac{v^2}{cg}\psi\right) .$$

In the case of small deflections, this equation represents the disturbed motion of the bicycle, relative to its steady motion along the stationary straight line M_1M_2 with a constant velocity v. Here d is the distance of the center of gravity G of the bicycle and load from the line M_1M_2, C is the distance between M_1 and M_2, b is the distance between the projection of the center of gravity G on the line M_1M_2 and the point M_1 of tangency of the rear wheel, Figure 14.

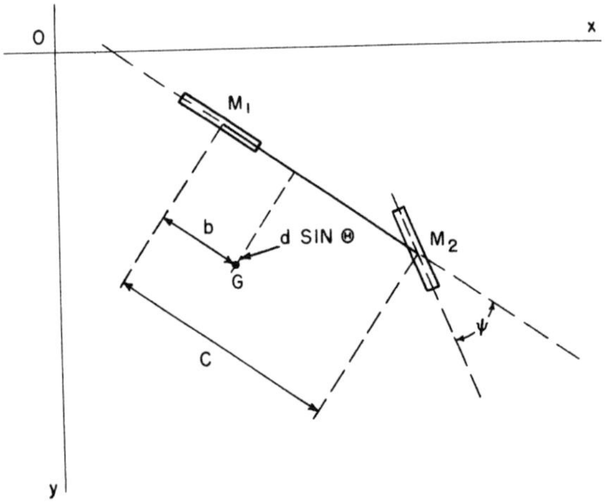

FIGURE 14. Coordinates of the Bicycle

* L. G. Loitsenskiĭ, A. I. Lur'e Teoreticheskaya Mekhanika (Theoretical Mechanics), Volume III, 1934, Gostekhizdat.

We assume that the front wheel is automatically controlled by means of a servomotor, whose equation of motion is

$$\dot{\psi} = w\dot{\theta} + f^*(\sigma)$$

$$\sigma = a\theta + E\dot{\theta} + G^2\ddot{\theta} - \frac{1}{\ell}\psi - N\dot{\psi} .$$

Here the constant $w > 0$ characterizes the magnitude of the gyroscopic moment of the forward wheel, the stabilizing effect of which is well known*, in the case of a bicycle rolling naturally.

Let us introduce the following symbols for the constants:

(12.37) $$m = \frac{g}{d} > 0 \quad n = \frac{v^2}{cd} > 0 \quad p = \frac{bv}{cd} > 0 .$$

We obviously have

(12.38) $$\begin{aligned} \ddot{\theta} &= m\theta - pw\dot{\theta} - n\psi - pf(\sigma) \\ \dot{\psi} &= w\dot{\theta} + f(\sigma) \\ \sigma &= (a + mG^2)\theta + [E - w(N + pG^2)]\dot{\theta} + \\ &\quad + (1/\ell + nG^2)\psi - (N + pG^2)f(\sigma) . \end{aligned}$$

We first transform these to the normal Cauchy form. Using the notation

(12.39) $$\theta = \eta_1, \quad \frac{d\theta}{d\tau} = \sqrt{s}\,\eta_2, \quad \psi = \eta_3, \quad \tau = \sqrt{s}\,t ,$$

where the new variables are dimensionless, and the constant $s > 0$ still remains arbitrary, we obtain

(12.40) $$\begin{aligned} \dot{\eta}_1 &= \eta_2 \\ \dot{\eta}_2 &= b_{21}\eta_1 + b_{22}\eta_2 + b_{23}\eta_3 + h_2 f(\sigma) \\ \dot{\eta}_3 &= b_{32}\eta_2 + h_3 f(\sigma) , \\ \sigma &= p_1\eta_1 + p_2\eta_2 + p_3\eta_3 - (N + pG^2)f(\sigma) . \end{aligned}$$

Here we denote

---

* Dr. R. Grammel, Der Kreisel, Seine Theorie und Seine Anwendungen, Zweiter Band, 1950.

$$\frac{m}{s} = b_{21}, \quad -\frac{pw}{\sqrt{s}} = b_{22}, \quad -\frac{n}{s} = b_{23}, \quad -\frac{p}{s} = h_2, \quad w = b_{32}, \frac{1}{\sqrt{s}} = h_3 \quad ,$$

(12.41)

$$p_1 = a + mG^2, \quad p_2 = [E - w(N + pG^2)] \sqrt{s} \ ,$$

$$p_3 = - \frac{1}{2} + nG^2 \quad .$$

Attention should be called to the expression for σ. If the bicycle equation includes a moment with a constant p, the use of the acceleration $\ddot{\theta}$ of the fall of the bicycle for control purposes reinforces the control exerted by a tachometric-feedback channel. Let us transform (12.40) to the canonical form. For this purpose, we introduce the canonical variables

$$x_s = C_1^{(s)}\eta_1 + C_2^{(s)}\eta_2 + C_3^{(s)}\eta_3 \qquad (s = 1, 2, 3) \ .$$

Let $\lambda_1, \lambda_2, \lambda_3$ be the roots of the equation

(12.42) $$\lambda(\lambda^2 - b_{22}\lambda - b_{21} - b_{22}b_{23}) = 0$$

and let the transformation constants $C_\alpha^{(s)}$ be subjected to the relations

$$\lambda_s C_1^{(s)} = C_{21}C_2^{(s)} \ , \qquad \lambda_s C_2^{(s)} + b_{22}C_2^{(s)} + b_{32}C_3^{(s)} \ ,$$

$$\lambda_s C_3^{(s)} = b_{23}C_2^{(s)} \ , \qquad 1 = h_2 C_2^{(s)} + h_3 C_3^{(s)} \ ,$$

then the sought transformation will be of the form

(12.43)
$$\begin{aligned} H(\lambda_1)x_1 &= b_{21}\eta_1 + \lambda_1\eta_2 + b_{23}\eta_3 \ , \\ H(\lambda_2)x &= b_{21}\eta_1 + \lambda_2\eta_2 + b_{23}\eta_3 \ , \\ h_3 x_3 &= - b_{32}\eta_1 \qquad + \eta_3 \ , \\ H(\lambda) &= h_2\lambda + b_{23}h_3 \ . \end{aligned}$$

This transformation is not singular, if $\lambda_1 \neq \lambda_2$, and therefore it can be inverted and we obtain

(12.44)
$$\eta_1 = - \frac{H(\lambda_1)}{\lambda_1(\lambda_2-\lambda_1)} x_1 + \frac{H(\lambda_2)}{\lambda_2(\lambda_2-\lambda_1)} x_2 + \frac{b_{23}h_3}{\lambda_1\lambda_2} x_3 \ ,$$

$$\eta_2 = \frac{H(\lambda_1)}{\lambda_2 - \lambda_1} x_1 + \frac{H(\lambda_2)}{\lambda_2 - \lambda_1} x_2 \qquad \eta_3 = \frac{b_{32}H(\lambda_1)}{\lambda_1(\lambda_2-\lambda_1)} x_1 + \frac{b_{32}H(\lambda_2)}{\lambda_2(\lambda_2-\lambda_1)} x_2 - \frac{b_{21}h_3}{\lambda_1\lambda_2} x_3 \ .$$

The canonical equations in the new variables will be of the form

$$\dot{x}_1 = \lambda_1 x_1 + f(\sigma) \;,$$
$$\dot{x}_2 = \lambda_2 x_2 + f(\sigma) \;, \tag{12.45}$$
$$\chi\dot{\sigma} = \beta_1 x_1 + \beta_2 x_2 - Rf(\sigma) \;.$$

Here we denote

$$\lambda_1 + \lambda_2 = b_{22} < 0, \; \lambda_1\lambda_2 = - b_{21} - b_{23}b_{32} > 0 \;,$$

$$\beta_1 = \frac{H(\lambda_1)}{\lambda_2 - \lambda_1}[p_1 + \lambda_1 p_2 + b_{32}p_3],$$
$$\beta_2 = \frac{H(\lambda_2)}{\lambda_2 - \lambda_1}[p_1 + \lambda_2 p_2 + b_{32}p_3] \;, \tag{12.46}$$

$$R = - h_1 p_2 - h_3 p_3 \;. \tag{12.47}$$

The equation $\dot{x}_3 = f(\sigma)$ for the canonical variable x_3 drops out from the system (12.45).

It is easily seen that the stability relative to the variables x_1, x_2, and σ guarantees the stability also relative to the variables η_1, η_2, and η_3. Let us consider the positive definite function

$$V = \frac{a_1^2}{2\lambda_1} x_1^2 - \frac{2a_1a_2}{\lambda_1 + \lambda_2} x_1x_2 - \frac{a_2^2}{2\lambda_2} x_2^2 + \int_0^\sigma \chi f(\sigma)d\sigma \;.$$

Its total derivative, calculated according to (12.45), will be

$$\dot{V} = - (a_1x_1 + a_2x_2 + \sqrt{R}\, f(\sigma))^2 \;.$$

If the following relations are satisfied,

$$\beta_1 + 2\sqrt{R}\, a_1 - \frac{2a_1a_2}{\lambda_1 + \lambda_2} - \frac{a_1^2}{\lambda_1} = 0, \; \beta_2 + 2\sqrt{R}\, a_2 - \frac{2a_1a_2}{\lambda_1 + \lambda_2} - \frac{a_2^2}{\lambda_2} = 0 \;, \tag{12.47}$$

then

$$R > 0 \;. \tag{12.48}$$

Relations (12.47) can be resolved with respect to the constants a_1 and a_2, if the following inequalities are satisfied:

$$r^2 + R + \frac{\beta_1}{\lambda} + \frac{\beta_2}{\lambda} > 0 \quad ,$$

(12.49)

$$D^2 = (\lambda_1^2 + \lambda^2)R + \lambda_1\beta_1 + \lambda_2\beta_2 \pm 2\lambda_1\lambda_2\sqrt{R}r^2 > 0 \quad .$$

Consequently the stability criterion reduces to limiting the choice of the regulator parameters to inequalities (12.48) and (12.49), and also to the inequality

(12.50) $$\lambda_1\lambda_2 > 0 \quad .$$

Let us examine these inequalities.

Inequalities (12.48) and (12.50) are of the form

(12.51) $$E + \frac{1}{p}\left(\frac{1}{\ell} + nG^2\right) > w(N + pG^2) \quad ,$$

(12.52) $$nw > m \quad .$$

We next have

$$\frac{\beta_1}{\lambda_1} + \frac{\beta_2}{\lambda_2} = -\frac{h_3b_{23}}{\lambda_1\lambda_2}p_1 + h_2p_2 - \frac{b_{23}b_{32}h_3}{\lambda_1\lambda_2}p_3 \quad ,$$

$$\lambda_1\beta_1 + \lambda_2\beta_2 = (h_2b_{22} + h_3b_{23})p_1 + [h_2(b_{22}^2 + b_{21} + b_{23}b_{32}) + h_3b_{22}b_{23})]p_2 +$$
$$+ (h_2b_{22} + h_3b_{23})b_{32}p_3 \quad .$$

Consequently

$$r^2 = \frac{h_3(a\ell n-m)}{\ell(nw-m)}$$

from which we should get

(12.53) $$a\ell n > m \quad .$$

The inversion of the inequality is impossible, by virtue of (12.52).

Finally, using (12.47) and (12.41) and (12.37), we find the last stability condition

$$D^2 = (p^2 w - n)a + pnw\,[E - w(N + pG^2)] + \tag{12.54}$$

$$+ m(\frac{1}{\ell} + wp^2G^2) + 2\sqrt{(wn - m)(a\ell n - m)[1/\ell + nG^2 + pE - pw(N + pG^2)]} > 0 .$$

The conditions obtained permit of mechanical interpretation.

If the parameters of the wheel are specified, then $w = kv$, where k is a proportionality factor, and inequality (12.52) becomes

$$kv^3 > cg \ . \tag{12.55}$$

This is certain to be satisfied if the speed of the bicycle is sufficiently high. In this case inequality (12.53) is restricted from below by the choice of the parameter "$a\ell$" of the control system, and we have

$$a\ell v^2 > cg \ . \tag{12.56}$$

Inequalities (12.51) and (12.54) can be satisfied by suitable choice of the parameters E and G^2.

AUTHOR INDEX

SUBJECT INDEX